NEUROMETHODS

For other titles published in this series, go to
www.springer.com/series/7657

Calcium Measurement Methods

Edited by

Alexei Verkhratsky

The University of Manchester, Manchester, UK

and

Ole H. Petersen

University of Liverpool, Liverpool, UK

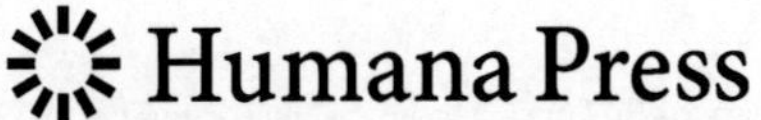

Editors
Alexei Verkhratsky
The University of Manchester
Manchester
UK
alex.verkhratsky@manchester.ac.uk

Ole H. Petersen
University of Liverpool
Liverpool
UK
o.h.petersen@liverpool.ac.uk

ISSN 0893-2336 e-ISSN 1940-6045
ISBN 978-1-60761-475-3 e-ISBN 978-1-60761-476-0
DOI 10.1007/978-1-60761-476-0
Springer New York Dordrecht Heidelberg London

Library of Congress Control Number: 2009939115

Printed on acid-free paper

Humana Press is part of Springer Science+Business Media (www.springer.com)

Preface to the Series

Under the guidance of its founders Alan Boulton and Glen Baker, the Neuromethods series by Humana Press has been very successful since the first volume appeared in 1985. In about 17 years, 37 volumes have been published. In 2006, Springer Science + Business Media made a renewed commitment to this series. The new program will focus on methods that are either unique to the nervous system and excitable cells or that need special consideration to be applied to the neurosciences. The program will strike a balance between recent and exciting developments like those concerning new animal models of disease, imaging, in vivo methods, and more established techniques. These include immunocytochemistry and electrophysiological technologies. New trainees in neurosciences still need a sound footing in these older methods in order to apply a critical approach to their results. Careful application of methods is probably the most important step in the process of scientific inquiry. In the past, new methodologies led the way in developing new disciplines in the biological and medical sciences. For example, Physiology emerged out of Anatomy in the nineteenth century by harnessing new methods based on the newly discovered phenomenon of electricity. Nowadays, the relationships between disciplines and methods are more complex. Methods are now widely shared between disciplines and research areas. New developments in electronic publishing also make it possible for scientists to download chapters or protocols selectively within a very short time of encountering them. This new approach has been taken into account in the design of individual volumes and chapters in this series.

Wolfgang Walz

Preface

In 1981, Roger Tsien introduced a fundamentally new class of cellular probes, the fluorescent calcium indicators, which revolutionized cellular physiology. These indicators allowed for the first time real-time monitoring and imaging of calcium movements in the intact cells. The immediate result of this technical breakthrough was a detailed characterisation of the calcium signaling system. This system is truly omnipresent and pluripotent, being involved in the regulation of a wide range of cellular reactions. Furthermore, calcium signals, which occur in cells in response to stimulation, can be used as universal reporters of cellular activity.

This volume of the *Neuromethods* series is dedicated to calcium imaging in neural cells. When composing this volume, we tried to balance the main principles of calcium imaging with specific applications of the technique to neural tissues. The wealth of microscopic and imaging technologies, which rapidly developed during the last 25 years, dramatically increased the versatility and power of calcium imaging. The synthetic calcium fluorescent probes are now represented by an extended range of indicators that allow precise measurements of Ca^{2+} concentrations within the entire physiological range, from nanomolar to millimolar. Calcium-sensitive fluorescent probes could be loaded into intracellular organelles, single cells, or cellular networks. The family of fluorescent calcium probes was recently extended by the development of a fundamentally new class of indicators based on proteins, which can be specifically targeted into organelles or cells of interest both in vitro and in vivo. Combinations of different indicators and recently developed microscopic techniques lead to spectacular successes in imaging of neural cells in the living brain in both physiological and pathophysiological models.

We therefore hope that this collection of chapters written by a team of recognized experts will provide references and help to the many scientists worldwide engaged in calcium imaging of the nervous system.

Manchester, UK *Alexei Verkhratsky*
Liverpool, UK *Ole H. Petersen*

Contents

Contributors

MARÍA TERESA ALONSO • *Instituto de Biología y Genética Molecular (IBGM), Universidad de Valladolid y Consejo Superior de Investigaciones Científicas (CSIC), Valladolid, Spain*

ROBERT BLUM • *Physiologisches Institut, Physiologische Genomik, Ludwig-Maximilians-Universität, München, Germany*

MARIO BORTOLOZZI • *Department of Physics, University of Padua, Padua, Italy Foundation for Advanced Biomedical Research, Venetian Institute of Molecular Medicine, Padua, Italy*

MARISA BRINI • *Department of Biochemistry and Department of Experimental Veterinary Science, University of Padova, Padova, Italy*

JOACHIM W. DEITMER • *Abteilung für Allgemeine Zoologie, FB Biologie, TU Kaiserslautern, Germany*

GERHARD EICHHOFF • *Institute of Physiology II, University of Tübingen, Tübingen, Germany*

LAURA FEDRIZZI • *Department of Biochemistry and Department of Experimental Veterinary Science, University of Padova, Padova, Italy*

OLGA GARASCHUK • *Institute of Physiology II, University of Tübingen, Tübingen, Germany*

JAVIER GARCÍA-SANCHO • *Instituto de Biología y Genética Molecular (IBGM), Universidad de Valladolid y Consejo Superior de Investigaciones Científicas (CSIC), Valladolid, Spain*

JEREMY GRAHAM • *Cairn Research Ltd, Faversham, Kent, UK*

OLIVER GRIESBECK • *Max-Planck-Institute of Neurobiology, Martinsried, Germany*

KNUT HOLTHOFF • *Department of Neurology, Friedrich-Schiller-Universität Jena, Jena, Germany*

SHIN HYE KIM • *Department of Physiology, Sungkyunkwan University School of Medicine, Suwon, Korea*

YURY KOVALCHUK • *Institute of Physiology II, University of Tübingen, Tübingen, Germany*

CHRISTIAN LOHR • *Interdisziplinäres Zentrum für Klinische Forschung, Institut für Physiologie I, Westfälische Wilhelms-Universität Münster, Münster, Germany*

FABIO MAMMANO • *Department of Physics, University of Padua, Padua, Italy Foundation for Advanced Biomedical Research, Venetian Institute of Molecular Medicine, Padua, Italy*

LUCÍA NÚÑEZ • *Instituto de Biología y Genética Molecular (IBGM), Universidad de Valladolid y Consejo Superior de Investigaciones Científicas (CSIC), Valladolid, Spain*

MYOUNG KYU PARK • *Department of Physiology, Sungkyunkwan University School of Medicine, Suwon, Korea*

OLE H. PETERSEN • *MRC Group, Physiological Laboratory, School of Biomedical Sciences, University of Liverpool, Liverpool, Merseyside, UK*

ROGER C. THOMAS • *Department of Physiology, Development and Neuroscience, University of Cambridge, Cambridgeshire, Cambridge, UK*

EMIL C. TOESCU • *School of Experimental Medicine, College of Medical and Dental Studies, University of Birmingham, Birmingham, West Midlands, UK*
ZSUZSANNA VARGA • *Institute of Physiology II, University of Tübingen, Tübingen, Germany*
ALEXEI VERKHRATSKY • *Faculty of Life Sciences, The University of Manchester, Manchester, Lancashire, UK*
CARLOS VILLALOBOS • *Instituto de Biología y Genética Molecular (IBGM), Universidad de Valladolid y Consejo Superior de Investigaciones Científicas (CSIC), Valladolid, Spain*

Chapter 1

Principles of the Ca^{2+} Homeostatic/Signalling System

Alexei Verkhratsky and Ole H. Petersen

Abstract

Calcium ions are the most ubiquitous and pluripotent signalling molecules, which regulate a wide array of physiological and pathological reactions. The specific system, controlling cellular Ca^{2+} homeostasis appeared very early in the evolution, being initially survival system preventing Ca^{2+}-mediated cell damage. Subsequently, the steep Ca^{2+} gradients maintained by Ca^{2+} homeostatic molecular cascades became the basis for Ca^{2+} signalling. This signalling system utilises Ca^{2+} channels and transporters localised in plasmalemma and intracellular membranes to create highly organised and compartmentalised cytosolic Ca^{2+} fluctuations occurring within the spatial and temporal domains. Changes in cytosolic Ca^{2+} concentrations regulate a multitude of Ca^{2+}-dependent proteins, which serve as "Ca^{2+} sensors" and thus the effectors of Ca^{2+} signalling system.

Key words: Calcium, Ca^{2+} homeostasis, Ca^{2+} signalling, Neurones, Glia, Nervous system

1. Early Evolutionary Roots of Ca^{2+} Signalling

It is a truth universally acknowledged that Ca^{2+} ions are the most ubiquitous and pluripotent signalling molecules, which regulate a wide array of physiological and pathological reactions. Indeed, Ca^{2+}-dependent regulation occurs in very different temporal and spatial domains, ranging from extremely rapid and localised events like, for example, exocytosis to long-lasting adaptive reactions, which may take days, months or even years to develop (e.g. learning and memory). The Ca^{2+} homeostatic/signalling system is present in virtually every tissue and cell and encompasses the whole of the kingdom of life from bacteria to humans.

Cellular ion homeostasis is essential for life. The ionic composition of the cell interior must be compatible with the biochemical reactions needed for cell survival and proper functioning. As a consequence, survival of early life forms ultimately depended on

A. Verkhratsky and Ole H. Petersen (eds.), *Calcium Measurement Methods*, Neuromethods, vol. 43, DOI 10.1007/978-1-60761-476-0_1, © Humana Press, a part of Springer Science+Business Media, LLC 2010

the control of ion movements between the extracellular milieu and the cytosol. The development and maintenance of transplasmalemmal ion gradients required specific homeostatic tools, represented by ion transport systems. Failure of such homeostatic cascades triggered universal cell death routines, which were firmly conserved throughout evolution (1). Compromised ion homeostasis also plays a leading role in tissue damage induced by environmental stress (i.e. in oxygen deprivation or toxic attack) in all cells and tissues.

With regard to ion compartmentalization, Ca^{2+} acquired a specific and unique role (2), most likely from the moment when ATP was selected as the main source of biological energy. This evolutionary choice ultimately required very low concentrations of cytosolic Ca^{2+}, since otherwise insoluble Ca^{2+}-phosphates would preclude cell energetics. In addition, Ca^{2+} has several specific properties (flexible coordination chemistry, high affinity for carboxylate oxygen, which is the most frequent motif in amino acids, rapid binding kinetics, etc. – see Jaiswal (3), which greatly promote Ca^{2+} binding and interaction with numerous biological molecules. At high concentrations, however, Ca^{2+} causes aggregation of proteins and nucleic acids, instigates precipitation of phosphates and damages the integrity of lipid membranes. A long-lasting and massive increase in the cytosolic Ca^{2+} concentration is therefore incompatible with life; at all phylogenetic stages, from the most ancient bacteria to the most specialised eukaryotes, a large sustained increase in the cytosolic Ca^{2+} concentration is invariably cytotoxic (4–8).

Ca^{2+} was readily available from the very beginning of life; Ca^{2+} is the fifth most abundant element in the earth's crust; only oxygen, silicon, aluminium and iron have greater mass. As a consequence, establishing an extremely low cytosolic Ca^{2+} concentration became a daunting task, which could only be fulfilled by several families of Ca^{2+} homeostatic molecules. These Ca^{2+} transporters, responsible for the up-hill transport of Ca^{2+} against steep concentration gradients appeared very early in phylogeny: Ca^{2+} pumps and exchangers, which are structurally similar to eukaryotic analogues, are present already in bacteria and other primitive life forms (9–12).

At the very same time, huge Ca^{2+} concentration gradients were utilised by evolution as one of the most versatile and omnipresent intracellular signalling systems. Indeed, the existence of a great "Ca^{2+} pressure" allows massive Ca^{2+} influx into the cytosol upon relatively small changes in the Ca^{2+} permeability of the plasma membrane. This Ca^{2+} influx can rapidly increase the cytosolic Ca^{2+} concentration ($[Ca^{2+}]_i$), which acts as a Ca^{2+} signal. The Ca^{2+} homeostatic mechanisms, designed for keeping $[Ca^{2+}]_i$ low, assumed responsibility for termination of Ca^{2+} signals.

Finally, early eukaryotes developed a broad family of Ca^{2+}-binding (EF-hand) proteins, which became both Ca^{2+} buffers and Ca^{2+} sensors, through which Ca^{2+} signalling systems govern the cellular processes (13).

2. Cellular Physiology of Ca^{2+} Homeostasis and Ca^{2+} Signalling

The life of every cell occurs within a confined space delineated by the plasma membrane; in complex cells, this space is further divided (also by membranes) into several compartments, which have unique purposes and properties. Conceptually, the Ca^{2+} homeostatic/signalling system operates within these cellular compartments, which have different modes of Ca^{2+} handling, thus creating Ca^{2+} concentration gradients within the cell and between the cell and the extracellular environment (14–18). Translocation of Ca^{2+} between these compartments, either due to passive diffusion or by active transport, produces rapid and spatially segregated changes in the free Ca^{2+} concentration that interact with "Ca^{2+} sensors". The latter are represented by a multitude of Ca^{2+}-sensitive enzymes. Binding of Ca^{2+} to these enzymes affects their activity, hence regulating diverse biochemical processes, which underlie various cellular reactions (19–21). The "Ca^{2+} sensors" have different affinities to Ca^{2+} and different intracellular localization, thus providing for amplitude- and space-coding as well as for the specificity of the Ca^{2+} signalling system. In addition, the kinetics of $[Ca^{2+}]_i$ fluctuations, determined by the incoming activating signals control the timing of Ca^{2+}-binding/unbinding to Ca^{2+}-sensitive proteins thus forming the basis for temporal coding of Ca^{2+} signalling events (22).

The three main intracellular compartments, in which Ca^{2+} signals emerge and dissipate are the cytoplasm, the endoplasmic reticulum (ER) and the mitochondria (although other intracellular organelles, such as the Golgi complex, lysosomes and secretory granules may also participate in intracellular Ca^{2+} homeostasis – (23, 24)).

In the cytosol, the Ca^{2+} concentration is kept at the very low level of ~50–100 nM, resulting in a constant "Ca^{2+} pressure" from the extracellular milieu and several intracellular organelles. Cell stimulation opens plasmalemmal or intracellular Ca^{2+} channels, which produce rapid and substantial Ca^{2+} influx increasing $[Ca^{2+}]_i$. These increases in $[Ca^{2+}]_i$ can be local or global depending on the nature of the stimulus, the degree of activation of the Ca^{2+} channels and the buffering of Ca^{2+} in the cytosol. This Ca^{2+} buffering is provided by specialised proteins (such as e.g. calbindin 28), which have high affinity for Ca^{2+} (25, 26). The concentration of cytosolic buffers varies between different cells; due to their high Ca^{2+} affinity, these buffers are instrumental in localising Ca^{2+} signalling events and in producing microdomains of high $[Ca^{2+}]_i$ concentration, which control many rapid and spatially confined cellular reactions such as, for example, exocytosis.

Intracellular organelles, the ER and mitochondria, process Ca^{2+} in a different fashion. The ER is arguably the largest intracellular organelle formed by the endomembrane that in turn defines the internal continuity of the ER structure (27–32). The ER is involved in many vital cellular functions, including protein synthesis, protein maturation and folding, intracellular transport, regulation of cell survival etc. (33). As far as the Ca^{2+} homeostasis/signalling system is concerned, the ER acts as the largest dynamic Ca^{2+} store, which establishes the spatial and temporal organization of Ca^{2+} signalling events. A special complement of endomembrane-resident Ca^{2+} channels and endomembrane-resident intracellular Ca^{2+} pumps of the SERCA type as well as the relatively low Ca^{2+} affinity of ER lumen-resident Ca^{2+} buffers (13, 17, 33) determine the Ca^{2+} signalling function of the ER. The intra-ER concentration of free Ca^{2+} varies between 200 and ~1,000 μM (34–37), thus creating a steep Ca^{2+} concentration gradient between the ER lumen and the cytosol. In fact, in many cells the ER acts as a major source of Ca^{2+} for production of cytosolic Ca^{2+} signals, as indeed the ER assumes the leading role in Ca^{2+} signal generation in all electrically non-excitable cells, and provides ~90–95% of the Ca^{2+} required for the contraction of muscle (22).

The ER lumen contains several classes of Ca^{2+}-binding proteins (such as e.g. calsequestrin, calreticulin or calumenin), whose affinity to Ca^{2+} is rather low; the K_D being in the range of 0.5–1 mM (38). As a result, the diffusion of Ca^{2+} within the ER lumen is much easier than in the cytosol (39). As a consequence Ca^{2+} within the ER lumen can rapidly equilibrate, allowing ER to act as a "Ca^{2+} tunnel" capable of rapidly moving Ca^{2+} in the polarised cells (30, 40).

The second important intracellular Ca^{2+} store is represented by the mitochondria, which are able to accumulate, store and release Ca^{2+}. The electronegativity of the mitochondrial matrix with regard to the cytosol creates an electro-driving force for Ca^{2+} uptake across the inner mitochondrial membrane whenever $[Ca^{2+}]_i$ rises above 300–400 nM (41, 42). Thus, local or global $[Ca^{2+}]_i$ rises lead to Ca^{2+} influx into the mitochondria; in physiological conditions this Ca^{2+} influx increases mitochondrial ATP production therefore creating an activity-energy production coupling loop. The overload of mitochondria with

Ca^{2+} is detrimental and often is a key step in triggering various death programmes (43).

3. Molecular Physiology of Ca^{2+} Homeostasis and Ca^{2+} Signalling: Ca^{2+} Channels and Transporters

Control of Ca^{2+} movements through cellular membranes represents the core of the Ca^{2+} homeostatic and Ca^{2+} signalling system. Membrane Ca^{2+} transport is a function of two broad groups of proteins: the membrane Ca^{2+} channels and the membrane Ca^{2+} transporters.

3.1. Plasmalemmal Ca^{2+} Channels

Ca^{2+} channels are essentially transmembrane protein-made aqueous pores endowed with selective filter and gating mechanisms, which allow Ca^{2+} diffusion driven by an electro-chemical gradient. These Ca^{2+} channels can be broadly subdivided into plasmalemmal and intracellular channels, which in turn are represented by several families distinct in their gating, selectivity and function. Plasmalemmal Ca^{2+} channels divide into voltage-gated channels, ligand-gated channels (or ionotropic receptors), store-operated channels and nonselective cationic channels that are partially permeable to Ca^{2+}. The classic voltage-gated channels and store-operated ($I_{CRAC/STIM1/ORAI}$) channels are almost exclusively permeable to Ca^{2+}; in contrast, Ca^{2+} permeability for all other types of channels varies quite widely. The highest Ca^{2+} permeability for ionotropic receptors (when $P_{Ca}/P_{monovalent}$ approaches 7–12) is reported for NMDA-type glutamate receptors, for certain types of P2X purinoceptors and for brain type acetylcholine receptors (44). The cationic channels, which are mostly represented by an extended family of transient receptor potential (TRP) channels, have very different permeation properties and hence have very different Ca^{2+} permeabilities (45).

The activation mechanisms of plasmalemmal Ca^{2+} channels differ markedly, depending on the cell type. Nevertheless, opening of these channels invariably results in massive Ca^{2+} entry into the cytosol, which, depending on the duration, may have both physiological and pathophysiological relevance. For excitable cells (nerve, muscle and endocrine) the prevalent mechanism of plasmalemmal Ca^{2+} entry is associated with depolarization and opening of voltage-gated channels or (especially in neurones) with direct activation of ionotropic receptors by relevant transmitters. In electrically nonexcitable cells (for example, exocrine) Ca^{2+} entry is predominantly a consequence of the opening of store-operated and/or nonselective cationic channels, which almost invariably involves the activation of metabotropic receptors with subsequent recruitment of intracellular signalling cascades.

3.2. Intracellular Ca^{2+} Channels

Intracellular Ca^{2+} channels dwelling in the membrane of the ER are represented by several subfamilies (46–49): the Ca^{2+}-gated Ca^{2+} channels, generally referred to as ryanodine receptors (RyRs), inositol 1,4,5-trisphosphate ($InsP_3$)-gated channels, generally known as $InsP_3$ receptors ($InsP_3Rs$) and NAADP receptors. The NAADP receptors are not yet molecularly identified and could also be combinations of NAADP binding proteins, intermediary linking proteins and RyRs (48).

The RyRs and $InsP_3Rs$ are tetramers homomerically assembled from primary subunits; the assembly having a four-cloverleaf-like structure when observed by electron microscopy. Each of these two families further comprises three sub-types of the receptors, designated as RyRs1,2,3 and $InsP_3R1,2,3$. Activation of RyR1, which is generally expressed in skeletal muscle cells, is triggered by depolarization-induced opening of sarcolemmal Ca^{2+} channels that are directly coupled to the RyR1 via cytosolic "foot" extensions of the channel proteins (50). Activation of RyR2,3 (which are abundantly expressed in the heart muscle, in the nervous system, in secretory cells and in some types of non-excitable cells) requires an initial increase in $[Ca^{2+}]_i$ attained either by plasmalemmal Ca^{2+} entry though voltage- and/or ligand gated Ca^{2+} channels or (especially in non-excitable cells) by Ca^{2+} release through $InsP_3Rs$. The latter are classical ligand-gated channels, which are gated by the intracellular second messenger $InsP_3$ produced as a consequence of activation of plasmalemmal metabotropic receptors controlling (through G-proteins) the activity of phospholiase C. This signalling cascade is almost omnipresent in electrically nonexcitable cells, is also operative in endocrine secretory cells, in some types of muscle cells and in neurones. Both RyRs and $InsP_3Rs$ may co-exist in the same cell and, depending on their location and functional interactions, are involved in fine-tuning and shaping Ca^{2+} signals (51).

Mitochondrial Ca^{2+} uptake occurs through an extremely selective Ca^{2+} channel (52) known since Peter Mitchell as the uniporter. The molecular identity of this channel remains unknown although the biophysical characteristics of the uptake system have been characterised in detail (41, 53).

3.3. Ca^{2+} Transporters

Energy-dependent Ca^{2+} transport against concentration gradients is central for both Ca^{2+} homeostasis and Ca^{2+} signalling as it balances the diffusion-based Ca^{2+} movements between the intracellular compartments. There are two families of ATP-dependent Ca^{2+} pumps, the plasmalemmal Ca^{2+} ATPases (PMCAs) and sarco-(endo)-plasmic reticulum Ca^{2+} ATPases (SERCA), which transport Ca^{2+} from the cytosol to the extracellular space and into the ER lumen, respectively (54, 55). In addition, Ca^{2+} can be transported across the cellular and mitochondrial membranes by the sodium–calcium exchanger (NCX), which uses the energy stored in the form of the transmembrane Na^+ gradient (54). The NCX can rapidly deal with large Ca^{2+} loads, which, for example, accompany heartbeats (56). Another interesting property of the NCX is its ability to almost instantly go into the "reverse" mode (depending on the Na^+ gradient and membrane polarization), when it starts to bring Ca^{2+} into the cytosol while expelling Na^+. This "reverse" mode may have important implications not only for Ca^{2+} signalling but also for the regulation of Na^+ homeostasis, as for example happens in astroglial cells, experiencing Na^+ overload as a consequence of glutamate uptake (57). Another type of exchanger operative in mitochondrial membranes is the Ca^{2+}/H^+ exchanger (58).

Importantly, all components of the Ca^{2+} homeostatic/signalling system operate under tight control of $[Ca^{2+}]$; the latter establishing a number of feedback loops which maintain the balance of Ca^{2+} movements. Indeed, most of the plasmalemmal Ca^{2+} channels display a strong $[Ca^{2+}]_i$-dependent inactivation (59, 60). The intracellular Ca^{2+} channels are under the control of both $[Ca^{2+}]_i$ and the intra-ER Ca^{2+} concentration: conceptually, an increase in the ER Ca^{2+} concentration increases the availability of intracellular Ca^{2+} channels for activation and *vice versa* depletion of Ca^{2+} stores inhibits both RyRs and $InsP_3Rs$ (61). The intra-ER Ca^{2+} concentration also controls the velocity of SERCA pumps: a decrease in the ER free $[Ca^{2+}]$ increases the SERCA pump activity, whereas, replenishment of the store inhibits Ca^{2+} accumulation (61, 62). Finally, the mitochondrial uniporter also displays Ca^{2+}-dependent inactivation (63, 64). Taken together all these feedbacks ascertain the versatility and sturdiness of cellular Ca^{2+} homeostasis.

4. Ca^{2+} Signalling in Neural Cells

Dynamic ensembles of neurones and glia represent the substrate for integration in the nervous system. These two types of cells form interdependent and constantly communicating cellular networks, which although employ distinct mechanisms for information transfer, work in concert to provide an unparalleled cognitive power. Conceptually, the electrically excitable neuronal networks are embedded into the astroglial syncytia, which provides functional compartmentalisation of the grey matter, regulation of brain homeostasis and, most likely, participate in information processing (65).

The integrative processes, which constantly occur in neuronal–glial circuits, ultimately depend on cellular Ca^{2+} signalling. In the neuronal networks $[Ca^{2+}]_i$, microdomains triggered by depolarisation of synaptic terminals assume the sole responsibility for exocytotic release of neurotransmitters (66–68). The Ca^{2+} signals, which regulate neurotransmitter release may be further amplified and shaped by both the ER and mitochondria (69, 70). Furthermore, Ca^{2+} plays a vital role for postsynaptic integration by controlling membrane excitability, synaptic plasticity and gene expression (71–74).

Ca^{2+} signalling is even more important for glial cells, as it provides for the special form of glial excitability. Indeed, glial membranes, which are in close contact with synaptic terminals, are endowed with the full complement of receptors, which permit them to perceive neuronal activity (75, 76). In most of the cases these receptors are coupled to the $InsP_3$-signalling cascade, which controls Ca^{2+} release from the ER store (77). The $InsP_3$-induced Ca^{2+} release, which occurs in glia following the activation of metabotropic receptors, triggers both Ca^{2+} oscillations and intracellular Ca^{2+} waves, which result from Ca^{2+}-assisted recruiting of $InsP_3Rs$ along the ER membrane. This wave of ER excitation crosses the cellular borders, and spreads through the astroglial syncytium acting therefore as a long-range signalling system (78). Astroglial Ca^{2+} signals directly control communications in neuronal–glial loops, as they trigger vesicular release of gliotransmitters, which act upon both neighbouring astrocytes and closely associated neurones (79). In conclusion, Ca^{2+} signals are instrumental for acquisition and processing of information within the astroglial syncytium as well as for a wide range of neuronal–glial communications.

5. Experimental Concepts of Probing for Ca^{2+} Signalling: Measuring Fluxes and Concentrations

All experiments aimed at investigations of cellular Ca^{2+} homeostasis and signalling are concerned with measuring either Ca^{2+} fluxes or Ca^{2+} concentrations or both of these simultaneously. This book is designed to provide a comprehensive and up-to-date accounts of these techniques within the specific context of neuroscience. In the first part of the book, general questions related to various types of Ca^{2+} probes and Ca^{2+} measuring set-ups are presented. In the second part, more specific accounts of Ca^{2+} measurements in cellular sub-compartments in single cells and in nervous tissues are discussed. We hope that this manual will provide the reader with concise yet comprehensive guidelines for Ca^{2+} measurements and practical approaches to experimental probing of the Ca^{2+} signalling system.

References

1. Nicotera P, Petersen OH, Melino G, Verkhratsky A (2007) Janus a god with two faces: death and survival utilise same mechanisms conserved by evolution. Cell Death Differ 14:1235–1236
2. Case RM, Eisner D, Gurney A, Jones O, Muallem S, Verkhratsky A (2007) Evolution of calcium homeostasis: from birth of the first cell to an omnipresent signalling system. Cell Calcium 42:345–350
3. Jaiswal JK (2001) Calcium – how and why? J Biosci 26:357–363
4. Choi YM, Kim SH, Chung S, Uhm DY, Park MK (2006) Regional interaction of endoplasmic reticulum Ca^{2+} signals between soma and dendrites through rapid luminal Ca^{2+} diffusion. J Neurosci 26:12127–12136
5. Orrenius S, Zhivotovsky B, Nicotera P (2003) Regulation of cell death: the calcium-apoptosis link. Nat Rev Mol Cell Biol 4:552–565
6. Paschen W, Mengesdorf T (2005) Endoplasmic reticulum stress response and neurodegeneration. Cell Calcium 38:409–415
7. Petersen OH, Sutton R (2006) Ca^{2+} signalling and pancreatitis: effects of alcohol, bile and coffee. Trends Pharmacol Sci 27:113–120
8. Trump BF, Berezesky IK (1995) Calcium-mediated cell injury and cell death. FASEB J 9:219–228
9. Berkelman T, Garret-Engele P, Hoffman NE (1994) The pacL Gene of *Synechococcus sp.* Strain PCC 7942 Encodes a Ca^{2+}-Transporting ATPase. J Bacteriol 176:4430–4436
10. Gambel AM, Desrosiers MG, Menick DR (1992) Characterization of a P-Type Ca^{2+}-ATPase from *Flavobacterium odoratum*. J Biol Chem 267:15923–15931
11. Shemarova IV, Nesterov VP (2005) Evolution of mechanisms of calcium signaling: the role of calcium ions in signal transduction in prokaryotes. Zh Evol Biokhim Fiziol 41:12–17
12. Shemarova IV, Nesterov VP (2005) Evolution of Ca^{2+} signaling mechanisms. Role of calcium ions in signal transduction in lower eukaryotes. Zh Evol Biokhim Fiziol 41:303–313
13. Petersen OH, Michalak M, Verkhratsky A (2005) Calcium signalling: past, present and future. Cell Calcium 38:161–169
14. Berridge M, Lipp P, Bootman M (1999) Calcium signalling. Curr Biol 9:R157–R159
15. Berridge MJ, Bootman MD, Roderick HL (2003) Calcium signalling: dynamics, homeostasis and remodelling. Nat Rev Mol Cell Biol 4:517–529
16. Kostyuk P, Verkhratsky A (1995) Calcium signalling in the nervous system. Wiley, Chichester
17. Petersen OH, Tepikin AV (2008) Polarized calcium signaling in exocrine gland cells. Annu Rev Physiol 70:273–299
18. Petersen OH, Petersen CC, Kasai H (1994) Calcium and hormone action. Annu Rev Physiol 56:297–319
19. Carafoli E (2002) Calcium signaling: a tale for all seasons. Proc Natl Acad Sci USA 99: 1115–1122
20. Carafoli E (2004) Calcium-mediated cellular signals: a story of failures. Trends Biochem Sci 29:371–379
21. Carafoli E, Santella L, Branca D, Brini M (2001) Generation, control, and processing of cellular calcium signals. Crit Rev Biochem Mol Biol 36:107–260
22. Toescu EC, Verkhratsky A (1998) Principles of calcium signalling. In: Verkhratsky A, Toescu EC (eds) Integrative aspects of calcium signalling. Plenum, New York/London, pp 2–22
23. Gerasimenko OV, Gerasimenko JV, Belan PV, Petersen OH (1996) Inositol trisphosphate and cyclic ADP-ribose-mediated release of Ca^{2+} from single isolated pancreatic zymogen granules. Cell 84:473–480
24. Michelangeli F, Ogunbayo OA, Wootton LL (2005) A plethora of interacting organellar Ca^{2+} stores. Curr Opin Cell Biol 17:135–140
25. Ikura M, Osawa M, Ames JB (2002) The role of calcium-binding proteins in the control of transcription: structure to function. Bioessays 24:625–636
26. Lewit-Bentley A, Rety S (2000) EF-hand calcium-binding proteins. Curr Opin Struct Biol 10:637–643
27. Jones VC, McKeown L, Verkhratsky A, Jones OT (2008) LV-pIN-KDEL: a novel lentiviral vector demonstrates the morphology, dynamics and continuity of the endoplasmic reticulum in live neurones. BMC Neurosci 9:10
28. Mogami H, Nakano K, Tepikin AV, Petersen OH (1997) Ca^{2+} flow via tunnels in polarized cells: recharging of apical Ca^{2+} stores by focal Ca^{2+} entry through basal membrane patch. Cell 88:49–55
29. Park MK, Petersen OH, Tepikin AV (2000) The endoplasmic reticulum as one continuous Ca^{2+} pool: visualization of rapid Ca^{2+} movements and equilibration. EMBO J 19:5729–5739
30. Petersen OH, Verkhratsky A (2007) Endoplasmic reticulum calcium tunnels integrate signalling in polarised cells. Cell Calcium 42:373–378

31. Subramanian K, Meyer T (1997) Calcium-induced restructuring of nuclear envelope and endoplasmic reticulum calcium stores. Cell 89:963–971
32. Terasaki M, Slater NT, Fein A, Schmidek A, Reese TS (1994) Continuous network of endoplasmic reticulum in cerebellar Purkinje neurons. Proc Natl Acad Sci USA 91:7510–7514
33. Verkhratsky A (2005) Physiology and pathophysiology of the calcium store in the endoplasmic reticulum of neurons. Physiol Rev 85:201–279
34. Alonso MT, Barrero MJ, Michelena P, Carnicero E, Cuchillo I, Garcia AG, Garcia-Sancho J, Montero M, Alvarez J (1999) Ca^{2+}-induced Ca^{2+} release in chromaffin cells seen from inside the ER with targeted aequorin. J Cell Biol 144:241–254
35. Mogami H, Tepikin AV, Petersen OH (1998) Termination of cytosolic Ca^{2+} signals: Ca^{2+} reuptake into intracellular stores is regulated by the free Ca^{2+} concentration in the store lumen. EMBO J 17:435–442
36. Solovyova N, Verkhratsky A (2002) Monitoring of free calcium in the neuronal endoplasmic reticulum: an overview of modern approaches. J Neurosci Methods 122:1–12
37. Tse FW, Tse A, Hille B (1994) Cyclic Ca^{2+} changes in intracellular stores of gonadotropes during gonadotropin-releasing hormone-stimulated Ca^{2+} oscillations. Proc Natl Acad Sci USA 91:9750–9754
38. Michalak M, Robert Parker JM, Opas M (2002) Ca^{2+} signaling and calcium binding chaperones of the endoplasmic reticulum. Cell Calcium 32:269–278
39. Mogami H, Gardner J, Gerasimenko OV, Camello P, Petersen OH, Tepikin AV (1999) Calcium binding capacity of the cytosol and endoplasmic reticulum of mouse pancreatic acinar cells. J Physiol 518:463–467
40. Petersen OH, Tepikin A, Park MK (2001) The endoplasmic reticulum: one continuous or several separate Ca^{2+} stores? Trends Neurosci 24:271–276
41. Nicholls DG (2005) Mitochondria and calcium signaling. Cell Calcium 38:311–317
42. Toescu EC (2000) Mitochondria and Ca^{2+} signaling. J Cell Mol Med 4:164–175
43. Nicotera P, Orrenius S (1998) The role of calcium in apoptosis. Cell Calcium 23:173–180
44. Pankratov Y, Lalo U, Krishtal OA, Verkhratsky A (2009) P2X receptors and synaptic plasticity. Neuroscience 158:137–148
45. Voets T, Janssens A, Droogmans G, Nilius B (2004) Outer pore architecture of a Ca2+-selective TRP channel. J Biol Chem 279:15223–15230
46. Bezprozvanny I, Ehrlich BE (1995) The inositol 1, 4, 5-trisphosphate (InsP3) receptor. J Membr Biol 145:205–216
47. Galione A, Ruas M (2005) NAADP receptors. Cell Calcium 38:273–280
48. Galione A, Petersen OH (2005) The NAADP receptor: new receptors or new regulation? Mol Interv 5:73–79
49. Hamilton SL (2005) Ryanodine receptors. Cell Calcium 38:253–260
50. Franzini-Armstrong C, Protasi F (1997) Ryanodine receptors of striated muscles: a complex channel capable of multiple interactions. Physiol Rev 77:699–729
51. Cancela JM, Van Coppenolle F, Galione A, Tepikin AV, Petersen OH (2002) Transformation of local Ca^{2+} spikes to global Ca^{2+} transients: the combinatorial roles of multiple Ca^{2+} releasing messengers. EMBO J 21:909–919
52. Kirichok Y, Krapivinsky G, Clapham DE (2004) The mitochondrial calcium uniporter is a highly selective ion channel. Nature 427:360–364
53. Graier WF, Frieden M, Malli R (2007) Mitochondria and Ca^{2+} signaling: old guests, new functions. Pflugers Arch 455:375–396
54. Guerini D, Coletto L, Carafoli E (2005) Exporting calcium from cells. Cell Calcium 38:281–289
55. Vangheluwe P, Raeymaekers L, Dode L, Wuytack F (2005) Modulating sarco(endo) plasmic reticulum Ca^{2+} ATPase 2 (SERCA2) activity: cell biological implications. Cell Calcium 38:291–302
56. Venetucci LA, Trafford AW, O'Neill SC, Eisner DA (2007) Na/Ca exchange: regulator of intracellular calcium and source of arrhythmias in the heart. Ann N Y Acad Sci 1099:315–325
57. Kirischuk S, Kettenmann H, Verkhratsky A (2007) Membrane currents and cytoplasmic sodium transients generated by glutamate transport in Bergmann glial cells. Pflugers Arch 454:245–252
58. Putney JW Jr, Thomas AP (2006) Calcium signaling: double duty for calcium at the mitochondrial uniporter. Curr Biol 16:R812–R815
59. Fryer MW, Zucker RS (1993) Ca^{2+}-dependent inactivation of Ca^{2+} current in *Aplysia* neurons: kinetic studies using photolabile Ca^{2+} chelators. J Physiol 464:501–528
60. Kasai H, Aosaki T (1988) Divalent cation dependent inactivation of the high-voltage-activated Ca-channel current in chick sensory neurons. Pflugers Arch 411:695–697

61. Burdakov D, Petersen OH, Verkhratsky A (2005) Intraluminal calcium as a primary regulator of endoplasmic reticulum function. Cell Calcium 38:303–310
62. Li Y, Camacho P (2004) Ca^{2+}-dependent redox modulation of SERCA 2b by ERp57. J Cell Biol 164:35–46
63. Moreau B, Parekh AB (2008) Ca^{2+} -dependent inactivation of the mitochondrial Ca^{2+} uniporter involves proton flux through the ATP synthase. Curr Biol 18:855–859
64. Moreau B, Nelson C, Parekh AB (2006) Biphasic regulation of mitochondrial Ca^{2+} uptake by cytosolic Ca^{2+} concentration. Curr Biol 16:1672–1677
65. Verkhratsky A (2009) Neuronismo y reticulismo: neuronal-glial circuits unify the reticular and neuronal theories of brain organization. Acta Physiol (Oxf) 195:111–122
66. Barclay JW, Morgan A, Burgoyne RD (2005) Calcium-dependent regulation of exocytosis. Cell Calcium 38:343–353
67. Burnashev N, Rozov A (2005) Presynaptic Ca^{2+} dynamics, Ca^{2+} buffers and synaptic efficacy. Cell Calcium 37:489–495
68. Katz B, Miledi R (1967) Ionic requirements of synaptic transmitter release. Nature 215:651
69. Llano I, Gonzalez J, Caputo C, Lai FA, Blayney LM, Tan YP, Marty A (2000) Presynaptic calcium stores underlie large-amplitude miniature IPSCs and spontaneous calcium transients. Nat Neurosci 3:1256–1265
70. Yang F, He XP, Russell J, Lu B (2003) Ca^{2+} influx-independent synaptic potentiation mediated by mitochondrial Na^{+}-Ca^{2+} exchanger and protein kinase C. J Cell Biol 163:511–523
71. Bading H (2000) Transcription-dependent neuronal plasticity the nuclear calcium hypothesis. Eur J Biochem 267:5280–5283
72. Bliss TV, Collingridge GL (1993) A synaptic model of memory: long-term potentiation in the hippocampus. Nature 361:31–39
73. Hardingham GE, Arnold FJ, Bading H (2001) Nuclear calcium signaling controls CREB-mediated gene expression triggered by synaptic activity. Nat Neurosci 4:261–267
74. West AE, Chen WG, Dalva MB, Dolmetsch RE, Kornhauser JM, Shaywitz AJ, Takasu MA, Tao X, Greenberg ME (2001) Calcium regulation of neuronal gene expression. Proc Natl Acad Sci USA 98:11024–11031
75. Araque A, Parpura V, Sanzgiri RP, Haydon PG (1999) Tripartite synapses: glia, the unacknowledged partner. Trends Neurosci 22:208–215
76. Verkhratsky A, Steinhauser C (2000) Ion channels in glial cells. Brain Res Brain Res Rev 32:380–412
77. Verkhratsky A, Kettenmann H (1996) Calcium signalling in glial cells. Trends Neurosci 19:346–352
78. Cornell Bell AH, Finkbeiner SM, Cooper MS, Smith SJ (1990) Glutamate induces calcium waves in cultured astrocytes: long- range glial signaling. Science 247:470–473
79. Volterra A, Meldolesi J (2005) Astrocytes, from brain glue to communication elements: the revolution continues. Nat Rev Neurosci 6:626–640

Chapter 2

Ca^{2+} Recordings: Hardware and Software (From Microscopes to Cameras)

Emil C. Toescu and Jeremy Graham

Abstract

From an early start, more than two decades ago, Ca^{2+} measurements have evolved from the use of simple systems, built around an epifluorescent microscope, a fluorescent lamp and a photomultiplier, into highly complex set-ups exploiting solid-state light sources and Electron Multiplied cameras to capture in five dimensions with time resolutions from milliseconds to days. In addition to these technological advances, a series of conceptual breakthroughs have enabled microscopy to move well beyond the classical diffraction limit into the realm of optical nanoscopy. Further developments into miniaturization of optical components and advances in optical fibres are bringing Ca^{2+} imaging nearer to intra-vital microscopy applications. In this chapter, we review the basic hardware requirements for Ca^{2+} imaging setups, discuss the latest optical and technological developments and look at the future directions that promise to bring Ca^{2+} imaging to a more central place in the medicine of the future.

Key words: Numerical aperture, Point spread function, Microscope objective, Fluorescence microscopy, Cameleon, Wide-field microscopy, Confocal imaging, Nipkow disks, Confocal line scan microscopy, Total internal reflection fluorescence microscopy, Fluorescence resonance energy transfer, Forster resonance energy transfer, Fluorescence life time microscopy, Time-correlated single-photon counting, Fluorescence recovery after photobleaching, Fluorescence loss in photobleaching, 4 Pi microscopy, Stimulated emission-depletion microscopy, Multiphoton confocal microscopy, Two photon confocal microscopy, Intravital microscopy, Fibre optic microscopy, Halogen lamps, Arc lamps, Light emitting diodes, Gas laser, Solid state laser, Dye lasers, Monochromators, Acousto-optical tunable filters, Photobleaching, Signal to noise ratio, Dark noise, Read noise, Statistical noise, Spatial resolution, Temporal resolution, Photon counting, Photomultiplier, Photodiode, Charge coupled device camera, CCD camera, Frame transfer, Interline transfer, Electron multiplication CCD camera, Image intensification, Microchannel plate, CMOS camera, Gradient index lens technology

A. Verkhratsky and Ole H. Petersen (eds.), *Calcium Measurement Methods*, Neuromethods, vol. 43,
DOI 10.1007/978-1-60761-476-0_2, © Humana Press, a part of Springer Science+Business Media, LLC 2010

1. Specifics and Dynamics of Ca^{2+} Measurements

Upon stimulation, many cells generate a signal, transducing the extracellular stimulus into an intracellular signal. Calcium ions, in addition to cyclic AMP, are typical such second messengers that can either activate directly various effectors (e.g. muscle contraction proteins or membrane bound ionic channels) or, indirectly, trigger various downstream intracellular signalling pathways, the third messengers, resulting in many instances in the activation of various kinases (1). An important functional feature of the second messenger signalling is that the signal is intrinsically transient – it either triggers or activates down-stream targets or processes, or is buffered away. What the cell cannot do is to preserve the signal as such, and from this follows an important requirement for live, immediate, real-time measurements. Transmission of information through the Ca^{2+} signalling pathways exhibits various modalities of coding. In the temporal domain, earlier studies indicated both frequency (FM) and amplitude modulation (AM) of the Ca^{2+} signals (2). In FM mode, $[Ca^{2+}]_i$ signalling modulates secretion (3), glycogen metabolism in hepatocytes (4), or neuronal axonal growth (5). An interesting observation reported a functional coupling between FM modulation of Ca^{2+} signals and CaM kinase II (a Ca^{2+}-calmodulin-dependent protein kinase with an important role in controlling the transmission of information in the post-synaptic element) (6). In lymphocytes, the AM mode, triggered by different agonists, is able to code for differential gene activation with a dramatic difference in functional outcomes (7). Another modality of information coding through Ca^{2+} signalling takes place in the spatial domain, with small, defined and localised releases of Ca^{2+} from a variety of intracellular Ca^{2+} release sites generating functional microdomains, with significant functional repercussions (2, 8, 9).

All these results just illustrate the importance and need for measuring the changes in Ca^{2+} ion concentration, ideally with a high temporal and spatial resolution. The tools best suited for such measurements, in real time and space, are the Ca^{2+} sensitive probes – either chemical compounds, derived from Ca^{2+} buffers, or photoproteins. They differ in the modality of light emission – the former are fluorescent, while the latter are light emitting (luminescent). A more recent event has been the development of genetic tools to produce truly multifunctional proteins that can bind Ca^{2+} and fluoresce and that can also be targeted specifically to various regions and compartments of cells. For the development and application of one such family of protein probes, the green-fluorescent protein (GFP), Roger Tsien was awarded (together with Osamu Shimomura and Martin Chalfie) the 2008

Nobel Prize for Chemistry (10). To indicate, maybe, the versatility of these probes, this family of proteins was called the "cameleon" (11). It is also important to note that the Ca^{2+}-sensitive proteins (both the photoproteins and the variants of chameleon) have an important advantage over the chemical compounds: the latter are based on chemical Ca^{2+} buffers, and are thus interfering with the actual signal they are supposed to monitor. Further specifics about the types of Ca^{2+} sensitive probes and their specific properties can be found in the other chapters in this book.

2. Microscopy and Objectives

The resolving power of a microscope is limited by the wave-like nature of the electromagnetic radiation (light) and, as originally pointed out in 1873 by E. Abbe, a collaborator of C. Zeiss in Jena, the resolution of optical microscopy is limited, in principle, by diffraction of the exploring, incident light. Every microscope objective is characterised by its *numerical aperture* (NA), which describes the amount of light, coming from the focal plane that the objective can collect. This parameter is defined as the product of the refractive index (n_i) of the medium interposed between lens and specimen (1 for air, 1.333 for water, 1.47 for glycerol and up to 1.5 for immersion oil, etc) and the sine of the half-angle (α) of the lens aperture.

$$\mathrm{NA} = n_{\mathrm{i}} \sin(\alpha) \tag{2.1}$$

Since $\sin(\alpha)$ will always be less than 1 (the practical upper limit for an objective is about 144°, resulting in an α of 72°, with a sine value of 0.95), the NA will always be lower than the refractive index of the medium. This NA value is one of the most important criteria to be considered when selecting an objective.

Any optical device, such as a microscope objective, used for visualising a point source is also characterised by a *point spread function (PSF)*, which measures the degree of blurring in the image of a point source. As a result of the spreading, a point source becomes a disc, when viewed through a lens; if two point sources of light in the object are too close together, their overlapping discs will prevent their identification as independent points. This process determines the limit of resolution of an objective, which is quantified by the Rayleigh criterion (1879), such that the lateral resolution (in the *xy* plane) is given by:

$$R_{(\mathrm{x,y})} = 0.61\,\lambda\,/\,\mathrm{NA} \tag{2.2a}$$

where λ is the wavelength of the light. Thus, for a fluorescence measurement with a water immersion objective with a good NA, such as the Olympus PlanApochromat WI60x, with a NA 1.2, for fura-2 signals (emission λ 500 nm), the resulting lateral resolution would be, in ideal optical alignment and conditions, about 250 nm.

Clearly, what applies in the *xy* plane, should apply also in the axial (z plane), with the corresponding resolution power given by

$$R_{(z\ \text{axial})} = 2\lambda / \text{NA}^2 \tag{2.2b}$$

This would result, for the same objective as above, a theoretical axial resolution of 700 nm.

Apart from the optical resolution issues, modern objectives used for microfluorescence deal with other types of optical aberrations that can affect images. Some of these are subject to differences in the way that light of different wavelength interacts with the lens material, resulting in chromatic aberrations which depend on the nature of that material. There are also monochromatic aberrations, that result either from failures of a point source to form a point image (e.g. spherical aberrations) or distortions, whereby objects, with an optical plane perpendicular to the principal axis of the lens, fail to generate a corresponding perpendicular image plane, resulting in field curvature and shape distortions.

Chromatic aberrations are caused by a given lens or prism material refracting (i.e. bending) light of one wavelength (colour) to a different extent than for other wavelengths. The effect is to disperse white light into its constituent wavelengths (colours), a process first described in by Newton's optics following his prism experiments. For lenses, the differential refraction causes colour separation and colour fringes in the images. The simplest type of, achromat lenses correct for axial chromatic aberration at just two wavelengths (blue and red, 480 and 650 nm), bringing these extremes into focus in a common plane. These lenses are acceptable when using black-and white detection systems, and their performance can be increased by using a green interference filter, but they show their significant limitations when used for colour visualisation and photography. Intermediate in quality are the semi-apochromatic lenses (e.g. fluorite objective, produced from special glasses that contain fluorspar (fluoride, or Ca fluoride, which has one of the lowest natural light dispersion coefficients) or equivalent synthetic materials). The highest specification objectives for chromatic aberrations are the apochromat series; these are corrected for all three major colours (blue, green and red) and have the highest NA values. These are now available with a fourth correction point at 405 nm (Nikon CFI Plan Apochromat VC Series or Olympus super plan).

Another basic aberration of any optical system is the *spherical aberration* that results from the fact that the light rays which pass through the edges of a perfectly spherical lens are bent to a greater extent than the near-axial rays, and so they come to a focus which is closer to the lens than the focus of the near-axial rays. This out-of-focus effect can be corrected by a judicious use either of combinations of convex and concave lens elements or of aspheric lenses, resulting in the elaborated apochromat objectives.

A modern objective used for high-end fluorescence microscopy will also be able to handle the *field curvature* problems. Field curvature is the failure of a plane object perpendicular to the principal axis to create an image in a plane. Instead, the image is in focus on a curved surface (Petzval's surface, named after the discoverer of this aberration, in the 1840s, working on improvements on the early photographic equipment of Daguerre). The practical result of this aberration is that the edges of the image are fuzzy, particularly at lower magnifications. The effect can be reduced by restricting the area of the field of view with an iris. Optical correction for field curvature is difficult and requires addition of several lens groups in the objective, with a resultant significant decrease in the length of the working distance and major increase in the prices of such objectives (plan apochromat).

In epifluorescence (i.e. reflected fluorescence, requiring the use of a dichroic mirror), the measured emitted fluorescence light is only a fraction of the incident excitation light, and thus the light transmission properties of the objective become very important. In general, in transmitted light mode, image intensity is proportional to the square of NA of the objective and condenser and inversely proportional to the square of the objective's magnification (M). In epifluorescence, the objective acts also as a condenser for the excitation light and thus the image intensity becomes proportional to NA^4 while remaining inversely proportional to M^2. An important practical consideration emerging from this relationship is the need to choose the appropriate magnification for the process under investigation. This is because too powerful an objective lens, despite its having the optimal NA value, could result in an unnecessary decrease in the measured light intensity.

Another important issue to consider is the autofluorescence of various types of glass. With Ca^{2+} measurements dependent to a large extent on the ratiometric properties of fura-2, excitable in the UV range (340 and 380 nm), there is a stringent requirement for low fluorescence glass or quartz, specialised optical cements and anti-reflection coatings, to minimise the background fluorescence "noise" and to maximise light transmission. This, in turn, requires low autofluorescent glass or quartz, specialised optical cements and anti-reflection coatings.

A limitation of these high-performance fluorescence objectives is that many of them do not provide the space required to house the optical elements needed to correct the field curvature, resulting in images that do not have uniform focus throughout the entire field of view. This becomes a major issue when the objective is utilised for conventional illumination techniques such as brightfield, darkfield, and differential interference contrast.

However, the continuous development of Ca^{2+} probes based on fluorescent proteins that fluoresce in the visible spectrum, allows for the use of more conventional objectives.

The increased performance of the current generation of objectives as well as their capacity for multi-tasking (e.g. microfluorimetry and differential interference contrast (DIC) or phase contrast optics) has been facilitated by the adoption of infinity-corrected optics. In such systems, the objective generates a flux of parallel lightwaves, imaged at infinity (in contrast to the traditional fixed focal optics, in which the objective generates light trains that converge towards the image plane). These parallel lightwaves, travelling in the "infinity space" are then collected by a tube lens that is placed within the body of the microscope and focused directly on to a detector or to an intermediate plane, which is then visualised by the eyepiece oculars. (side and bottom ports on inverted microscopes usually produce a primary image, trinocular heads on upright microscopes use an intermediate plane). The parallel, rather than converging nature of the lightwaves in these objectives allows the introduction in the "infinity space", between the objective and the tube lens, of complex optical components without the introduction of new optical aberrations or modifications of the objective's working distance. Furthermore, such optical accessories can be designed so that they produce a 1× magnification without altering the alignment between the objective and tube lens.

3. Techniques of Fluorescent Measurements (with Emphasis on Ca^{2+} Measurements)

The various techniques for measuring fluorescence signals, using either wide field or confocal microscopy, are by now extremely well established and routine in many laboratories in the world. As a result, a description of the general concepts and principles of both fluorescence measurements and confocal microscopy are now better found in monographs rather than that in journal reviews (12). At the same time, at the sharp end, these methodologies are continuously developing on all levels, pushed by technological advances for various components (optical, mechanical or electronic), significant developments in the conceptual approaches and by the development of new fluorophores with chemical properties which are specific for focussed applications. The combination of various technologies with new ideas has led to the development of specific new modalities of imaging. It is beyond the scope of this chapter to address systematically and in detail all these various combinations, and many will be covered in the other chapters of this book. Here, we would like to attempt to provide a very broad categorisation of all these various imaging approaches, mainly on the basis of the type of information or scientific questions that are asked.

3.1. 3D Information, Morphological or Functional

This group of techniques is probably the largest, accounting for most of the published applications of fluorescence imaging. At one end of the spectrum is a *wide-field* "classical" microscopy, in which all the field of view is illuminated and the resulting signal is captured. In the very first implementations, in the early 1980s, the fluorescence signal was captured by photomultiplier detectors, but later, the development of sensitive cameras allowed capturing of spatial information. Such camera-based systems were starting to enter the labs interested in Ca^{2+} signalling by late 1980s, and this technology generated many thousands of papers. Its principles, the means and variation of implementation and advantages and limitations have been reviewed extensively in book format (e.g. (13–15)), including detailed comparisons with other imaging techniques, such as confocal imaging (e.g. (16)). If the experimental model is simple, using monolayers of cells (acutely isolated or cultured), if the interest is in the general spatial localization at the level of large "regions of interest" (ROIs) in the field of view (rather than subcellular localization of signals, at the level of microdomains or individual organelles), then the wide-field imaging is an extremely robust and powerful technique. Because it requires only a basic setup (good quality fluorescence microscope (upright or inverted), light source with the possibility of control of the excitation wavelength, filter-controlled emission pathway and a good quality digital CCD camera), it is much less expensive and easier to implement than other imaging approaches. It is also flexible enough to allow integration with a variety of other physiological techniques (e.g. electrophysiology) and all these characteristics explain why it still forms a very large majority, probably more than 75%, of the Ca^{2+} imaging setups used in the world.

Confocal microscopy (CLSM) is a relatively old technology (the first patent on it, by Marvin Minsky, was applied for in 1957 and granted in 1961), but which has entered the range of standard imaging technologies only in the last 20 years. Its optical principle is the use of point illumination and of a pinhole in an optically conjugate plane placed in front of the detector. Only the light within the focal plane can be detected with the out-of-focus information discarded, so that image quality is much better than that of the wide-field images. An important consequence of this approach is that only one point is illuminated at a time, and thus the generation of images requires scanning over a regular raster (i.e. a rectangular pattern of parallel scanning lines) in the specimen, a procedure that introduces obvious time delays. Spinning-disk approaches (based on the use of Nipkow-disks with appropriately spaced array of pin-holes) can be faster (17), and some have incorporated this technology for use in *fluorescent speckle microscopy (FSM)* (18). Advances in the conceptual approach to the photon activation of the fluorescent target molecule led to the development of *two-photon* and *multi-photon excitation microscopy*, which will be discussed in further detail in the last section of the chapter.

3.2. Collecting Information from (Quasi) Planar Level, 2D Information

There are several approaches that can be used to investigate fluorescent signals generated from point sources or from various cellular surfaces. Some of these methods allow also investigations in very thin cellular volumes. *Total internal reflection fluorescence microscopy (TIRFM)*, was a result of challenging the traditional way of performing microscopy using light paths perpendicular to the object plane. If the sample is illuminated at a very shallow angle, the excitation light will be completely reflected off the interface between the coverslip and the sample. No photons penetrate this interface, but a thin evanescent field of excitation light is generated that excites fluorophores within only 100 nm or less from the surface of the object investigated. Using such techniques (as discussed in a recent methods review from Rutter's group (19), it has been possible not only to study the dynamics of exocytosis but also to develop "optical patch-clamp recording" that probes Ca^{2+} channel activity and gating properties, while also visualising the generation of various functional Ca^{2+} microdomains (20).

Another method that allows capturing information from small regions or surfaces is the *fluorescence* (or Förster) *resonance energy transfer (FRET)* microscopy. In the late 1940s, the German physicist Thomas Förster described the non-radiating transfer of energy from one dye to another over the range of molecular distances (less than 10 nm). The final acceptor molecule could then release the energy in the form of a photon, at a specific wavelength, that can be monitored (a typical example involves a derivatives of green fluorescent protein, cyan FP (CFPs), that will absorb excitation at 440 nm and pass on, under the right chemical and stereological conditions, the energy to a yellow FP (YFPs) that will eventually fluoresce at around 530 nm) (21). Although many of the current biological applications of FRET (or its bioluminescent equivalent BRET) study protein structure, protein–protein or protein–membrane interactions (22), including studies in Alzheimer disease (23), this technique had been used in one of the many seminal contributions of R. Tsien to intracellular Ca^{2+} signalling in developing the cameleons, chimeric proteins consisting of a blue or cyan mutant of green fluorescent protein (GFP), calmodulin (CaM), a glycylglycine linker, the CaM-binding domain of myosin light chain kinase (M13), and a green or yellow version of GFP (24).

3.3. Collecting Information in the Temporal Domain (High Temporal Resolution)

The approach used by these methods touches on the fundamentals of the fluorescence process. Starting from the fact that each fluorescent dye has its own specific lifetime in an excited state, *fluorescence life-time imaging (FLIM)* measures the exponential decay of this fluorescence from which a specific time constant can be determined (for many fluorescent dyes used in biology, these lifetimes are in the nanosecond range). Since the rate of fluorescence decay is affected by a variety of factors, FLIM measurements can provide important information about such local conditions and factors that include: ion activity, hydrophobic properties, oxygen concentration, molecular binding, and molecular interaction (such as the energy transfer possible between two proteins approaching each other). At the same time, lifetime decay is independent of dye concentration, photobleaching, light scattering and excitation light intensity (25).

FLIM measurements can be performed in two different modes: in the time-domain and in the frequency-domain. Time-domain FLIM uses time-correlated single-photon counting (TCSPC), in which a laser with an ultra-short pulse duration of a few hundred picoseconds and a nanosecond-level shutter are coupled with highly sensitive detectors that record photon arrivals precisely timed against the excitation pulse time. With such approaches, rapid quantitative measurements of Ca^{2+} signals are possible, even using single wavelength Ca^{2+} dyes (26), while new Ca^{2+} specific dyes, suitable for FLIM measurements continue to be produced (27). The other modality of lifetime measurements, the frequency-domain FLIM, uses the effect on the fluorescence decay of phase shift modulation in the excitation light.

4. Generating "Light Input" (Light Sources)

Calcium fluorescence measurements require the delivery of photons of a defined wavelength to a biological preparation in order to excite a specific indicator to emit photons at a lower energy (longer wavelength).

When choosing a suitable photon source, many of the considerations are obvious: studying fast dynamics or small signal changes require a light source that is both intense and stable; whereas high resolution imaging demands a uniform field. There follows a discussion of the principal criteria in choosing a light source and wavelength selection device followed by an overview of the currently available equipment and how well it meets the demands.

4.1. Key Considerations

4.1.1. Intensity

In calcium studies, the excitation of fluorescence dyes is usually achieved episcopically with incident illumination through the microscope objective. This means that the intensity at the specimen is dependent on the number of photons that can be captured by the objective lens from the light source. Fluorescence objectives have a high NA (see above) for efficient capture of the light emitted from the specimen, but conversely, they have a low NA as seen by the light source. Consequently, the light entering the back aperture of the objective needs to be relatively well collimated in order to reach the sample. A subtler point regarding light sources is the difference between radiance, which defines the amount of light being emitted in a particular direction, at a given angle and intensity, which defines the amount of light being generated from the total surface of the light source. In practical terms, the radiance (point intensity) of the light source is more important than its total optical power (intensity).

4.1.2. Spatial Uniformity

If detailed spatial information is required, then the specimen needs to be evenly illuminated; at least across the field of view captured by the detector. The light source is usually optically coupled to provide Kolher illumination, wherein a magnified image of the source is formed at the back aperture of the objective and consequently defocussed at the specimen (i.e. the filament of the light source is not visible in the sample plane). If the emitting source itself is uniform (e.g. a fibre-optic with multiple bends (28)), then the so-called "critical illumination" (i.e. the simpler, pre-Kohler, method of illumination) can be used, with the light source in focus at the specimen plane. In practice, the nature of the light source itself will still have an effect on the evenness of illumination and in demanding applications, especially 3D deconvolution, great care must be taken to optimise the field uniformity.

4.1.3. Stability

If the fluorescence measurements are quantitative, especially if the anticipated changes in fluorescence are small, then the stability of the light source becomes paramount. It is important to consider the temporal resolution of the system when assessing the acceptable noise levels. With slow capture rates and gradual changes in biological fluorescence, any high frequency noise is likely to be filtered by the detection system. Noise is worse if it has a similar frequency to real biological changes, which in many cases makes any 50 Hz bleedthrough from poorly filtered mains a serious problem.

4.1.4. Wavelength and Bandwidth

Light sources are either broad spectrum, such as arc lamps, or have discrete wavelengths as is the case with Light Emitting Diodes (LEDs) and Lasers. Even with discrete emitters, it is usually necessary to "clean up" the spectral output of the source using either a difraction grating or more commonly an interference filter.

Choosing the appropriate wavelength and bandwidth of the optical filtering will depend on the excitation and emission spectra of the calcium indicator used, and in some cases on the choice of light source itself. The spectra of most dyes can be found on the Internet sites of the providers such as Molecular Probes (Invitrogen).

When selecting the excitation wavelength and bandwidth, the trade-off is between using a broad bandwidth to maximise the signal (more photons) and a narrow bandwidth to maximise the dynamic range of any changes in fluorescence (using the dye as close as possible to its peak emission wavelength). This process is further complicated by any overlap between the excitation and emission spectra of the dye, which sometimes necessitates using an artificially narrow excitation band so that the emission band can be kept as broad as possible. Keeping the bandwidth low also helps to reduce the relative autofluorescence, which is unlikely to have the same spectral peak as the indicator in question.

4.1.5. Shuttering and Switching

The detectable signal from a fluorescence dye will tend to decrease with exposure to excitation light as the fluorophore loses its ability to fluoresce. In order to prevent or minimise this chemical photobleaching process, it is important to minimise the exposure of the specimen to incident photons (29). To achieve this, the illumination source will usually be fitted with a shutter. Electromechanical shutters, where the blades physically move into and out of the light path, have the advantage of 100% contrast ratio, but they can introduce vibrations, and will have a time delay as the blades move. Solid state shutters based on acousto-optical or liquid crystal technology can be arbitrarily fast and are vibration free, but will have reduced contrast ratios and may not be spectrally neutral as the transparent state of the shutter is still in the light path when the shutter is "open".

In addition to switching the light source on and off, it may be necessary to change the spectral properties of the illumination during the experiment. This is usually achieved with an electromechanical wheel fitted with optical filters, but in some cases solid-state devices are used, especially in conjunction with lasers. The same considerations of vibration, speed of response, and transmission efficiency apply as do with shutters.

4.1.6. Minimising or Eliminating Ambient Light

The low light levels associated with fluorescence measurements mean that experiments often need to be carried out in a dark environment. A laboratory with blackout curtains or a Faraday cage with solid panels is generally sufficient, but in some cases, more stringent precautions need to be taken. If using bioluminescence indicators such as aequorin, it is vital that there are no stray photons, and the experimental apparatus should be mounted in a light-tight chamber.

4.2. Light Sources

4.2.1. Halogen Lamps

The majority of research microscopes are fitted with a halogen light source for transmitted illumination, configured for modalities such as DIC, phase-contrast and simple brightfield imaging.

Despite being bright, reliable, inexpensive and stable, these lamps are not routinely used for fluorescence excitation for two reasons. Firstly, they have relatively poor transmission in the ultraviolet and blue sections of the electromagnetic spectrum, where most calcium indicator dyes are used. Also, the light emitting area of a halogen lamp is relatively large and thus it is not always possible to couple it efficiently into the microscope through the objective. This second consideration is more important for high magnification and high NA lenses. Consequently, halogen lamps are mostly used for low maginfication micro and macro imaging.

4.2.2. Arc Lamps

Historically, most epifluorescence microscopes have used mercury or xenon arc lamps as an excitation source. These light sources produce photons by forming an electrical (voltaic) arc between two electrodes, bridged by ionised gas. A high voltage spark is used to ionise the gas and the flow of charge between the electrodes is then maintained with a low voltage DC source. The strength of such lamps is that the emitting area is highly concentrated on one of the electrodes, making them approximate to point sources (the bright point of the arc can be a fraction of a millimeter long). They produce light across the electromagnetic spectrum, with xenon lamps having a relatively flat output and mercury lamps having concentrated emission bands, in particular at 365, 405 and 436 nm, which can be useful for specific fluorophores (see Fig. 2.1). Although mercury lamps have higher point intensity, it is xenon lamps that are preferred for quantitative measurements, since they are more stable (the path that the arc forms between the two electrodes is less prone to wander) and have a longer useful life.

Several manufacturers now offer light sources with metal halide and mixed mercury xenon gases. These lamps offer the peaks of the mercury spectrum, coupled with the stability and life span xenon arc lamps.

The disadvantage of all types of arc lamps is that they produce a lot of heat, they are potentially dangerous and they require high voltage electronics to ignite. For this reason, it is preferable not to have the light source close to the biological preparations, and in most cases, the lamp is coupled to the microscope using a silica or liquid filled optical fibre.

4.2.3. Light Emitting Diodes

Recent developments in light emitting diode (LED) technology have yielded devices that have sufficient radiance to be suitable sources for fluorescence light microscopy. LEDs are electroluminescent devices which emit photons when an electrical current is passed

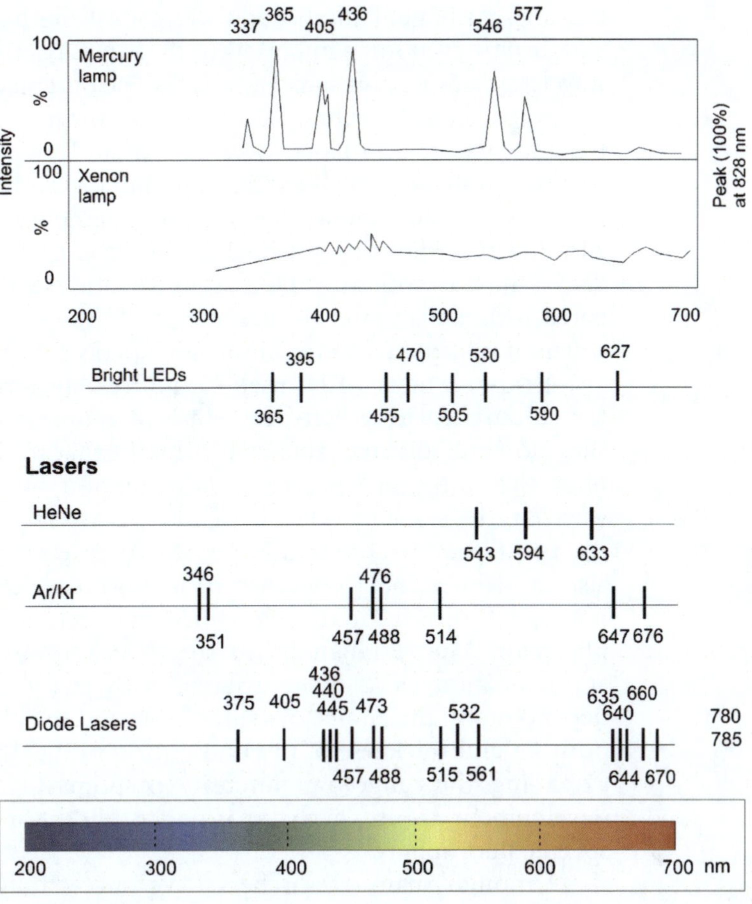

Fig. 2.1. Excitation wavelengths for various light sources. A diagrammatic representation of the emission spectrum of various light sources plotted against a spectrum in which the various wavelengths are colour coded. On top, the emission spectrum of the two main type of arc lamps is illustrated. Below, the emission lines of the most frequently used light emitting diodes (LEDs) is represented. For the lasers, one line is dedicated to illustrating the emissions spectrum for each of the two major types of gas lasers (see text for details). The last line shows the range of individual wavelengths covered by the diode lasers.

through them. They are extremely efficient, converting the applied current into visible or ultraviolet light rather than heat, as the other light sources tend to do. High intensity LEDs are available with small emitting areas of approximately 1 mm square, so they are easy to couple to the microscope objective. They are inexpensive and do not require sophisticated electronics, so are ideal for lab-built and custom applications. The intensity can be adjusted by varying the applied current and they can be rapidly switched off, so there is no need for a separate shutter. Added to this, they have a very long life span and do not suffer from the inherent safety and stability problems of arc lamps.

So, why have they not taken over completely from the other light sources? The main reason for this is that they are only available at a limited number of discrete wavelengths (see Fig. 2.1) and, maybe even more importantly, the intensity at some of these wavelengths is not yet sufficient for fast measurements.

For calcium signalling, there are high power LEDs available to excite fluorescein-based dyes and Indo 1. Although bright LEDs are still not available at 340 nm, Fura 2 can be excited near its isosbestic point and at its longer ratiometric wavelength, thus allowing recordings of fast dual-excitation Ca^{2+} signalling (30). At the time of writing, LEDs tend to perform poorly when compared with arc lamps in the green area of the spectrum, but this situation is likely to change in the not too distant future.

Although most LEDs used for fluorescence measurements are coloured, with a half-bandwidth of approximately 30 nm, they do emit sufficient out-of-band light that it is advisable to filter the output to prevent contamination of the emission wavelengths.

In addition to coloured LEDs, the so-called white LEDs are also available. These comprise of a short-wavelength blue or ultraviolet LED coated with a predominately green emitting phosphor. The resultant broad spectrum output produces an approximation of a white source, with the colour balance depending on the choice of source LED and the chemical composition and thickness of the phosphor coating. In addition to extending the range and intensity of primary LEDs, future development is also likely to produce phosphors tailored to specific fluorophores.

Although many researchers have successfully built their own LED illuminators, there are also commercial systems available from companies such as CoolLED and Cairn Research. These systems are designed to combine multiple LEDs, to couple them efficiently to the microscope and to provide fast and stable control of intensity. Bright (aka, Power) LEDs are available at the following wavelengths 365, 385, 455, 470, 505, 530, 590 and 627 nm (Fig. 2.1).

4.2.4. Lasers

Lasers are the default light source for confocal fluorescence microscopy and are also used for widefield applications where very high intensity is needed. They are distinct from other sources in that they produce light which is coherent (in phase), monochromatic and highly collimated. This coherence can introduce problems for widefield microscopy as the light scattering and diffraction that occurs at all optical surfaces (including apertures) can manifest itself as "laser speckle", and other undesirable artifacts (these effects are averaged out with an incoherent source, and are largely rejected in confocal systems).

A laser acts as a point source and is thus capable of illuminating small areas (diffraction limited, in principle) with extremely high optical power density. This makes it an ideal choice for scanning and spinning disk confocal systems and TIRF applications (see previous sections) where the range of beam angles produced by a non-coherent light sources prevents efficient coupling. The high point intensity and tightly collimated beam mean that lasers should be used with great care as they are capable of causing blindness or burns, and in most cases, the microscope must be made "laser safe" so that it is impossible to expose the naked eye to the beam. The required power range for microscopy is typically between 10 and 200 mW, with most lasers specified in the 25–50 mW range.

There are four main types of laser used in fluorescence microscopy, gas, dye, solid-state and diode lasers, all of which are available within the appropriate power range.

Gas Lasers

There are a variety of gas lasers available, the ones most commonly used in microscopy are Argon-ion and He–Ne (Helium–Neon mix) (Fig. 2.1).

The Argon-ion laser is capable of producing a variety of wavelengths (approximately 25 individual lines between 409 and 686 nm). However, it is most efficient when used for 488 and 514.5 nm, and both of these lines are valuable for fluorescence illumination.

He–Ne is the most common of all the gas lasers, but produces significantly less power than the Argon-Ion type. Usually constructed to produce light at 632.8 nm, they can also be configured to emit at 543.5 nm and also in the infra-red at 1,532 nm. Other common types of gas lasers include, Krypton (647.1 nm), Nitrogen (337.1 nm) as well as various metal vapour lasers which can produce useful wavelengths, e.g. Helium–cadmium lasers, that emit at 442 or 325 nm.

The drawbacks of gas lasers are that they have a fairly limited life time and that they generate a lot of heat, such that cooling and placement of the lasers in the lab environment requires careful consideration (in some instances, a separate laser room has to be built to accommodate them).

Dye lasers

These use a gain medium in a liquid state and are available in a wide range of wavelengths. Unfortunately, many of the dyes used are toxic or carcinogenic and careful handling is required. The dyes themselves have a limited life and generally will require replacing after a few days. In addition to this, the dye cells themselves must be cleaned out weekly to prevent the risk of damaging the cell. This type of laser is not in common use for microscopy, as in general, the support requirements out-weigh the usefulness of having variable wavelengths.

Solid-state lasers:

As the name suggests, these lasers have a gain medium in the solid state which makes them much easier to maintain and handle than either gas or liquid (dye) lasers. They are only available at a limited number of wavelengths with the most commonly used lines being at 460, 488 and 532 nm.

Semiconductor "Diode" lasers

These are fast becoming the most common type of laser used in microscopy. Over the last few years, the range of available wavelengths, the output power and the stability of this type of laser have improved immeasurably. Currently, the available wavelengths are 375, 405, 436, 440, 445, 457, 473, 488, 491, 515, 532, 561, 635, 640, 644, 660, 670, 780 and 785 nm with output energies, from 10 to 100 mW (Fig. 2.1).

Recently, two useful new lasers have been introduced at 355 and 594 nm (manufactured by Cobolt AB). Diode lasers offer many advantages over gas lasers. Not only is there a wide range of wavelengths at useful output power, but they are compact, which means that several can be fitted into a "laser combiner" with a single "multi-mode" fibre optic output. They generate relatively little heat which means there are no special placement requirements and they have very good lifetime, measured in several years of use. Many of the diode lasers can be attenuated directly which allows for simple equipment design; and the remaining types can be passed through an acusto-optical tunable filter (AOTF; see below) which gives rapid wavelength selection, as well as intensity attenuation when required.

4.3. Wavelength selection

Regardless of the choice of light source, it is usually necessary to limit the spectral output so as to prevent contamination of the fluorescence emission and to minimise phototoxicity. In most cases, this is achieved using optical filters, but in some instances a monochromator or acousto-optical tunable filter (AOTF) is used instead.

4.3.1. Optical Filters

Optical filters are available with longpass, shortpass, or bandpass characteristics and function by either reflecting or absorbing the unwanted wavelengths. The most important filters for defining the excitation wavelength are bandpass interference filters, which are designed and manufactured by companies such as Chroma Technology, Semrock and Omega, specifically for use with fluorescence indicators. There are two distinct families of filters, the traditional multi-layer interference filters and the newer "hard" ion beam sputtered (IBS) filters (31); the latter are more robust and can achieve transmission efficiencies of >90%, compared with up to 85% for traditional filters. The multi-layer filters are also still extensively used, since they are available in a wider range of wavelengths and can be manufactured with extremely high flatness for sub-pixel registration of images at different wavelengths. Both types of filters are available with single or multiple spectral bands, allowing them to be used with a variety of fluorescence probes.

In epifluorescence microscopy applications, the excitation light is diverted into the lightpath by a chromatic beamsplitter (dichroic mirror) mounted at 45° (this mirror reflects the excitation light and transmits the longer wavelength emission light). These mirrors are available from the same suppliers as the filters and their spectral properties are quoted, based on their use at 45°.

4.3.2. Wavelength Switching

Most wavelength switching devices work by moving different interference filters into and out of the light path. Stepper motor based instruments manufactured by companies such as Prior Scientific, Sutter Instrument and Ludl Electronic Products can switch wavelengths in approximately 50 ms. Other devices are available from Sutter and IonOptix Corporation which divert the light beam through alternative filters using galvanometer scanners. These have the benefit of allowing much faster switching times in the region of 1 ms. All these devices have the potential to introduce vibrations and so are typically coupled to the microscope using an optical fibre system.

4.3.3. Monochromators

As an alternative to optical filters, monochromators can provide a fast and convenient means of controlling the illumination wavelength (and in some cases bandwidth). Commercial instruments are broadly based on the Czerny–Turner configuration (32), where the light from an arc lamp is focused on an input slit and is reflected by a concave mirror onto a diffraction grating, which disperses the light beam so that all wavelengths leave the grating at different angles. From the grating, the light is focused by a second mirror onto an exit slit, which for convenience is located at the input to an optical fibre. By rotating the grating (using a galvanometer), only the desired wavelength of the spectrum is focused on the exit slit. With this arrangement, changes in the wavelength can be reduced to 1 or 2 ms, and all the commercially available monochromators (TILL Photonics, Photon Technology International, Cairn Research) are capable of this.

A monochromator offers several benefits, in particular fast switching, good ultraviolet output, and the ability to run spectral scans to optimise wavelength and bandwidth for the specific experiment. The drawback is that in most cases, the intensity level will be lower than for a well designed optical filter and the rejection of out of band light is less effective.

4.3.4. Acousto-Optical Tunable Filters and Solid State Shutters

For applications using laser illumination, AOTFs provide an extremely fast means of switching between different wavelengths. The AOTF comprises of a birefringent crystal with difraction properties that are dependent on interaction with an acoustic wave. By changing the acoustic frequency, the wavelength of the first order diffracted beam can be adjusted to pass a particular laser line into a single or multi mode optical fibre (33). The transmission is comparable to that of optical filters and can achieve 85% with switching speeds in the microsecond domain. Fine tuning of the intensity is also possible on the same timescale and the operation is entirely vibration free. The drawback of AOTFs is that they have relatively small apertures (typically 3–5 mm) and their acceptance angle of around 5° makes them only suitable for use with lasers.

4.4. Maximizing Effective Light Signal

The fluorescence emission obtained from a biological signal needs to be captured by a digital or analogue detection system for further assessment and analysis. Maximizing the amount of effective information extracted from the detector involves not only increasing the brightness of the fluorescent image, but also a reduction to a minimal level of the noise in the system. This could be achieved by manipulating the parameters of the light capturing device (see next section), but also by reducing the incoming noise (i.e. extraneous light). When a sample is excited, emitting fluorescence will spread in all directions, which means that signals generated from cells marginal to the field of excitation could contribute to the background light noise signal (the haze). With modern objectives, with a high NA, the increased light captured also means a potential increase in stray light, particularly in wide-field microscopy. One effective solution to this problem is the use of a field diaphragm in the excitation light pathway.

The fact that dichroic beamsplitters and other interference filters are designed for optimal performance on-axis, makes it important to avoid scatter within the microscope as this can lead to out-of-band off-axis rays passing through the system. Ensuring that the microscope optics are free from dirt, and that there are no extraneous optical surface will help to ensure this.

In a discussion about maximising the light signal, photobleaching deserves a special mention, since it can negatively affect the quality of the image, particularly in experiments using confocal techniques that use powerful incident light. As a note, one could argue that the development of two-photon confocal imaging implicitly addressed this problem. The atomic mechanisms of photobleaching are still not entirely clear, but the process manifests itself by a fading of fluorescence over time, when a sample is exposed to the excitation light. It is presumed that this happens because a small, but significant number of the already excited molecules undergo a further photochemical reaction that results in the production of a new molecule which either is not fluorescent, or is not sensitive to the excitation wavelength used (34). A variant of this process is photo-oxidation, which involves chemical reactions between the fluorescent molecule and oxygen, and which is one of the most common mechanisms for the fading of samples during storage. The chemistry of photobleaching or photo-oxidation can generate, as a function of the specific dye used, toxic compounds, leading to photo-toxicity which is typical of the dyes used for mitochondrial staining (35). While an array of antifading agents can be used on fixed specimens, prepared for immunocyto/histochemistry, reduction of photobleaching in physiological imaging experiments on live cells will depend mainly on a judicious assessment of the illumination requirements (light intensity, frequency of exposure, exposure/integration time).

The process of photobleaching has been harnessed, on the other hand, to provide important information on the dynamics of intracellular compartmentalisation, connectivity and diffusional kinetics. One such technique is the Fluorescence Recovery After Photobleaching (*FRAP*), in which a sharply defined region of the specimen is exposed to an intense burst of laser light, with associated rapid photobleaching, accompanied by the subsequent observation of the rates and pattern of fluorescence recovery in the photobleached area (36). Using such techniques, important questions about the heterogeneity of the mitochondrial compartment (37), or the continuity of the sarco(endo)plasmic reticulum with the nuclear envelope (38) have been addressed successfully. A related technique, known as Fluorescence Loss In Photobleaching (*FLIP*), can be used to monitor the decrease of fluorescence in a defined region lying adjacent to a photobleached area. Similar to FRAP, the latter technique is useful in the investigation of molecular mobility and dynamics in living cells, and has been used, for example, to demonstrate the continuity between the dendritic shaft and spine ER (39). As a note for those quick scanners of the literature, this imaging meaning of FLIP should not be confused with the other meaning of FLIP, as in Fluorescent Indicator Proteins (FLIPs) which consist, typically, of a ligand-sensing domain, which is allosterically coupled to a pair of green fluorescent protein (GFP) variants with properties making them suitable for Fluorescence Resonance Energy Transfer (*FRET*). This allows very interesting imaging assessments of carbon-dependent metabolomics of glutamate in brain slices (40).

5. Capturing the Light Signal

5.1. General Characteristics of Photodetectors

The choice of a photodetector for optical calcium measurements needs to be considered with careful reference to the scientific questions being posed. To address this issue, it is useful to consider the fundamental properties of the detector itself. The following discussion is mostly concerned with camera technology, but reference is also made to single point detectors, where such considerations differ.

5.1.1. Sensitivity: Quantum Efficiency and Amplification

The most fundamental measure of sensitivity of a photodetector is its Quantum Efficiency (QE), defined, in a photomultiplier or image intensifier, as the chance of any one incident photon generating a photoelectron (e^-) at the photocathode. In a silicon detector, such as a charge coupled device (CCD) or photodiode, the QE is the chance of an incident photon producing an electron-hole pair, but this is usually also referred to as a photoelectron. Quantum efficiency, expressed as a percentage, is dependent on the wavelength of the incident photon, and thus should ideally be specified by the manufacturer in the form of a spectral graph so that the detector can be chosen to be the best match that emitting the wavelength of the fluorophore (Fig. 2.2). Detectors such as photomultipliers, image intensifiers and electron multiplied (EM) cameras have low-noise amplification stages which also contribute to the effective sensitivity and will be discussed in more detail in the appropriate sections.

5.1.2. Signal to Noise Ratio

When assessing the noise of a photodetector in low-light applications, it is important to bear in mind that the best-case scenario is when the Signal to Noise Ratio (SNR) is *shot noise* limited. Shot noise is the noise signal associated with the uncertainty in the actual number of photons captured from a given intensity of signal. This has a Poisson distribution, and yields a statistical noise (standard deviation) equal to the square root of the average number of photons captured (*N*). In optimal conditions this gives a SNR:

$$\mathrm{SNR} = N / \mathrm{SQRT}(N) = \mathrm{SQRT}(N)$$

where *N* is the number of photons hitting the detector, multiplied by the quantum efficiency.

The easiest way to improve the SNR of an optical signal is to increase the number of photons captured, by either using a

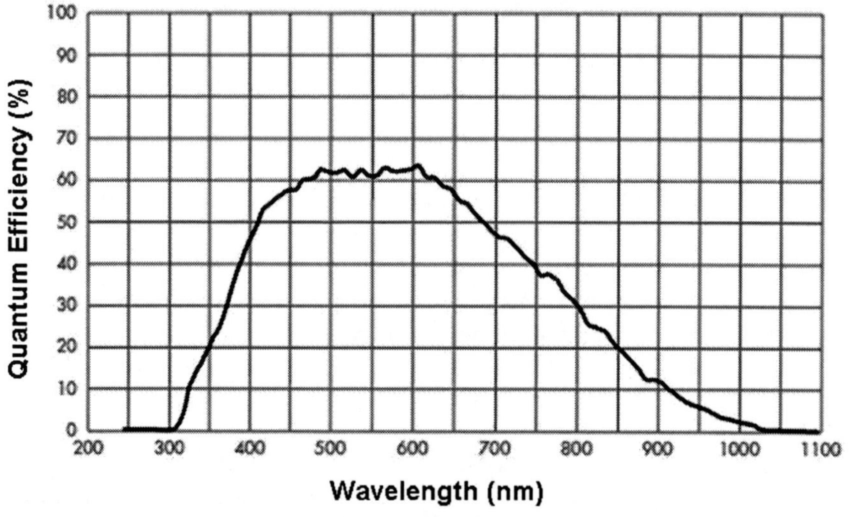

Fig. 2.2. Graph of quantum efficiency as a function of the incident light wavelength. As an example, the image shows the sensitivity of the CoolSNAP Monochrome HQ2 camera produced by Photometrics (and reproduced with permission, from the camera's specification sheet).

brighter illumination source or increasing the acquisition interval. This explains why cheap colour detectors (relatively noisy and insensitive), used in consumer products, can produce sharper images than expensive scientific CCD cameras: they are operating at orders of magnitude higher illumination levels.

To get as close as possible to the shot noise limit, the three major noise sources that should be minimised are: (1) the dark noise (dark current or background signal), (2) the read noise (electronic noise associated with transferring and pre-amplifying the charge) and (3) the statistical noise introduced by gain stages in intensification or electron multiplication.

The *dark noise* (or thermionic noise in a photomultiplier) is caused by vibrational energy in the silicon or photocathode substrate causing detectable photoelectron events in the absence of incident photons. Because this noise source builds up over time and is independent of the illumination level, it presents most of the problems when the long acquisition periods are used at low light levels (e.g. for aequorin luminescence). Fortunately, dark noise is relatively easy to eliminate by cooling the detector. With a silicon detector, the dark current is reduced by an order of magnitude for every 20°C reduction in temperature (41). For calcium fluorescence measurements, cooling to –30°C is usually sufficient to reduce the dark noise to negligible levels, even with long exposures, and for acquisition intervals of <1 s much less intense cooling is required. Photomultipliers vary from tube to tube and so should be selected for low dark current based on the manufacturer's data. Because ambient heat radiation is significant in the infrared region of the electromagnetic spectrum, the dark noise is likely to increase with the sensitivity of the sensor to red light.

Read noise is not usually a problem for single-point detectors such as photomultipliers, but is often significant for CCD cameras, especially at fast readout rates where the system electronics tend to be noisier. Most of the noise is introduced in the preamplifier as the charge carriers are amplified to produce the voltages required for digitisation. The read noise cannot be specified with certainty, but instead it is expressed as an average number of photoelectrons per pixel using the root mean square (RMS) value. Like dark noise, read noise is only a problem at low light levels, but, as it is applied once per acquisition interval (not as a function of time), it is dominant at fast acquisition rates rather than over prolonged exposures, for which dark noise is the main consideration.

A third type of noise, the *statistical noise*, is associated with technologies which multiply each detection event prior to digitisation. In photomultipliers, image intensifiers, and electron multiplied cameras, one or more multiplication stages are applied to each detection event. Gains of 10^5 or higher can be imposed, making such devices extremely sensitive. Although the electronic process is remarkably quiet, there is still a statistical, multiplicative noise associated with the uncertainty of whether or not each intended amplification event actually occurs. As a rule of thumb, photoelectron multiplication should be applied only to increase the pre-digitisation signal levels to the point where read noise is effectively eliminated. Further multiplication may make very dim signals qualitatively easier to see, but will actually reduce the SNR. Due to its statistical nature, the multiplicative noise is often treated in tandem with the shot noise and serves to amplify the shot noise by a factor between SQR(2) and 2. (42)

5.1.3. Dynamic Range

The dynamic range of a photodetector is defined as the ratio between the largest and smallest detectable signals and is probably the most misunderstood property of the device. The term refers both to the range of intensity values that can be recorded within a single captured image (The intrascence dynamic range), as well as range of intensities that can be recorded between images by changing the detector settings (Interscene dynamic range). The interscene dynamic range is usually much higher as the gain, multiplication or integration interval can be altered between acquisitions. As an example, the human eye is adapted to a wide range of illumination levels and has a vast interscene dynamic range, of perhaps 10^{10}, but can differentiate fewer than 256 intensity levels (8-bits) at any one time. Scientific cameras have much higher intrascene dynamic range and signals are typically digitised to 10, 12, 14 or 16 bits (up to 65,536 intensity levels) , but it is important to understand a number of considerations related to this concept.

The dynamic range is only meaningful if it is considered in conjunction with the noise of the detector and, for cameras, this is usually defined by dividing the full well capacity (the photoelectron storage capacity of each pixel) by the pixel read noise. This gives an indication of the number of discrete levels that can be discriminated between a minimum signal, which is just above the noise threshold and a maximum of which will lead to saturation. The problem with this simplification is that the detector will tend to become non-linear when the wells are close to capacity and a signal close to the read noise floor will have a sufficiently poor SNR to be useless for quantitative measurements. Taking the data from a typical camera used for calcium measurements (e.g. Photometrics CoolSnap HQ2), the well-depth of a 2×2 binned pixel is 30,000 photoelectrons (e^-) and the minimum read noise is 4.5e^- RMS giving a dynamic range of 6,667. This camera is available with either 12-bit (4,096 level) or 14-bit (16,384 level) digitisation. However, even using the 12-bit digitiser will not reduce significantly the useful dynamic range, particularly bearing in mind that the target (shot limited) SNR, assuming a nearly saturated signal, would be SQRT(30,000) or 173. With 12-bit digitisation (4,096 levels), the precision for the brightest signal levels would thus only be 4,096/173 or 23 intensity levels. So, why is it useful to digitise to a higher precision than the SNR would seem to justify? First, the dynamic range of these experiments can be very high, and having a large number of discrete signal levels allows weak signals to be analysed in the same field as brighter regions. Although the relative shot noise is higher for weaker signals the absolute value is lower, so the precision becomes more important to avoid posterization or other digitisation artefacts. Second, it is usually not necessary to monitor calcium transients on a pixel by pixel basis. The signal can usually be spatially averaged by plotting the mean of all pixels corresponding to a particular cell or part of cell. Temporal averaging is also sometimes appropriate if the time resolution of calcium change is sufficiently slow. Both these forms of averaging will increase the effective well size and thus make a high bit-depth more useful.

If small measurements are to be recorded (as is the case for Voltage-Sensitive Dyes (VSDs)), then the photoelectron store has to be huge in order to allow acceptable SNR. For example, to measure a 1% change in a signal with 100:1 SNR requires 10^8e^-.

A final consideration while using signals digitised to high bit depths is that due to the limitations of both the human eye and most monitor technology (8 bit or 256 levels), it is not possible for the researcher to actually visualise the dynamic range as a gray-scale intensity. This problem can be overcome by changing the Look Up Table (LUT) to view the light and dark areas of the images independently, or applying a pseudo-colour LUT so that different intensity ranges are displayed with different colour palettes.

5.1.4. Spatial Resolution

Unless used in conjunction with a scanning confocal microscope, a single-point detector has, by definition, no spatial resolution, so this section will concentrate on CCD camera and related technologies.

The starting point for defining the useful resolution is the difraction limit of the optical system, defined by Eq. 2.2 above.

With a 1.3 NA 40× oil immersion objective lens at 510 nm (green emission for Fura 2) the difraction limited resolution is approximately 240 nm, which with 40 times magnification corresponds to 9.6 μm at the detector plane. In order to ensure that the detector is not limiting, then the pixel size should be half of this or 4.8 μm. For static histology slides, getting as close as possible to the difraction limit makes sense, but for dynamic calcium measurement, this high degree of spatial resolution is rarely useful and the number of photons captured by such a small pixel is, in any case, too small to achieve acceptable SNR in any case.

Most scientific cameras have pixel sizes from 6.4 to 24 μm, with the trade-off being between resolving power and the greater full well depth (higher dynamic range) offered by the larger pixels.

Of more relevance to calcium researchers is the field of view that the camera images, and the number of vertical and horizontal pixels that the image can be divided into (also referred to as "resolution", generating possible confusion). There are no hard and fast rules about what these should be, but when configuring an imaging system, it is valuable to have a concept of the actual size and detail required at the biological level so that magnification, sensor size and sensor resolution can be optimised. In digital imaging, it is trivial to increase the effective pixel size and reduce the resolution in software (or even in hardware), so an ideal detector would have a very large number of small pixels. Unfortunately, this can have implications on other properties of the sensor, in particular on the temporal resolution.

5.1.5. Temporal Resolution

How many images or data points per second are required to track the ion concentration or morphological change under investigation? If changes occur slowly, over several minutes or hours, then few demands are placed on the photodetector and software, and a relatively simple (inexpensive) acquisition system may suffice. For changes that occur in the millisecond (ms) domain, care should be taken to ensure that the photodetector can generate sufficient data sets to resolve these changes.

Taking the example of a mid duration physiological event, such as the contraction of the cardiac myocyte which occurs over the course of around 100 ms. The highest frequency of interest within this event might be 100 Hz (10% of the overall transient). According to the Nyquist–Shannon sampling theorem (43), the sampling rate should be more than twice the maximum frequency of interest. Therefore, in this example, the photodetector sampling frequency should be >200 Hz equating to a period of <5 ms (in practice, 5 ms would suffice). If a slower time resolution is used, then this may result in temporal aliasing whereby a higher frequency component causes interference at lower frequencies depending on its instantaneous intensity at each sampling period. In calcium fluorescence, this is most likely to occur if the light source is not adequately filtered to remove ripple from the 50 Hz (60 Hz in some countries) alternating mains supply.

For a point detector, producing a single value per sampling interval, the maximum data rates will far exceed those required for biological fluorescence and the main consideration is to filter off high frequency components to reduce noise. Cameras and other spatial detectors may require a million or more discrete data points to be captured per time interval so the speed of the detector is critical. Most scientific cameras have clock speeds between 10 and 35 MHz (Mpixel/s), which, assuming a 500×500 array (250,000 pixels) will give theoretical maximum rates of between 40 and 140 frames/s. Due to unavoidable overheads, these rates will never quite be matched, but with good camera design and the appropriate software then it is possible to get close to these limits.

5.2. Single Point and Low Resolution Detectors

If the researcher is only interested in the total ion concentration within a specific cell or group of cells, then single-point and low resolution detectors are adequate, or, in some cases, preferable. Despite the lack of direct spatial discrimination, it is useful to be able to determine where in the biological preparation the fluorescence is coming from. The usual method to achieve this is to have a user adjustable aperture in the primary image plane of the micro- or macroscope and to relay this mask on to both the sensitive detector and to an inexpensive auxiliary camera for alignment purposes. The "viewing" camera can be illuminated with deep red or infrared transmitted light to avoid interfering with the fluorescence emission. Integrated systems to achieve this are available from companies such as Rapp Optoelectronic, Till Photonics and Cairn Research Ltd.

Both photomultipliers and photodiodes usually take the form of single detectors, but are also available in square and linear array format (e.g. 8×8, 16×16 or 32×1 elements) which allow low resolution *optical mapping*. All types tend to have large photosensitive elements and hence high dynamic range. Due to their low spatial resolution, they are also arbitrarily fast in the context of biological optical measurements.

5.2.1. Photomultipliers

The classic photomultiplier tube (PMT) was the workhorse photodetector for many decades and is still used today; especially, where there is a requirement for high temporal and low spatial resolution. When optical measurements are combined with electrophysiology (e.g. in patch clamping), there are significant logistical and cost benefits to using a PMT to produce analogue data which can be captured and processed by the same software as the electrical signals. To facilitate this, the output of the PMT should either be low-pass filtered and processed by a current-to-voltage converter, or integrated over discrete time intervals (as with a digital camera). For luminescence and other very low light measurements, they can also be used in a *photon counting* mode where photon events are discriminated and counted in the frequency domain.

A PMT consists of a photocathode which generates electrons upon being hit by photons, and a series of metallic electrodes (dynodes) held at progressively higher electrical potential, and which are coated with light-sensitive phosphors. An incident photon strikes the photocathode thereby releasing a photoelectron. This electron is accelerated in vacuum to the first dynode by a potential difference striking the dynode with high kinetic energy. This photoelectron releases many new photoelectrons in a process known as *secondary emission*, which will each in turn accelerate to the next dynode in the chain. Each dynode collision increases the total number of electrons by a factor of up to about 100 and there are typically between 5 and 11 stages. This results in a large pulse of over one million electrons producing a negative current at the anode. The interval between the arrival of the incident photon and the time of the maximal anode current is in the range of 20 ns allowing data rates of up to 50 MHz. This is several orders of magnitude faster than that is required for single point detection, but is important in laser scanning confocal microscopy where an image is constructed by producing a faster scan at each time point and consequently required data rates need to be multiplied by the number of spatial points (virtual pixels).

The dark noise (dark current) in PMTs is caused by ***thermionic emissions*** emitted by the cathode and/or dynodes, ***ionisation of residual gases*** due to the presence of high velocity electrons and also ***field emissions*** which appear when the PMT is operated at too high a voltage. Selecting tubes with inherently low dark current and using low-noise electronics can ensure that dark and read noise are minimised so that the signal is dominated by the statistical noise both from the photocathode and the dynodes. The major drawback with PMTs is that they tend to have only QE in the order of 20–30%, so the shot noise at low light levels is much higher than the ideal case. Scientific quality PMTs are produced by a variety of manufacturers, including Hamamatsu and Electron Tubes.

5.2.2. Photodiodes

Photodiodes typically have higher QEs than PMTs, but they lack the intrinsic gain (with the exception of avalanche photodiodes, which are sometime used as an alternative to PMTs) and unless the photosensitive area is very small, they also have high dark current. For this reason, they are rarely used in low-light applications. However, because they can be designed with arbitrarily large photoelectron storage capacity, they are useful for recording small changes in experiments which are not light limited. Photodiode arrays are typically used for optical mapping of calcium or voltage-sensitive dyes in sections of tissue or whole organs.

5.3. Cameras

Although point detectors are still valuable for some measurements (especially where time resolution is critical), the majority of systems are now configured with a camera detector. This usually takes the form of a so-called digital or Electron-Multiplied (EM) Charge Coupled Device (CCD) camera, but it is useful to at least be aware of frame-rate and intensified options, and the competing Complementary Metal-Oxide Semiconductor (CMOS) technology.

Common to all the different types of camera available for scientific imaging are the concepts of digitisation, binning and sub-array acquisition.

Digitisation is the conversion of the number of photoelectrons stored in each pixel into a numerical value which can be processed by the host computer. This usually occurs in a serial fashion, so in order to process a frame each pixel in turn has to pass through an analog-to-digital converter, usually within the camera head. The digital signal is then passed to the computer using either a bespoke digital interface card or a generic interface such as Cameralink, Firewire or USB 2.

Binning is simply the combining of groups of adjacent pixels prior to digitisation. For example, a 512×512 array could be "2×2 binned" to give a resolution of 256×256 with an effective pixel size of four times the area (i.e. each pixel comprises of a 2×2 array). The advantage of this is that each resultant "pixel" has 4 × the photon capture area and the digitiser has 4 × fewer signals to process (and is hence faster). The obvious disadvantage is the loss in resolution.

Sub-array acquisition refers to capturing a specific region of interest from within the array. If the specimen only takes up a section of the field of view of the camera, then limiting the acquired area can be used to increase frame rates and reduce the required storage capacity in the computer.

5.3.1. Charge Coupled Device Cameras

A CCD chip comprises of a two dimensional array of individual "picture" sensing elements (pixels) on a thin silicon substrate. The pixels absorb energy from incident photons leading to the formation of electron-deficient sites or holes in a crystal lattice. These pixels can be considered an array of electrically isolated silicon photodiodes of definite dimensions (commonly 4.8–24 μm across), which respond to incident illumination by producing a voltage.

Scientific CCD cameras used in calcium signalling will typically have an array size of between 80×80 pixels for fast transient measurements and 1,392×1,040 for detailed spatial information. Many devices are peltier cooled (rendering dark current negligible) so the differentiation between cameras is usually based on the quantum efficiency, read noise, pixel size and the speed and flexibility with which images can be captured. These properties are, to a large extent, related to the design of the CCD chip itself so it is important to understand the two principal architectures that are used.

Frame-Transfer

A frame transfer CCD chip comprises a photoactive silica array alongside a second masked array which acts as a storage area (Fig. 2.3). Each frame can be rapidly transferred from the active area to the masked area allowing the next image to be acquired whilst the previous frame is digitised by the host computer. The parallel transfer of charge across the array takes in the order of 1 μs per pixel (faster for some cameras such as the Andor Ixon+), so the

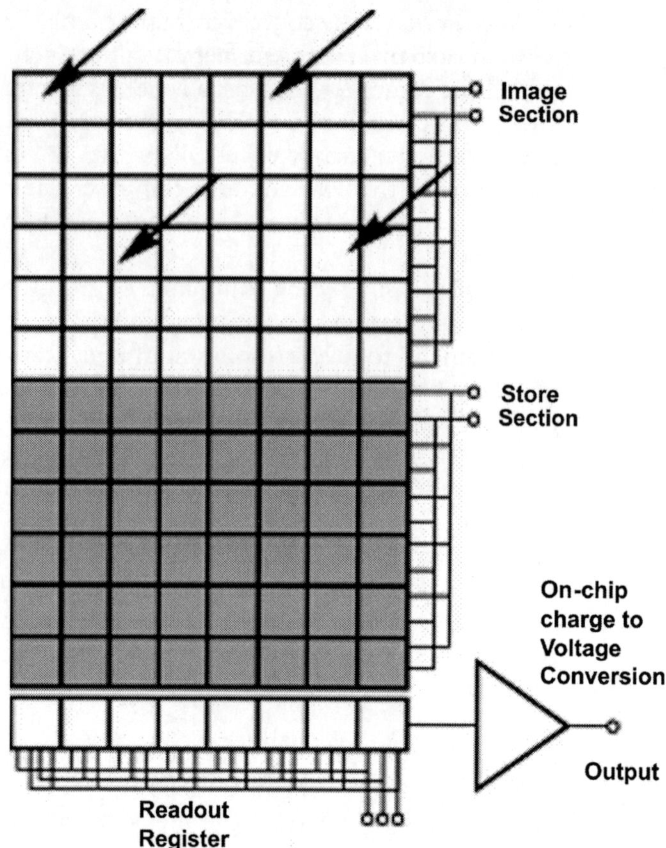

Fig. 2.3. Schematic representation of the frame-transfer CCD camera. This type of camera uses a two-part sensor in which one-half of the photosensitive parallel array is used as a storage region and is protected from light by a light-tight mask. Incoming photons are allowed to fall on the uncovered portion of the array and the accumulated charge is then rapidly transferred into the masked storage region for charge transfer to the serial output register. While the signal is being integrated on the light-sensitive portion of the sensor, the stored charge is read out (Adapted, with permission, from Andor Technologies).

frame transfer time for a 512 × 512 element chip is approximately 0.5 ms. If possible, the illumination source should be shuttered off during this interval to prevent smearing, as the the chip continues to absorb photoelectrons during the transfer process.

For a typical camera with a clock speed of 10 MPixel/s, the digitisation of the signal from the masked area will take around 30 ms. Provided that the interframe interval is greater than this digitisation time plus a small allowance for the frame-transfer time, then the camera will be able to continuously image the so-called "frame-transfer mode". For shorter integration times, the camera will need to skip alternate frames so that the active area can be cleared before the next exposure.

In addition to being able to image at full resolution with minimal deadtime, frame transfer chips are extremely flexible in terms of binning and sub-array acquisition. Data rates can be increased further by choosing sub-arrays adjacent to the masked area or by using part of the photoactive area for storage (if the illumination can be masked with high contrast). It is also possible to expand on this architecture so that two or more masked areas and digitisers are used in parallel (multiple taps) to increase acquisition rates without needing a faster clock speed.

The quantum efficiency of frame transfer cameras is reduced by photons having to pass through the gate structure of the device. For this reason, many scientific frame transfer cameras use back-thinned chips where the reverse side of the chip is etched away so that it can be used for sensing directly, producing a mirror image of the equivalent front-illuminated chip. This yields sensors that can have quantum efficiencies of >90% making them suitable for the most sensitive measurements.

Most of the frame transfer cameras used for calcium measurements are also either intensified or electron multiplied as being discussed in the later session.

Interline

Interline CCD chips are commonly used in scientific, security and consumer cameras. The charge from the photoactive pixels is stored in linear masked arrays adjacent to each column of pixels (Fig. 2.4). The problem with this arrangement is that the fill

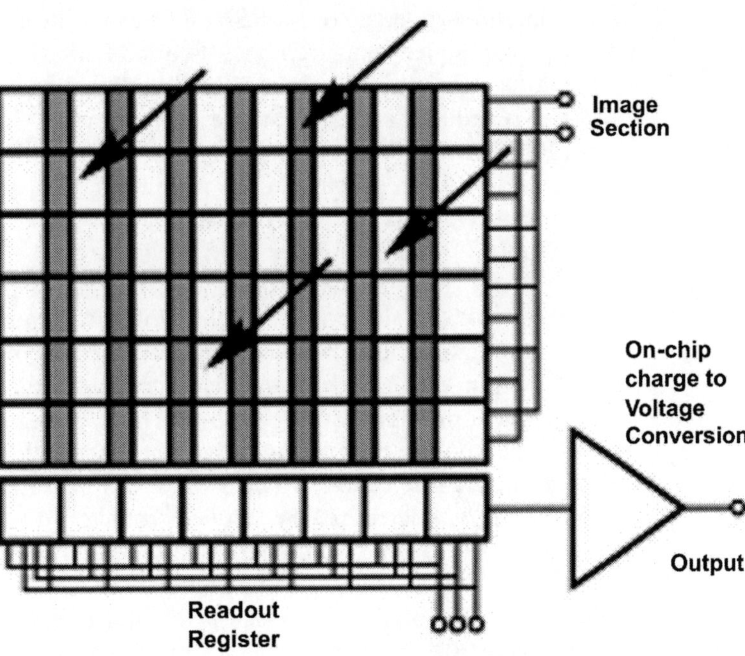

Fig. 2.4. The interline-transfer CCD camera. This type of camera incorporates charge transfer channels called *Interline Masks* (*greyed* in the image). These are immediately adjacent to each photodiode so that the accumulated charge can be rapidly shifted into the channels after image acquisition has been completed. The very rapid image acquisition virtually eliminates image smear. Altering the voltages at the photodiode so that the generated charges are injected into the substrate rather than shifted to the transfer channels can electronically shutter interline-transfer CCDs (Adapted, with permission, from Andor Technologies).

factor of the detector is poor, with much of the image hitting areas of the chip that are not photoactive (by comparison a frame transfer camera has a 100% fill factor). This shortcoming is addressed in scientific cameras by having a microlens array in front of the chip to project more of the light onto the photoactive areas. This works surprisingly well and interline cameras can achieve quantum efficiencies of up to 70% with no discernible optical aberrations introduced by the lenses.

The advantage of an interline camera is that the transfer from photoactive area to the store only requires a single step and is thus effectively instantaneous (sub microsecond), so there is no need to shutter the light source between frames. The drawback is that the architecture is less flexible and the speed benefits of binning and sub-array acquisition are often less significant than they are with frame transfer devices. The majority of interline CCD cameras used for calcium measurements are based on the Sony ICX285 chip, with cheaper cameras using the Sony ICX205 or Kodak KAI-20200 sensors. These are available from many suppliers including Q Imaging, Photometrics, Hamamatsu and PCO.

5.3.2. Frame-Rate "Analogue" Cameras

Before the advent of inexpensive digital cameras and digital interfaces such as USB2 and firewire, it was common to use the so-called frame rate cameras for calcium imaging. These cameras are sometimes referred to as analogue because the camera outputs an analogue signal which can be subsequently viewed on a monitor or digitised on a frame grabber card within the host computer. The two standards used for these cameras are CCIR in Europe and RS170 in the United States and they run at a fixed frequency of 50 and 60 Hz respectively. The main limitation, apart from the fixed frame rate, is that noise is seldom good enough to justify digitising to more than 8 bits (256 levels).

5.3.3. Electron Multiplication

In recent years, electron multiplication (EM) technology has set a new standard for fast low-light fluorescence measurements. As shown in Fig. 2.5, the frame transfer architecture is extended to include a multiplication register where clock voltages of up to 50 V are applied to generate secondary photoelectrons via an impact-ionisation process. The extent of the resultant gain is exponentially related to the applied voltage and can reach in excess of 1,000 times. Compared with conventional intensified cameras (ICCDs), the difference is that the gain is applied after the silica detector allowing small pixels and quantum efficiencies in excess of 90%, but with the disadvantage that the noise is amplified along with the signal.

Used with care, EM cameras are the best detectors for most low-light calcium fluorescence applications, but certain factors need to be taken into consideration. Because the multiplication process is applied to both signal and noise, it is vital that the

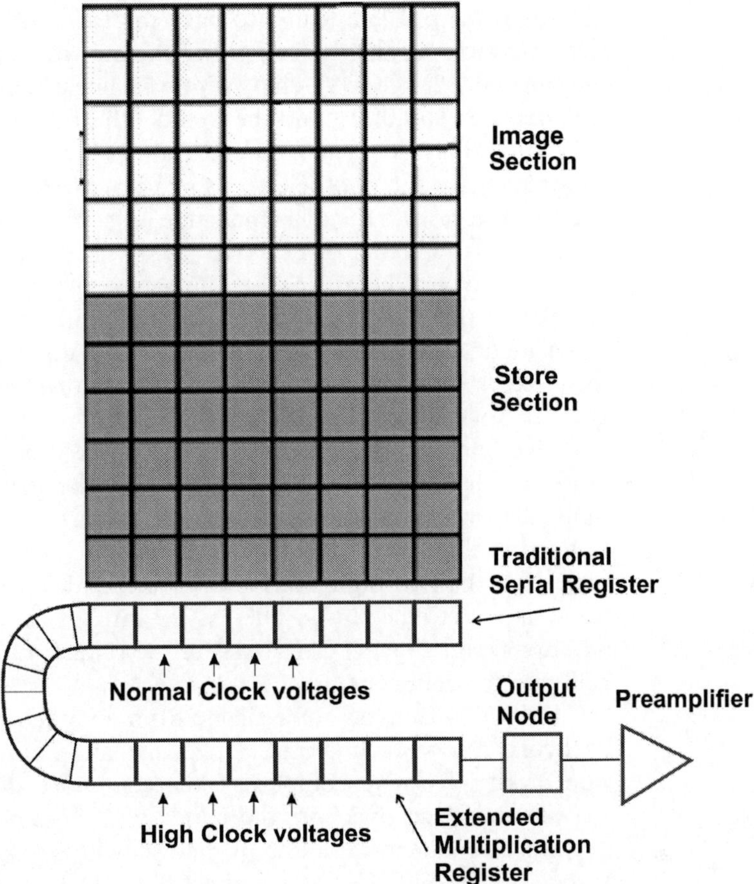

Fig. 2.5. Electron-multiplying CCD camera. The distinguishing feature of EMCCDs (also called "on chip multiplication" cameras) is the incorporation of a specialised extended serial register on the sensor chip that generates multiplication gain through the process of *impact ionization* within the silicon sub-structure. The photon-generated charge is elevated above the read noise even at high frame rates and is applicable to any of the current CCD sensor configurations.

camera is cooled to reduce dark current and that any spurious charges (sometimes referred to as *clock induced charge* (cic)) are minimised. Because the multiplication process amplifies the signal sufficiently to effectively remove read noise, the dark and cic noise sources tend to be limiting. When comparing cameras, it is informative to acquire frames in total darkness with appropriate gain and integration intervals to assess the background signal (the probability of a given pixel having a spurious event). Some of the early EM CCDs did not fully address these issues leading to unfavourable comparisons with ICCDs.

Another consideration is that the EM gain is temperature dependent with higher gain achieved at lower temperatures. Cameras are typically operated at between −30 and −85°C and this temperature must be stable to avoid fluctuations in gain.

Applying EM gain can also have the effect of reducing the dynamic range of the camera. Using the industry standard back-illuminated E2V CCD97 chip (within its linear range), the pixel well depth is 160,000e$^-$ and the well depth of the gain register is only 400,000e$^-$. If a gain of over 2.5 times is applied, then there is potential for the photoelectrons in a pixel to saturate the gain register and restrict the dynamic range (e.g. If a gain of ten times is applied to a signal of 50,000e$^-$ this will produce 500,000e$^-$ which is beyond the capacity of the gain register and will hence saturate at this stage. For this reason high-end EM cameras will often incorporate an alternative, conventional, amplifier with a lower clock speed (and hence reduced read noise) for recordings that are not photon limited.

Two final points to consider are that the EM process is inherently non-linear and over and above this, the gain can reduce gradually over time (thought to be the result of electrons being trapped in the silicon – silicon dioxide interface in the gain register), particularly if high gains and/or intensities have been routinely applied to the sensor. These issues can both be addressed by additional (intelligent) electronics in the camera to linearise and calibrate the gain response.

EM CCD cameras are available from a range of suppliers in particular Andor Technology, Hamamatsu and Photometrics and due to their relative complexity, there is more differentiation between products than with the standard CCD cameras. Most of the available cameras use one of the following sensors manufactured by E2V or Texas Instruments (TI):

E2V	CCD97	512×512	16 μm pixels	QE at 510 nm = 90%
E2V	CCD60	128×128	24 μm pixels	QE at 510 nm = 90%
E2V	CCD201	1,024×1,024	13 μm pixels	QE at 510 nm = 90%
TI	TC885	1,004×1,002	8 μm pixels	QE at 510 nm = 52%

5.3.4. Image Intensification

An ICCD camera comprises of an image intensifier optically coupled to the sensor of a CCD camera. Image intensifiers are only suitable for low-light applications and can be damaged by exposure to excessive photons even when the device is switched off.

The intensifier itself is made up of a photocathode, *a micro-channel plate* (MCP) and a phosphor screen enclosed in a high-vacuum environment. The photocathode is a thin layer of material onto which the image is focused and which generates electrons following the photon impact. The most efficient gallium arsenide phosphide (GaASP) photocathodes are well matched to fluorescence dyes and have a QEs of >40% at 510 nm. The electrons generated by the photocathode are accelerated by a strong electric field towards the MCP. Depending on the chosen voltage at the photocathode, the electrons are either accelerated towards the MCP (negative voltage) or remain in the photo cathode (positive voltage). Consequently, the image intensifier can be used as an extremely fast (5 ns) and efficient electronic shutter which is of particular value in fluorescence lifetime imaging (FLIM).

The MCP is made out of leaded glass and includes one to ten million channels with a characteristic diameter of 6–10 μm and a length of approx. 0.5 mm. Each channel represents a photomultiplier and each time an electron hits the wall of the channel, it will generate further electrons by secondary emission. The degree of intensification depends mainly on two parameters: a) the length and diameter of the channels (increasing the number of hits) and b) the applied voltage. Therefore, the degree of intensification can be influenced by setting the acceleration voltage (gain). Finally, the electrons leaving the MCP are accelerated at a voltage of several kV towards the anode, which is a phosphor screen on which they generate the photon emission. The choice of phosphor is a compromise between its optical efficiency and its time response. Of the most commonly used materials, P43 is efficient at converting electrical power into optical power, but it takes 10 ms to decay to 1% of its maximum after being hit by electrons (phosphor lag), whereas P46 is less efficient, but decays to 1% of its maximum within 2 μs.

The final consideration is the coupling of the phosphor screen to the CCD sensor. Because the phosphor emits photons in all directions, this makes lens coupling inefficient (typically 12% with a well designed system). Using a tapered fibre-optic bundle directly bonded onto both the screen and the CCD sensor can significantly increase this efficiency, but the manufacturing process is expensive and requires great skill.

In conclusion, a top of the range ICCD camera may in practice perform as well as an EM camera for fast imaging at very low light levels. However, the increased cost, complexity, risk of damage, lower dynamic range and inherently reduced quantum efficiency mean that they are now mainly used for applications where fast gating is required or in the special case of photon counting cameras described below.

5.3.5. Complementary Metal Oxide Semiconductor Cameras

Complementary Metal Oxide Semiconductor sensors compete with CCD sensors in the high-volume consumer and security camera markets, but are not (yet?) routinely used for calcium microfluorescence measurements. Unlike a CCD, which transfers charge to be converted into voltage in a single serial readout register, the CMOS chip has individual charge-to-voltage conversion for each pixel. This has the advantage of making the CMOS architecture far more parallel and capable of being rapidly addressed on an individual pixel level. Unfortunately, the additional on-chip electronics and the multiple charge converters have a negative effect on the fill-factor and the uniformity respectively. The relatively high noise and low QE of currently available devices means that they are rarely used for low-light measurements, however, because they can be built with arbitrarily deep pixel wells and very high speed addressing, they are used for fast measurements where high illumination levels are achievable especially macroimaging of tissue or whole organs.

5.3.6. Colour

The discussion of photodetectors so far has made the implicit assumption that the optical system has a single defined optical bandwidth of interest for the sensor to detect. This is often sufficient for calcium indicators such as Fura 2 or Fluo3 that emit at one wavelength, but does not address other ratiometric emission dyes such as Indo 1 with multiple emission wavelengths, or the numerous cases where several fluorophores of different colours are used simultaneously.

There are three broad approaches to multi colour detection: (1) separate the colours sequentially in the time domain using an electronic filter to acquire at one wavelength and then the next; (2) divide the light using spectral beamsplitters and use single or multiple detectors and finally, (3) specific camera solutions that allow simultaneous acquisition using a single detector.

If the task in hand is to detect blue, green and red fluorophores simultaneously, why not simply use a colour camera? To answer that, we need to first consider what a digital colour camera actually is. The digital cameras used in conventional photography have a matrix of red, blue and green colour filters in front of the pixels. Figure 2.6 shows the most commonly used Bayer matrix named after its inventor Bryce Bayer. There are twice as many green pixels as red and blue as green is the dominant colour detected by the human eye. Algorithms are applied to the raw (RAW format) image captured by the sensor to interpolate the colours from the mask and produce the TIFF and JPEG images used by computers.

The problems with using such sensors for scientific imaging are numerous; and include the facts that the colour mask is unlikely to be a good match for the fluorophores, the true spatial information of each colour is offset by the grid, binning pixels loses all colour information and perhaps the greatest drawback is that only a maximum of 50% of the pixels are used for any given colour. Despite this, it is possible to capture good colour images using these camera if the sample is bright and static.

A second type of colour camera uses 3 CCD chips incorporated into a single camera head with a prism to divert the three colour channels to the appropriate sensor. These so called 3-chip

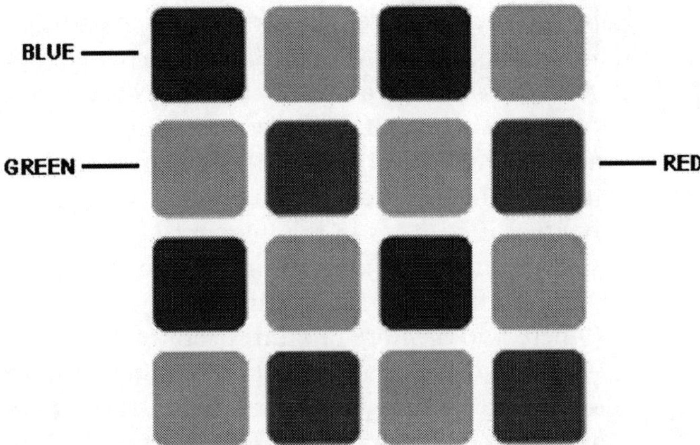

Fig. 2.6. Bayer filter. The figure illustrates the arrangement of colour filters on the pixel array of an image sensor.

cameras are not surprisingly, difficult to build and expensive, but they avoid most of the problems associated with Bayer mask cameras. The remaining drawback is that the RGB filters are still fixed although at least one manufacturer, Hamamatsu, will supply a version with the colour channels matched to popular fluorophores rather than the response of the human eye.

The most flexible method of multicolour imaging on a single sensor is to use a dedicated image splitter device. These devices use bespoke optics to spectrally separate the image from the microscope into up to four different colours and to focus these onto different sections of a single sensor, which can then be recombined in software. These devices have the significant advantage that it is straightforward to change filters to match with specific fluorophores, and that they can be used with any of the wide range of available monochrome cameras. The disadvantage is that they require careful alignment and that they reduce the effective size of the sensor (as only part of the sensor can be used for each colour channel). Commercial image splitters are available from Hamamatsu, Photometrics and Cairn Research.

5.4. Photon Counting

The light levels associated with most optical calcium measurements are sufficiently high that it is more convenient to digitise an analogue signal at the output stage of the camera or other detector. For bioluminescence measurements and some very low light fluorescence, it can be necessary to discriminate individual photon events in order to achieve acceptable signal-to-noise.

A PMT may be operated in a pulse(photon)-counting mode where each pulse of electrons is registered as "an event". This concept of discrimination allows threshholds to be set such that an event below (or above) a certain predetermined level is rejected as being a spurious or dark noise and anything within the preset range is recorded as single "photon" count, regardless of its amplitude. In this mode, the PMT measures incident light not in the amplitude domain but in the frequency domain, i.e. counting events per unit of time. Each "count" is the consequence of an incident photon and the intensity (amplitude) of the incident light is coded by the number of such counts integrated over the measuring time. Because it operates in the frequency domain, care must be taken to set appropriate intervals or deadtime between counts so that as many photons as possible can be counted and linearity is not lost (where two or more photons produce a single count). The pulse width of commercially available counting heads (from Hamamatsu and Electron Tubes) is in the order of 10 ns, but to ensure reliable and linear counts it can be useful to limit the bandwidth to several million counts per second.

Sufficiently cooled digital cameras can also be used in a photon counting mode, but this would usually be for astronomy where it is possible to have frame integration times of several minutes or even hours.

Some electron multiplied and intensified cameras can be operated in a photon counting mode to effectively eliminate the multiplicative noise. The camera integration time is set so that there is a significantly greater chance of a single photon event occurring than a spurious or dark event, but that there is minimal chance of a second photon event occurring in the same pixel during the frame. This generates a binary image, where every pixel is either white to indicate an event or black if no event has occurred. By summing several frames, an image can be constructed with an improved signal-to-noise ratio over the equivalent image that would have been produced by a single prolonged exposure.

For fast aequorin and and luciferase imaging, there are also dedicated photon counting cameras incorporating up to three microchannel plates to achieve gains far higher than the 1,000× provided by EM and standard intensified CCDs. These systems are available from specialist manufacturers such as Photek Ltd.

6. Analysing the Output

Image acquisition is the first and most important step in Digital Imaging Systems. High quality images require not only a good microscope, properly aligned optics, an appropriate camera with enough sensitivity for collecting images, but also proper integration of various hardware components: microscope and stage (*z* axis, but also *xy* location drives), light source, capturing devices (cameras or PMT), image frame grabbers, data capturing modules, filter drivers and for more sophisticated approaches, means of integration with other equipment (e.g. electrophysiology).

As seen in the previous sections, there is now a huge range of options and technologies available for investigating the variety of processes and mechanisms associated, triggered or resulting from changes in intracellular Ca^{2+} and it is most likely that the task of setting together, hand-shaking and controlling in an integrated manner all this hardware is beyond the expertise of most laboratories. As a result, image analysis packages are usually now supplied as integrated, almost turn-key systems.

All this hardware setting and integration is required to capture the image, but, in the lab context, the actual science starts only after capturing the images and it is an often mentioned adage that an imaging experiment "takes 1 day to capture and 1 week (sometime month) to analyse". For performing morphological and intensity-levels image analysis on single images or on stacks of independent images, the choice of software available is wide starting from the freely-available *Image J* (http://rsb.info.nih.gov/ij/index.html) to powerful commercial software suites such as *Image-Pro* (from MediaCybernetics, at http://www.mediacy.com/index.aspx?page=IPP) and MetaMorph from MDS. Image J deserves a special mention, since it is a public domain program based on Java script language designed with an open architecture that provides free access to the code and thus allows continuous development through routines and macros written by users and made available from the Image J site. Various Ca^{2+} imaging-specific routines (particularly for time-course recordings of fluorescence ratiometric signals) are available through some of the heavy-weights of image analysis – *MetaFluor*, from the MetaMorph stable at MDS Analytical Technologies , and Volocity from Improvision/Perkin Elmer; or through dedicated imaging suites developed by various hardware manufacturers (e.g. the *Cell*^{R} from Olympus, IQ from Andor Technology, or *TILLvisION*, from the monochromator manufacturers TILL (Germany)). All these programs will contain a number of specific application modules, in addition to the ratiometric measurements, to include FRET or TIRFM measurements or routines allowing software deconvolution.

7. Enhancements in Fluorometric Microscopy Techniques and Future Developments

Despite these improvements in the quality of optical components, fluorescent imaging used either for tissue and cellular morphology and protein localization or for metabolic, real-time imaging of Ca^{2+} homeostasis still suffer from a number of important limitations. Some of these concern the optical characteristics of the process – such as limited spatial resolution or relatively modest tissue penetration, while other limitations are due to a restricted physical access, particularly important when attempting in-vivo or intra-vital microscopy (IVM). The advances in imaging techniques of the last decade are truly staggering and it is now more an issue of science catching up, in setting relevant questions and/or experimental paradigms with the opportunities opened by these technological developments. Not all these improvements in technology are strictly applicable to the field of intracellular Ca^{2+} imaging, but this is most likely just a temporary state of affairs as prices continue to decrease while enhancements take place in streamlining the available equipment.

Improvements in spatial resolution, particularly in the (z) plane resolution were initially brought by implementation of confocality. By challenging the conventional assumptions of traditional microscopy by either using different imaging geometries, by imposing structure upon the pattern of illumination, or by breaking the linearity of the excitation–emission process of fluorescence, spatial resolution of light microscopy was pushed down to an incredible 30 nm level in the far-field of optical nanoscopy (44).

Some of the improvements in resolution came from challenging the relatively simple geometry of traditional optical inspection. As discussed in a previous section, TIRF resulted from a change in the geometry of the light path used to excite the specimen – from the traditional perpendicular path to illumination at a very shallow angle. Another assumption of classical microscopy is that imaging is done with only one microscope lens. It is now more than 15 years since Stefan Hell then at Heidelberg University proposed the *4Pi microscopy* (45). This system uses the focussed light from two objectives facing each other. This generates a full-surround illumination while at the same time creating a constructive interference of the opposing wave fronts. This results in a significantly lower focal volume with the most dramatic effect on the axial resolution. The name 4Pi is derived from the solid angle (4π) of a complete spherical wave. To date, most of the applications of this method are in the field of exploration of the finer details of cell morphology (46). One such recent report using the model of mitochondrial dynamics in response to metabolic challenges illustrate the significant improvements in detail afforded by this technique in comparison even with the "classical" confocal microscopy (47).

A very successful method to increase the penetrative power of microscopy was the development of non-linear fluorescent processes such as multiphoton microscopy. The multiphoton approach is based on the idea that two or more photons of low energy can excite a fluorophore in a quantum event and thus generate a fluorescent emission. The probability of quasi-simultaneous absorption (within less than a femtosecond (10^{-15} s)) of two or more photons is very low, and thus this technique requires a large flux of excitation photons. This increased flux of photons can be generated both in the spatial and temporal domains. Spatial concentration of photons is achieved by focussing a laser beam to a small spot through an objective with a high NA objective as in normal confocal microscopy. The temporal concentration is accomplished by compressing photons from a continuous source into short (femtosecond) pulses. A typical Ti-sapphire oscillator exhibit high peak intensities but low average power (48). This combination of ultra-short high peak pulses with sharp spatial focussing elicits multiphoton excitation in a small focal plane eliminating a significant part of the background fluorescence.

A variant of multiphoton microscopy is *stimulated emission depletion (STED)* microscopy which involves the overlapping of two light beams in the focal regions. The first beam is just a conventional excitation beam that generates a fluorescent emission signal. The second beam of different wavelength acts as a damper, suppressing the initial fluorescence by relaxing the excited molecules of the fluorophore before they are able to emit the fluorescent photons. The further ingenious approach is that, in STED microscopy, the depletion beam is shaped as a doughnut containing a small hole (50–70 nm) in the centre, which thus produces a very sharply defined fluorescence signal surrounded by dark background (49). Using such approaches, significant insights into the dynamics of synaptic vesicles recycling and the fate of some of the vesicle-associated proteins (50) or of dendritic spines in live cells (51) have been obtained recently.

Although expensive and difficult to setup, multiphoton microscopy including the two-photon incarnation is moving strongly now from a lab specific tool (e.g. W. Denk's various setups, starting from the original versions described in 1990 (52)) towards turn-key systems, available currently from most of the major imaging systems manufacturers (e.g. BioRad/Zeiss, Leica, Olympus). One of the reasons for this continuous development of the two(multi)-photons microscopy is the significant range of improvements it brings, including a positive double whammy of decreased amount of light scattering coupled with greater depth of penetration (five-six times deeper) in comparison even with the simple confocal scanning laser microscopy (CLSM). The two-photon systems also provide better axial (z) resolution and less photodamage or toxicity. All these features dramatically increase the scope and opportunities for intravital microscopy (IVM) that attempts to explore cellular physiology and morphology deep within tissues in in vivo conditions. Such approaches also benefit from the recent developments in GRIN (gradient index) lens technology, which can be fabricated with submillimeter diameters and act either as objective lenses to focus the light or as relay lens to transfer an image at some distance from the object plane (53).

The use of GRIN lenses helped also in the significant development of another technology allowing for an even more powerful and flexible IVM – the use of fibre-optic imaging (54). Optic fibres have been used for some time in fluorescence imaging as simple, efficient and flexible light conduits both for excitation and emission light. Continuous developments in reducing the fibre size facilitating the incorporation of individual conduits in bundles of optic fibres, development of fibres that can act differently as spatial, spectral or polarization filters, together with the introduction of smaller control hardware, such as the micro electromechanical system scanners, all contributed to the expansion of fibre-optic imaging (55, 56). In fact, fibre-based imaging devices are now so widespread and used in so many variant configurations that some will classify them in three broad categories on the basis of their size and uses: microscopy, endoscopy and microendoscopy, with the latter allowing for further subdivisions as a function of the light configuration used: epifluorescence, confocal and multiphoton (57). As an example of the efficiency and explanatory potential of such techniques, particularly when applied intravitally, it is worth mentioning the recent work in the field of Alzheimer disease research that provided intriguing information about the dynamics of neuritic plaques and of their interactions with the local neurites (58, 59).

8. Conclusions

The interest in observing and monitoring the finest spatial and temporal characteristics of cellular physiology has maintained a constant pressure on the development of optical technologies to allow such endeavours. Ingenious modalities of circumventing the limitation of classical optics are opening new opportunities, covering the whole range from in vitro observations to in vivo measurement at subcellular level. The increased resolving power in all four dimensions (three spatial and one temporal) is generating a large increase in the amount and quality of information content of imaging experiments, allowing more and more refined modelling of data. One likely direction of future development is the further miniaturization of image capture devices, particularly through the use of fibre-optic technologies. All such enhancements will bring fluorescence imaging nearer the fields of drug development and therapeutic intervention. It is probably just a question of time until such cellular fluorescence technologies will be able to complement directly and significantly the coarser volume imaging approaches, such as magnetic resonance imaging or positron emission tomography and become a fully integrated component of translational medicine.

9. Note

For those interested in taking any of the theoretical points related to microscopy to a further level, two Internet sites maintained and sponsored by two of the major microscope manufacturers (Olympus and Nikon) are recommended: the Olympus Microscopy Resource Centre, at: http://www.olympusmicro.com/primer/index.html and the Nikon's Microscopy U at: http://www.microscopyu.com/. A similar educational site geared more towards camera technology is maintained by Hamamatsu at http://learn.hamamatsu.com/explore/.

References

1. Verkhratsky A, Toescu EC (eds) (1998) Integrative aspects of calcium signalling. Plenum, New York and London
2. Berridge MJ, Lipp P, Bootman MD (2000) The versatility and universality of calcium signalling. Nat Rev Mol Cell Biol 1:11–21
3. Petersen CCH, Toescu EC, Petersen OH (1991) Different patterns of receptor-activated cytoplasmic Ca2+ oscillations in single pancreatic acinar cells: dependence on the receptor type, agonist concentration and intracellular Ca2+ buffering. EMBO J 10:527–533
4. Woods NM, Cuthbertson KSR, Cobbold PH (1986) Repetitive transient rises in cytoplasmic free calcium in hormone-stimulated hepatocytes. Nature 319:600–602
5. Tang F, Dent EW, Kalil K (2003) Spontaneous calcium transients in developing cortical neurons regulate axon outgrowth. J Neurosci 23:927–936
6. De Koninck P, Schulman H (1998) Sensitivity of CaM kinase II to the frequency of Ca2+ oscillations. Science 279:227–230
7. Dolmetsch RE, Lewis RS, Goodnow CC, Healy JI (1997) Differential activation of transcription factors induced by Ca^{2+} response amplitude and duration. Nature 386:855–858
8. McCarron JG, Chalmers S, Bradley KN, MacMillan D, Muir TC (2006) Ca2+ microdomains in smooth muscle. Cell Calcium 40:461–493
9. Parekh AB (2008) Ca^{2+} microdomains near plasma membrane Ca2+ channels: impact on cell function. J Physiol 586:3043–3054
10. Miyawaki A (2008) Green fluorescent protein glows gold. Cell 135:987–990
11. Miyawaki A (2005) Innovations in the imaging of brain functions using fluorescent proteins. Neuron 48:189–199
12. Olympus (2008) Olympus Fluoview Resource Center: recommended books on confocal microscopy. In:
13. Mason WT (ed) (1993) Fluorescent and luminescent probes for biological activity – a practical guide to technology for quantitative real-time analysis. Academic, London
14. Petersen OH (ed) (2001) Measuring calcium and calmodulin inside and outside cells. Springer, Berlin
15. Stephens D (ed) (2006) Cell imaging: methods express series. Scion, Oxford
16. Jepson MA (2006) Confocal or wide-field? A guide to selecting appropriate methods for cell imaging. In: Stephens D (ed) Methods express: cell imaging. Scion, Oxford, pp 17–48
17. Nakano A (2002) Spinning-disk confocal microscopy – a cutting-edge tool for imaging of membrane traffic. Cell Struct Funct 27:349–355
18. Adams MC, Salmon WC, Gupton SL et al (2003) A high-speed multispectral spinning-disk confocal microscope system for fluorescent speckle microscopy of living cells. Methods 29:29–41
19. Ravier MA, Tsuboi T, Rutter GA (2008) Imaging a target of Ca2+ signalling: dense core granule exocytosis viewed by total internal reflection fluorescence microscopy. Methods 46:233–238
20. Demuro A, Parker I (2006) Imaging single-channel calcium microdomains. Cell Calcium 40:413–422
21. Jares-Erijman EA, Jovin TM (2003) FRET imaging. Nat Biotechnol 21:1387–1395

22. Piston DW, Kremers GJ (2007) Fluorescent protein FRET: the good, the bad and the ugly. Trends Biochem Sci 32:407–414
23. Munishkina LA, Fink AL (2007) Fluores-cence as a method to reveal structures and membrane-interactions of amyloidogenic proteins. Biochim Biophys Acta 1768:1862–1885
24. Miyawaki A, Llopis J, Heim R et al (1997) Fluorescent indicators for Ca2+ based on green fluorescent proteins and calmodulin. Nature 388:882–887
25. Suhling K, French PM, Phillips D (2005) Time-resolved fluorescence microscopy. Photochem Photobiol Sci 4:13–22
26. Wilms CD, Schmidt H, Eilers J (2006) Quantitative two-photon Ca2+ imaging via fluorescence lifetime analysis. Cell Calcium 40:73–79
27. Kim HM, Kim BR, An MJ, Hong JH, Lee KJ, Cho BR (2008) Two-photon fluorescent probes for long-term imaging of calcium waves in live tissue. Chemistry 14:2075–2083
28. Reitz FB, Pagliaro L (1994) Fibre optic scrambling in light microscopy: a computer simulation and analysis. J Microsc 176:143–151
29. Dyba M, Hell SW (2003) Photostability of a fluorescent marker under pulsed excited-state depletion through stimulated emission. Appl Opt 42:5123–5129
30. Fukano T, Shimozono S, Miyawaki A (2006) Fast dual-excitation ratiometry with light-emitting diodes and high-speed liquid crystal shutters. Biochem Biophys Res Commun 340:250–255
31. New Optical Filters – Improve high-speed multicolor fluorescence imaging (2006) (Accessed at http://www.semrock.com/Data/Documents/Biophotonics_March06_Article.pdf)
32. Rosfjord KM, Villalaz RA, Gaylord TK (2000) Constant-bandwidth scanning of the czerny-turner monochromator. Appl Opt 39:568–572
33. Overview of fiber optic cable (2008) (Accessed at http://www.arcelect.com/fibercable.htm)
34. Picciolo GL, Kaplan DS (1984) Reduction of fading of fluorescent reaction product for microphotometric quantitation. Adv Appl Microbiol 30:197–234
35. Lacerda SH, Abraham B, Stringfellow TC, Indig GL (2005) Photophysical, photochemical, and tumor-selectivity properties of bromine derivatives of rhodamine-123. Photochem Photobiol 81:1430–1438
36. Rizzuto R, Pinton P, Carrington W et al (1998) Close contacts with the endoplasmic reticulum as determinants of mitochondrial Ca^{2+} responses. Science 280:1763–1766
37. Collins TJ, Berridge MJ, Lipp P, Bootman MD (2002) Mitochondria are morphologically and functionally heterogeneous within cells. EMBO J 21:1616–1627
38. Wu X, Bers DM (2006) Sarcoplasmic reticulum and nuclear envelope are one highly interconnected Ca2+ store throughout cardiac myocyte. Circ Res 99:283–291
39. Toresson H, Grant SG (2005) Dynamic distribution of endoplasmic reticulum in hippocampal neuron dendritic spines. Eur J Neurosci 22:1793–1798
40. Dulla C, Tani H, Okumoto S, Frommer WB, Reimer RJ, Huguenard JR (2008) Imaging of glutamate in brain slices using FRET sensors. J Neurosci Methods 168:306–319
41. Introduction to Charge-Coupled Devices (CCDs) (2009) (Accessed at http://www.microscopyu.com/articles/digitalimaging/ccdintro.html)
42. Unravelling Sensitivity, Signal to Noise and Dynamic Range – EMCCD vs CC (2007) (Accessed at http://www.emccd.com/what_is_emccd/unraveling_sensitivity/Signal_to_Noise_in_CCDs/)
43. Luke HD (1999) The origins of the sampling theorem. IEEE Commun Mag 37:106–108
44. Hell SW (2007) Far-field optical nanoscopy. Science 316:1153–1158
45. Hell SW, Stelzer EHK, Lindek S, Cremer C (1994) Confocal microscopy with an increased detection aperture: type-B 4Pi confocal microscopy. Opt Lett 19:222–224
46. Gugel H, Bewersdorf J, Jakobs S, Engelhardt J, Storz R, Hell SW (2004) Cooperative 4Pi excitation and detection yields sevenfold sharper optical sections in live-cell microscopy. Biophys J 87:4146–4152
47. Plecita-Hlavata L, Lessard M, Santorova J, Bewersdorf J, Jezek P (2008) Mitochondrial oxidative phosphorylation and energetic status are reflected by morphology of mitochondrial network in INS-1E and HEP-G2 cells viewed by 4Pi microscopy. Biochim Biophys Acta 1777:834–846
48. Oheim M, Michael DJ, Geisbauer M, Madsen D, Chow RH (2006) Principles of two-photon excitation fluorescence microscopy and other nonlinear imaging approaches. Adv Drug Deliv Rev 58:788–808
49. Klar TA, Jakobs S, Dyba M, Egner A, Hell SW (2000) Fluorescence microscopy with diffraction resolution barrier broken by stimulated emission. Proc Natl Acad Sci USA 97:8206–8210

50. Willig KI, Rizzoli SO, Westphal V, Jahn R, Hell SW (2006) STED microscopy reveals that synaptotagmin remains clustered after synaptic vesicle exocytosis. Nature 440:935–939
51. Nagerl UV, Willig KI, Hein B, Hell SW, Bonhoeffer T (2008) Live-cell imaging of dendritic spines by STED microscopy. Proc Natl Acad Sci USA 105:18982–18987
52. Denk W, Strickler JH, Webb WW (1990) Two-photon laser scanning fluorescence microscopy. Science 248:73–76
53. Levene MJ, Dombeck DA, Kasischke KA, Molloy RP, Webb WW (2004) In vivo multiphoton microscopy of deep brain tissue. J Neurophysiol 91:1908–1912
54. Helmchen F (2002) Miniaturization of fluorescence microscopes using fibre optics. Exp Physiol 87:737–745
55. Piyawattanametha W, Barretto RP, Ko TH et al (2006) Fast-scanning two-photon fluorescence imaging based on a microelectromechanical systems two- dimensional scanning mirror. Opt Lett 31:2018–2020
56. Rochefort NL, Jia H, Konnerth A (2008) Calcium imaging in the living brain: prospects for molecular medicine. Trends Mol Med 14:389–399
57. Bullen A (2008) Microscopic imaging techniques for drug discovery. Nat Rev Drug Discov 7:54–67
58. Eichhoff G, Busche MA, Garaschuk O (2008) In vivo calcium imaging of the aging and diseased brain. Eur J Nucl Med Mol Imaging 35(Suppl 1):S99–S106
59. Skoch J, Hickey GA, Kajdasz ST, Hyman BT, Bacskai BJ (2005) In vivo imaging of amyloid-beta deposits in mouse brain with multiphoton microscopy. Methods Mol Biol 299:349–363

Chapter 3

Ca^{2+} Imaging: Principles of Analysis and Enhancement

Fabio Mammano and Mario Bortolozzi

Abstract

In this chapter, we review the theoretical and experimental foundations underling a quantitative approach to Ca^{2+} imaging, discuss equilibrium conditions and their violations and present a computational framework that can be used to estimate the spatial and temporal dynamics of Ca^{2+} signals based of fluorescence measurements with Ca^{2+} indicators.

Key words: Buffers, Law of mass action, Binding reactions, Fluorescent indicators, Optical measurement of Ca^{2+} concentration, Single wavelength and ratiometric imaging, Chemical equilibrium, Non-equilibrium conditions, Diffusion, Ion fluxes, Differential equations, Monte Carlo methods

1. Optical Measurement of Ca^{2+} Concentration

Optical measurement of the intracellular concentration of selected ion species is paramount to understanding cell physiology and function. Several molecular probes, namely fluorescent dyes, capable of sensing the local ion concentration with high selectivity have been developed over the last 20 years. These are based on BAPTA (1,2-bis(o-aminophenoxy)ethane-*N*,*N*,*N′*,*N′*-tetraacetic acid), a pH-insensitive evolution of the widely used Ca^{2+}-selective chelator EGTA (ethylene glycol tetraacetic acid) (Fig. 3.1). Chelation is the binding or complexation of a bi- or multidentate ligand with a single metal ion. The mechanism of Ca^{2+} chelation by BAPTA is shown in Fig. 3.2.

Chelation of Ca^{2+} by a buffer B to form a complex CaB is described by the reaction

A. Verkhratsky and Ole H. Petersen (eds.), *Calcium Measurement Methods*, Neuromethods, vol. 43, DOI 10.1007/978-1-60761-476-0_3, © Humana Press, a part of Springer Science+Business Media, LLC 2010

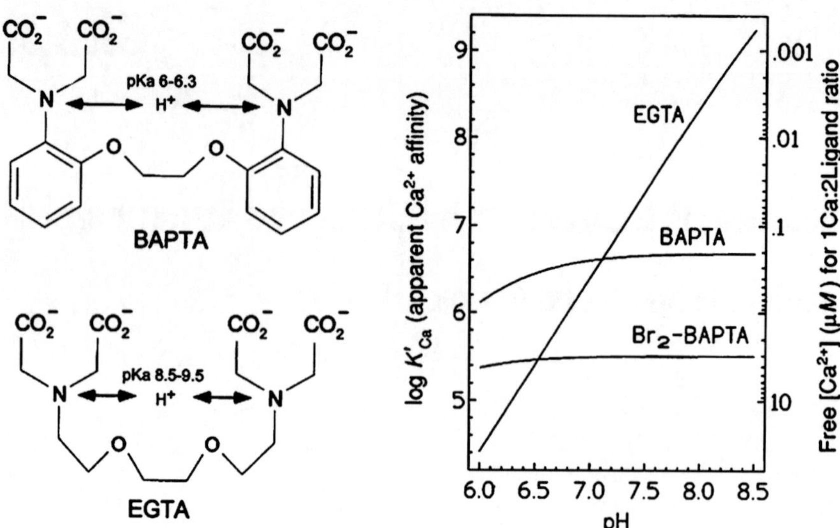

Fig. 3.1 Chemical structure and pH dependence of Ca^{2+} affinity of BAPTA and EGTA.

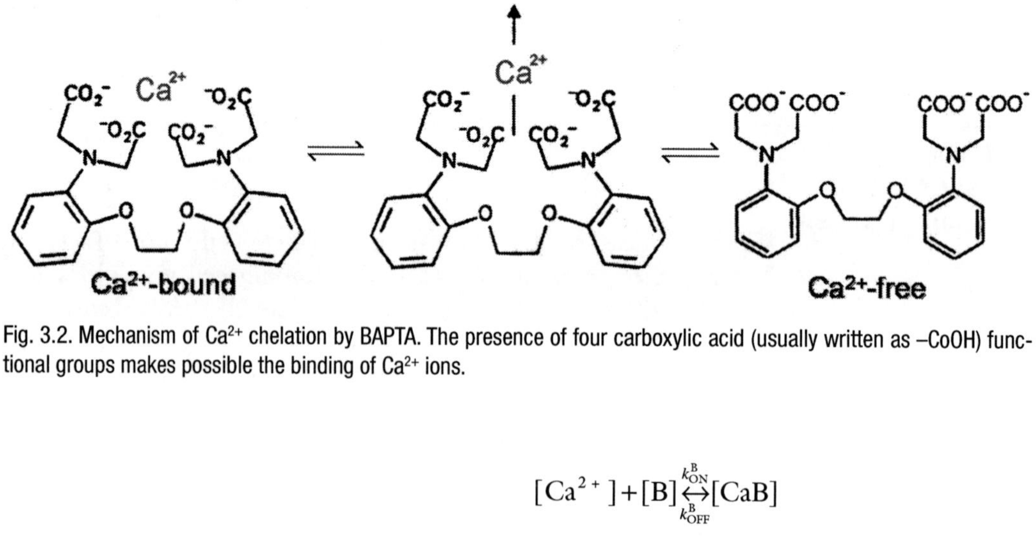

Fig. 3.2. Mechanism of Ca^{2+} chelation by BAPTA. The presence of four carboxylic acid (usually written as –CoOH) functional groups makes possible the binding of Ca^{2+} ions.

$$[\mathrm{Ca}^{2+}]+[\mathrm{B}] \underset{k_{\mathrm{OFF}}^{\mathrm{B}}}{\overset{k_{\mathrm{ON}}^{\mathrm{B}}}{\leftrightarrow}} [\mathrm{CaB}]$$

and the corresponding kinetic equation is

$$\frac{\mathrm{d}[\mathrm{CaB}]}{\mathrm{d}t} = k_{\mathrm{ON}}^{\mathrm{B}}[\mathrm{Ca}^{2+}][\mathrm{B}] - k_{\mathrm{OFF}}^{\mathrm{B}}[\mathrm{CaB}], \tag{3.1}$$

where square brackets are used to indicate concentration, $k_{\mathrm{ON}}{}^{\mathrm{B}}$ is the rate constant for Ca^{2+} binding to B and $k_{\mathrm{OFF}}{}^{\mathrm{B}}$ is the rate constant for Ca^{2+} dissociation. At chemical equilibrium

$$\frac{\mathrm{d}[\mathrm{CaB}]}{\mathrm{d}t} = 0$$

therefore

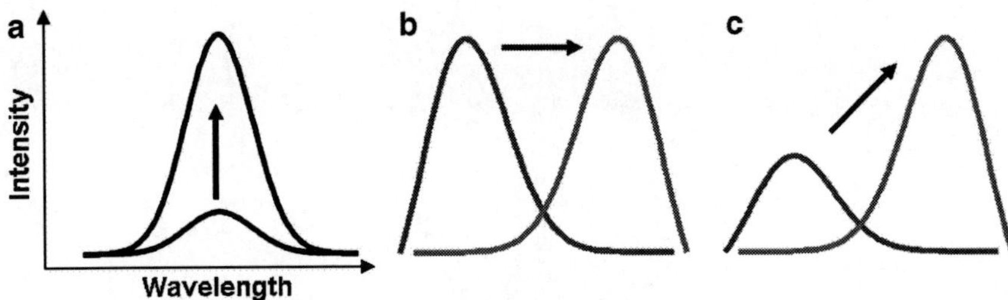

Fig. 3.3. Spectral properties that can be utilized to measure Ca^{2+} concentration.

$$\frac{[Ca^{2+}][B]}{[CaB]} = \frac{k_{OFF}^{B}}{k_{ON}^{B}} \equiv k_{D}^{B}. \tag{3.2}$$

In the above equation, which represents an instance of the law of mass action under equilibrium conditions, $k_D{}^B$ is the equilibrium or dissociation constant (for BAPTA: $k_D{}^B = 0.192\,\mu M$ (1); $k_{ON}{}^B = 500\,\mu M^{-1} s^{-1}$ (2); $k_{OFF}{}^B = k_D{}^B \times k_{ON}{}^B = 96\ s^{-1}$ (3)).

Ca^{2+}-selective fluorescent probes share a modular design consisting of a metal-binding site (or sensor) B covalently coupled to a fluorophore A therefore:

$$[A] = [B]. \tag{3.3}$$

In order for such a fluorescent probe to provide useful information about its environment, it is necessary that its spectral properties be altered in a suitable manner by the parameter to be measured. For most biological applications, any one of the following three property changes is appropriate (Fig. 3.3):

(a) A change in fluorescence yield

(b) A shift in the excitation or emission spectrum

(c) A combination of the two

2. Single Wavelength Indicators

Case (a) comprises the Fluo family of the so-called *single wavelength* fluorescent Ca^{2+} indicators (Fig. 3.4). The green-fluorescent emission (~525 nm) of Ca^{2+}-bound fluo-3 is conventionally detected using optical filter sets designed for fluorescein (FITC). Fluo-4 is an analog of fluo-3 with the two chlorine substituents replaced by fluorine atoms.

Fluo-3 is essentially non-fluorescent unless bound to Ca^{2+} and exhibits an at least 100-fold Ca^{2+}-dependent fluorescence enhancement (Fig. 3.5). The fluorescence quantum yields of Ca^{2+}-bound fluo-3 and fluo-4 are essentially identical (~0.14 at saturating Ca^{2+}).

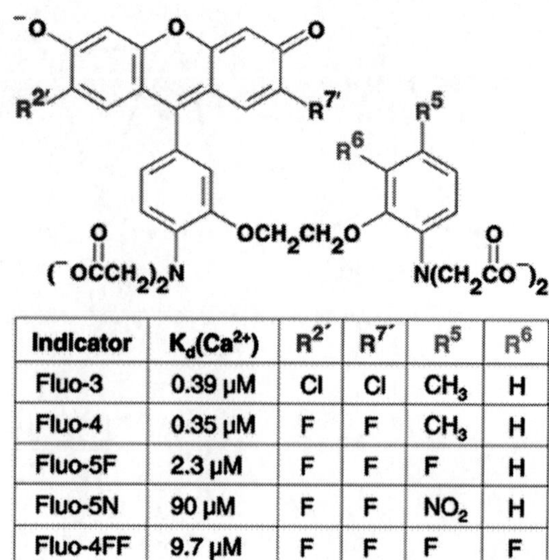

Indicator	$K_d(Ca^{2+})$	$R^{2'}$	$R^{7'}$	R^5	R^6
Fluo-3	0.39 μM	Cl	Cl	CH_3	H
Fluo-4	0.35 μM	F	F	CH_3	H
Fluo-5F	2.3 μM	F	F	F	H
Fluo-5N	90 μM	F	F	NO_2	H
Fluo-4FF	9.7 μM	F	F	F	F

Fig. 3.4. Chemical structure and dissociation constants for the Fluo family of Ca^{2+} indicators.

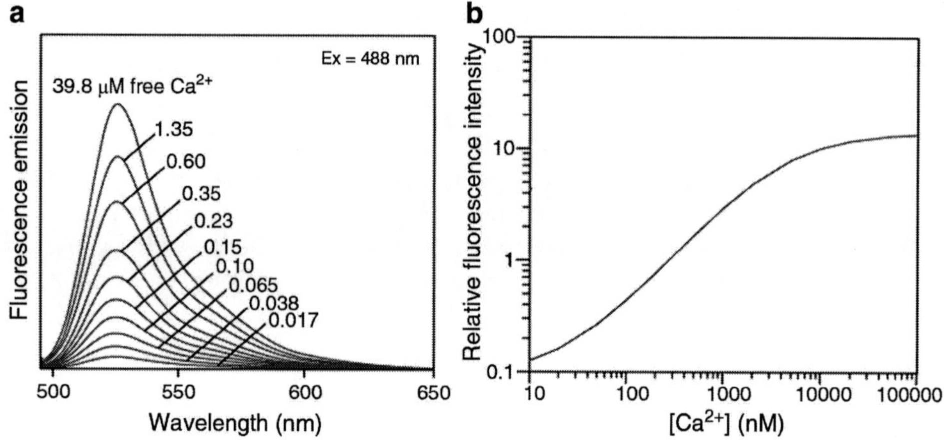

Fig. 3.5. (**a**) Fluo-3 spectra, excited by the 488-nm line of the Argon laser, are shown for different values of the free Ca^{2+} concentration ($[Ca^{2+}]$). (**b**) Relative fluorescence emission intensity, measured at the peak of each spectrum in (**a**), plotted against the corresponding $[Ca^{2+}]$.

The intact acetoxymethyl (AM) ester derivative of fluo-3 is also nonfluorescent. The absorption maximum of fluo-4 is blue-shifted about 12 nm when compared to fluo-3, resulting in an increased fluorescence excitation at 488 nm and consequently higher signal levels for confocal laser-scanning microscopy. When fluo-4 is substituted for fluo-3 (i.e., using identical loading protocols), fluorescence signals are at least doubled. The stronger fluorescence signals provided by fluo-4 are particularly advantageous in most cell types.

Also shown in Fig. 3.5b is the typical sigmoid dependence of the fluorescence emission on $[Ca^{2+}]$ (the free calcium concentration), which limits the useful detection range to approximately one log unit to the left and the right of the k_D^B. In particular, as is the case with other high affinity indicators, the upper limit for Fluo-3 and Fluo-4 is not exceeding 1–2 μM. Demonstrations that intracellular (i) free-Ca^{2+} concentrations ($[Ca^{2+}]_i$) can reach levels as high as 100 μM have sparked interest in fluorescent Ca^{2+} indicators (e.g. Fluo-5F, Fluo-5 N, Fluo-4FF), which can be used to measure calcium concentrations in the micromolar range (Fig. 3.4).

Suppose now that we have a system of fluorophores A at a total concentration c_T, which we excite with light of a given intensity and wavelength λ, (i.e. energy $h\nu$, where h is the Plank constant and ν is frequency). We can represent the excitation process as

$$\mathrm{A} + \text{photon} \xrightarrow{k_A} \mathrm{A}^*$$

where k_A is the excitation rate constant (with units of s^{-1}) and A^* represents fluorophores in the excited state. The system comes back down either non-radiatively (nr), with a rate k_{nr}, or radiatively (r), i.e. emitting a photon of longer wavelength (i.e. reduced energy $h\nu'$) with a rate constant k_r (Fig. 3.6).

The overall de-activation rate constant k_M is given by

$$k_M = k_r + k_{nr} = \frac{1}{\tau_{ex}}$$

(with units of s^{-1}) where τ_{ex} is the excited state lifetime (typically few ns). Under constant illumination conditions, a steady state is rapidly reached such that

$$k_A \alpha \left(c_T - [\mathrm{A}^*]\right) = k_M [\mathrm{A}^*], \tag{3.4}$$

where the dimensionless parameter α represents the fraction of absorbed photons. Therefore, the equilibrium (steady state) concentration of excited state fluorophores $[A^*]_{eq}$ is given by

$$[\mathrm{A}^*]_{eq} = \frac{\alpha c_T}{\alpha + k_M / k_A},$$

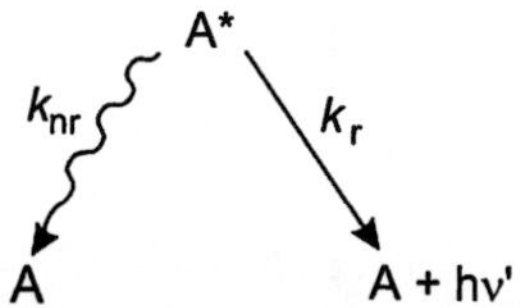

Fig. 3.6. Radiative and nonradiative decay from the excited state.

where, in general, $k_M \ll k_A$. The fluorescence emission intensity $F(t)$ is proportional to the number of photons emitted in the process of return from the excited state

$$A^* \xrightarrow{k_r} A + \text{photon}$$

therefore, its steady state value is

$$F = k_r[A^*]_{eq} = \frac{\alpha k_r c_T}{\alpha + k_M/k_A} \tag{3.5}$$

where the result is expressed in mols of photons emitted per unit time and per unit volume of dye solution.

For a given c_T, α is proportional to the product $\varepsilon(\lambda)l$ where $\varepsilon(\lambda)$ is the molar absorption coefficient (with units of $\mathrm{L\ mol^{-1}\ m^{-1}}$) and l is the length of absorbing medium traversed by the illuminating beam. The fluorescence quantum yield (sometimes incorrectly termed quantum efficiency) is a gauge for measuring the efficiency of fluorescence emission relative to all of the possible pathways for relaxation and is generally expressed as

$$\eta_F = \frac{k_r}{k_M} \tag{3.6}$$

We thus conclude that F depends on factors such as illumination intensity, molar concentration of fluorescent probes, fluorescence quantum yield, molar absorption coefficient, and path length.

Let us then assume that the concentration of Ca^{2+}-selective fluorescent probes is kept low enough that the relationship between fluorescence emission intensity and concentration is indeed linear, as predicted by (3.5). In general, the concentration [B] and [CaB] of the Ca^{2+}-free (f) and Ca^{2+}-bound (b) forms differ with respect to quantum yield and absorption. Therefore, we write F as a linear combination

$$F = S_f[B] + S_b[CaB], \tag{3.7}$$

where the proportionality constants S_f, S_b lump all (system-dependent) factors such as illumination intensity, η_F, $\varepsilon(\lambda)$and l. We are interested in measuring $[Ca^{2+}]$ in a closed system (e.g. the cell cytoplasm). Hence, we must also include the mass balance equation

$$c_T = [B] + [CaB]. \tag{3.8}$$

We then define

$$F_{max} = S_b c_T \tag{3.9}$$

as the fluorescence emission under Ca^{2+} saturation conditions and

$$F_{min} = S_f c_T \tag{3.10}$$

as the corresponding emission under Ca^{2+}-free conditions (see Fig. 3.5).[1] Thus, the expression for F can be re-written as

$$F = F_{\min} + (S_b - S_f)[\mathrm{Ca}B] = F_{\max} - (S_b - S_f)[B].$$

Therefore, we can then also write

$$F - F_{\min} = (S_b - S_f)[\mathrm{Ca}B],\ F_{\max} - F = (S_b - S_f)[B],$$

yielding

$$\frac{F - F_{\min}}{F_{\max} - F} = \frac{[\mathrm{Ca}B]}{[B]}.$$

At chemical equilibrium (3.2),

$$[\mathrm{Ca}^{2+}] = k_D^B \frac{[\mathrm{CaB}]}{[\mathrm{B}]}.$$

Therefore, we conclude that

$$[\mathrm{Ca}^{2+}] = k_D^B \frac{F - F_{\min}}{F_{\max} - F}. \quad (3.11)$$

Equation (3.11) expresses a quantitative relationship between the physiologically relevant equilibrium [Ca^{2+}], the dissociation constant $k_D{}^B$ and optically measurable quantities F_{max}, F_{min} and F for single wavelength Ca^{2+} selective probes such as Fluo-3 (Fig. 3.7). However, there are a number of caveats and problems with the practical use of (3.11).

First, we note that the denominator vanishes as $F \to F_{max}$. Consequently, even small fluctuations in the estimate of F (due, e.g. to instrumental noise) may cause unacceptably large fluctuations in the estimate of [Ca^{2+}] (unreliable zone in Fig. 3.7). Furthermore, (3.11) is difficult to apply to imaging experiments where F_{max}, F_{min} and F change rapidly over time due to photobleaching (see, e.g. Fig. 2 of (4)).

Provided that adverse effects are kept under control, (3.11) can be used to estimate the change in concentration

$$\Delta[\mathrm{Ca}^{2+}] \equiv [\mathrm{Ca}^{2+}] - [\mathrm{Ca}^{2+}]_{\mathrm{rest}}$$

caused, e.g., by a physiological stimulus, where

$$[\mathrm{Ca}^{2+}] = k_D^B \frac{F - F_{\min}}{F_{\max} - F}$$

and F_0 are, respectively, the pre-stimulus free Ca^{2+} concentrations and fluorescence emission intensity. For stimuli that keep [Ca^{2+}]

[1] Note that, because of (3.3), the c_T in (3.4) is the same as in (3.8)–(3.10).

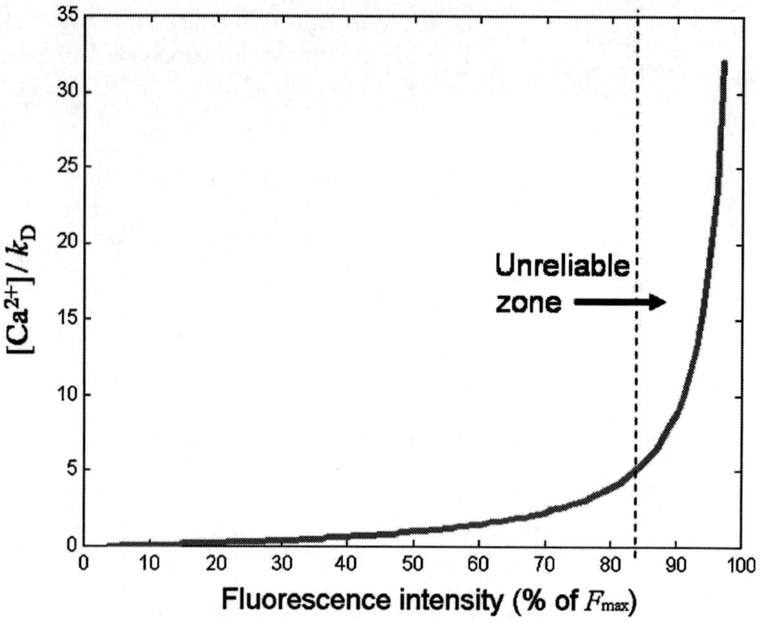

Fig. 3.7. Graphical representation of the single wavelength [Ca^{2+}]-measurement formula.

within the approximately linear region of (3.11) (i.e., outside the unreliable zone) we can approximate $\Delta[Ca^{2+}]$ as

$$\Delta[\mathrm{Ca}^{2+}] \cong \left.\frac{\mathrm{d}[\mathrm{Ca}^{2+}]}{\mathrm{d}F}\right|_{F_0}(F - F_0),$$

that is

$$\Delta[\mathrm{Ca}^{2+}] \cong k\frac{\Delta F}{F_0}, \tag{3.12}$$

where

$$k = k_\mathrm{D}^\mathrm{B}\frac{F_0\cdot(F_\mathrm{max} - F_\mathrm{min})}{(F_\mathrm{max} - F_0)^2}.$$

Thus, if imaging experiments are performed in such a way that bleaching is moderate and as long as the total Ca^{2+}-sensor/fluorescent dye concentration c_T and the path length l do not change during the measurement, the *pixel-by-pixel ratio*

$$\frac{\Delta F}{F_0} = \frac{F - F_0}{F_0} \tag{3.13}$$

yields a unique function of $\Delta[Ca^{2+}]$, i.e. of the stimulus-induce change in the free Ca^{2+} concentration.

3. Ratiometric indicators

Calibration problems associated with the estimate of F_{max}, F_{min} and F are alleviated by the use of ratiometric (or dual wavelength) Ca^{2+} probes (Fig. 3.3b, c) such as fura-2 for the high affinity case or mag-Fura-2 (furaptra), Fura-2FF, and BTC for the low affinity case (5). Consider, for example, the excitation spectrum of fura-2 (Fig. 3.8).

The most remarkable feature is the existence of an *isosbestic point* at ~360 nm for which fluorescence is independent of $[Ca^{2+}]$. Consequently, spectral amplitudes increase with increasing $[Ca^{2+}]$ to the left of the isosbestic point and decrease to the right. Consider then what happens to the florescence emission recorded, e.g., at the peak of the fura-2 emission spectrum (510 nm) in response to a hypothetical stimulus that causes a step-like increase in $[Ca^{2+}]$(Fig. 3.9).

The traces in Fig. 3.9 represent idealized fura-2 readout vs. time of the $[Ca^{2+}]$ averaged within a region of interest (ROI) where the step-like increase is spatially uniform (for a more appropriate description of kinetics, see Sect. 4). As bleaching affects similarly the readout at all wavelengths, its effects can be neutralised by ratioing, e.g., λ_1/λ_0 or λ_1/λ_2 (Fig. 3.10).

Figure 3.10 also shows that taking the signal ratios for wavelengths on opposite sides of the isosbestic point (e.g. λ_1/λ_2) maximises the dynamic range, and therefore (optics transmittance permitting) should be preferred over ratioing with respect to the (isosbestic point) λ_0 signal.

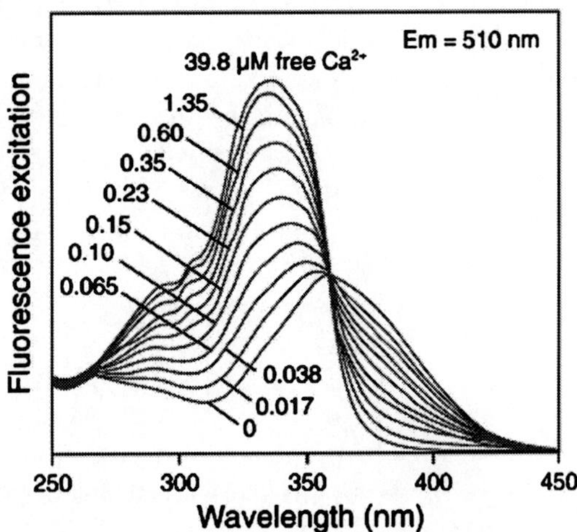

Fig. 3.8. Excitation spectra of fura-2 for the indicated values of the free Ca^{2+} concentration.

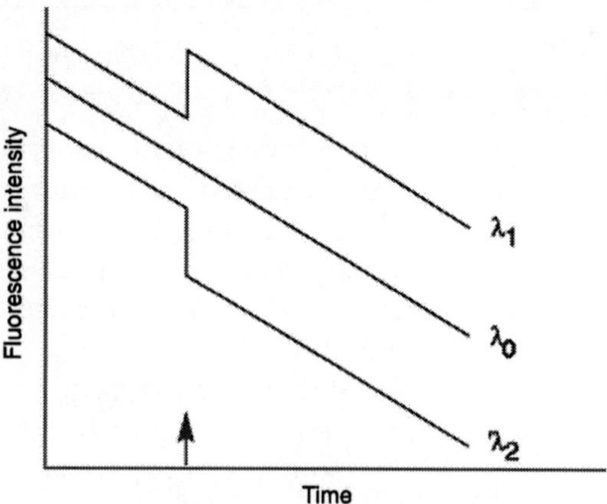

Fig. 3.9. Effect of a hypothetical step-like increase in $[Ca^{2+}]$(whose occurrence is marked by the *vertical arrow*) in the presence of substantial bleaching of fura-2. λ_0, excitation wavelength coincident with the isosbestic point;λ_1, excitation wavelength to the left of the isosbestic point; λ_3, excitation wavelength to the right of the isosbestic point.

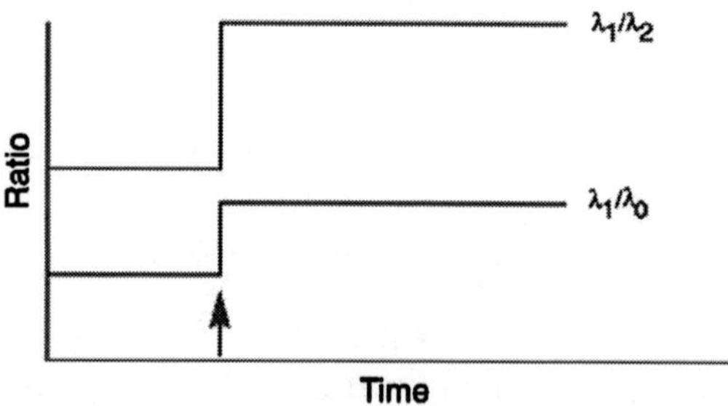

Fig. 3.10. Ratioing the signals acquired at different wavelengths cancels the effects of photobleaching.

We shall now derive a formula that allows quantifying the equilibrium $[Ca^{2+}]$ in terms of ratio signals.

Two excitation wavelengths λ_1,λ_2 and two dye species, free and bound, require four proportionality coefficients, hereby symbolized as S_{f1} for free dye measured at wavelength λ_1, S_{f2} for free dye at λ_2, and S_{b1}, S_{b2} for Ca^{2+}-bound dye at λ_1 and λ_2, respectively (typical values for fura-2 are λ_1 = 340 nm and λ_2 = 380 nm; Fig. 3.11).

Then, we may write

$$F_1 = S_{f1}[B] + S_{b1}[CaB], \; F_2 = S_{f2}[B] + S_{b2}[CaB]. \qquad (3.14)$$

Using the law of mass action at equilibrium (3.2), we can express [CaB] in (3.14) in terms of [B] and $[Ca^{2+}]$ as

$$[CaB] = [Ca^{2+}][B] / k_D^B.$$

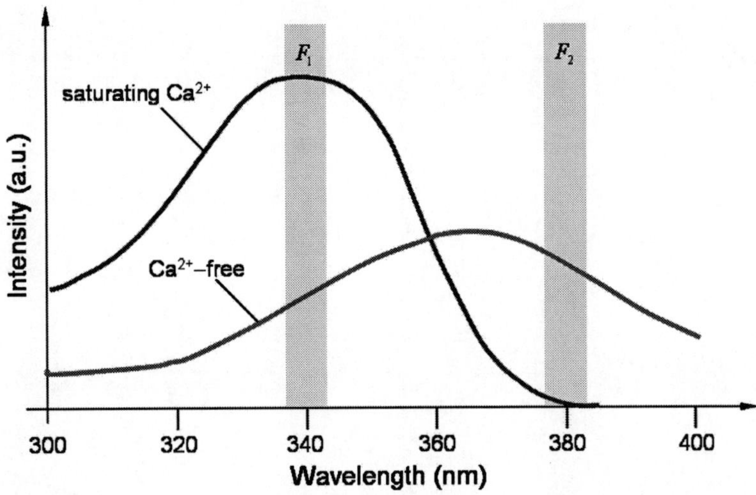

Fig. 3.11. Excitation spectra of fura-2 under Ca^{2+}-free and saturation Ca^{2+} conditions.

Thus, factoring out the [B] term from (3.14), and taking the ratio signal as

$$R = \frac{F_1}{F_2},$$

yields

$$R = \frac{S_{f1} + S_{b1}[Ca^{2+}]/k_D^B}{S_{f2} + S_{b2}[Ca^{2+}]/k_D^B}.$$

The minimum value of the above expression achieved under Ca^{2+}-free conditions is

$$R_{min} = \frac{S_{f1}}{S_{f2}}, \quad (3.15)$$

Whereas, the maximum value achieved under saturating Ca^{2+} conditions is

$$R_{max} = \frac{S_{b1}}{S_{b2}}. \quad (3.16)$$

We may thus re-express R in terms of R_{min} and R_{max} as

$$[Ca^{2+}] = k_D^B \left(\frac{R - R_{min}}{R_{max} - R} \right) \left(\frac{S_{f2}}{S_{b2}} \right), \quad (3.17)$$

which is known as the Grynkiewicz equation (6) (Fig. 3.12) and has a form closely analogous to (3.11). In particular, it displays a

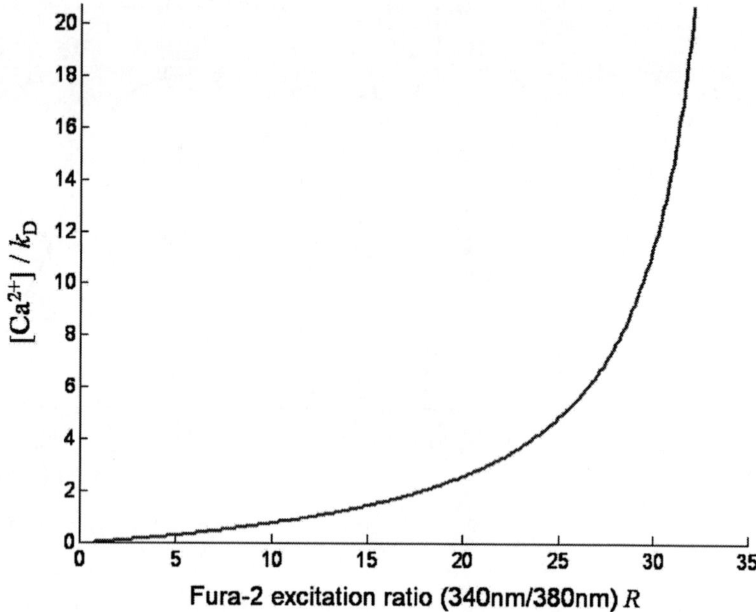

Fig. 3.12. Graphical representation of the Grynkiewicz equation for λ_1 = 340 nm and λ_2 = 380 nm, and reported values R_{min} = 0.768, R_{max} = 35.1, S_{f2}/S_{b2} = 2.01 and k_D = 0.250 μM (6)..

hyperbolic singularity for $R = R_{max}$ which makes the estimate of $[Ca^{2+}]$ problematic when $[Ca^{2+}]$ exceeds the k_D^B by more than ~5-fold (the actual viability range depends on the available signal to noise ratio). However, using ratios and (3.17), dye content and instrumental sensitivity are free to change between one ratio and another since they cancel out in each ratio. Of course, stability is still required within each individual ratio measurement. Also R, R_{min}, R_{max} and S_{f2}/S_{b2} should all be measured on the same instrumentation so that any wavelength biases influences all of them equally.

Thus, though not as large as that of fluo-3 (~100), the dynamic range of fura-2 is a remarkable R_{max}/R_{min} ~ 45. Detailed calibration procedures are described in (7).

Note that (3.17) applies equally well to emission ratios of dyes such as indo-1 (Fig. 3.13a). With minor modifications, the versatile (3.17) accommodates also other dual emission approaches to $[Ca^{2+}]$ measurement, e.g. dye mixtures such as fluo-3 and fura red (8) (Fig. 3.13b) or fluorescent resonance energy transfer (FRET)-based genetically encoded calcium indicators (GECIs), such as: cameleons (9–11); troponin-C based biosensors (12); or D3cpv (13,14).

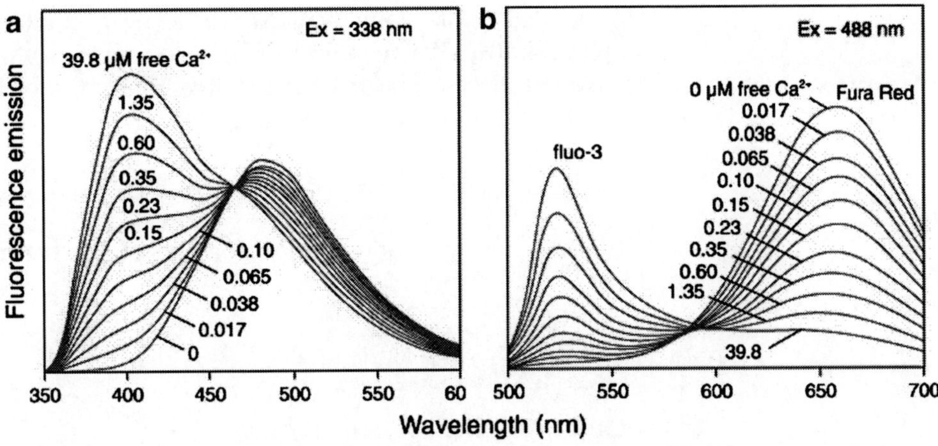

Fig. 3.13. Emission spectra of (**a**) indo-1 and (**b**) a mixture of fluo-3 and fura red, for the indicated values of the free Ca^{2+} concentration.

4. Displacement from Equilibrium

The endogenous buffers normally present inside the cell bind most of the Ca^{2+} that enters the cytoplasm through Ca^{2+} permeable channels. However, because these buffers are relatively immobile and may present with a lower affinity for Ca^{2+}, addition of a mobile high-affinity buffer such as fura-2 allows the exogenous buffer to out-compete these endogenous buffers and alter the nature of the *transient* intracellular free calcium ($[Ca^{2+}]_i$) signals (15). Dramatic examples of how buffering effects alter the time course of fluorescence emission of fura-2 are presented in Fig. 2A and Fig. 3A of (16). Indeed, the key factor in (3.11), (3.12), and (3.17), which are all equilibrium equations based on (3.2) is the dissociation constant k_D^B, i.e. the ratio k^B_{OFF}/k^B_{ON}. By contrast under dynamic, i.e. non-equilibrium conditions, the shape of the fluorescence signals evoked by $[Ca^{2+}]$ transients depends on each one of the factors k_{OFF}^B, k_{ON}^B and c_T. Thus, for a binding reaction between Ca^{2+} and the indicator the rate of approach to equilibrium can vary even for indicators with the same k_D^B. Therefore, a better description of kinetics is the relaxation time τ, which describes how fast a reaction comes to a new equilibrium when one or more of the reactants is suddenly shifted away from its equilibrium concentration. To define τ precisely, consider again the binding reaction

$$[Ca^{2+}]+[B]\underset{k^B_{OFF}}{\overset{k^B_{ON}}{\leftrightarrow}}[CaB]$$

with the initial equilibrium concentrations $[Ca^{2+}]_0, [B]_0, [CaB]_0$. If the system is perturbed, the transiently shifted concentrations

$[Ca^{2+}],[B],[CaB]$ will come rapidly to a new equilibrium $[Ca^{2+}]_1,[B]_1,[CaB]_1$. We therefore define a quantity Δx, which represents the difference between the shifted concentrations and the new equilibrium concentrations that is

$$\Delta x = [Ca^{2+}] - [Ca^{2+}]_1 = [B] - [B]_1 = [CaB]_1 - [CaB].$$

Next, consider the reaction equation for one of the reactants, e.g., $[Ca^{2+}]$:

$$\frac{d[Ca^{2+}]}{dt} = k^{B}_{OFF}[CaB] - k^{B}_{ON}[Ca^{2+}][B],$$

which can be re-written in term of Δx as

$$\frac{d\Delta x}{dt} = k^{B}_{OFF}([CaB]_1 - \Delta x) - k^{B}_{ON}\left([Ca^{2+}]_1 + \Delta x\right)([B]_1 + \Delta x)$$

or

$$\frac{d\Delta x}{dt} = -\left\{k^{B}_{OFF} + k^{B}_{ON}\left([Ca^{2+}]_1 + [B]_1\right)\right\}\Delta x + k^{B}_{OFF}[CaB]_1 - k^{B}_{ON}[Ca^{2+}]_1[B]_1 - k^{B}_{ON}\Delta x^2$$

But, $[Ca^{2+}]_1,[B]_1,[CaB]_1$ are equilibrium values, therefore,

$$k^{B}_{OFF}[CaB]_1 - k^{B}_{ON}[Ca^{2+}]_1[B]_1 = 0.$$

Furthermore, for sufficiently small changes

$$\Delta x \ll [Ca^{2+}]_1 + [B]_1$$

the quadratic term can be neglected and the rate of change of Δx can be approximated by the equation

$$\frac{d\Delta x}{dt} \cong -k_S \Delta x, \qquad (3.18)$$

where the rate constant of the reaction, k_S (with units of s^{-1}), is

$$k_S \equiv \frac{1}{\tau} = k^{B}_{ON}\left([Ca^{2+}]_1 + [B]_1\right) + k^{B}_{OFF}. \qquad (3.19)$$

For example:
fura-2 on-rate $6.0 \times 10^8\,M^{-1}s^{-1}$ off-rate $98\ s^{-1}$
furaptra on-rate $7.5 \times 10^8\,M^{-1}s^{-1}$, off-rate $26{,}760\ s^{-1}$

The prediction that furaptra should be much faster then fura-2 is verified experimentally (Fig. 3.14), lending support to the argument used in the derivation of (3.18) and (3.19). Consider then what happens in a typical experiment that rapidly and markedly displaces $[Ca^{2+}]$ from its equilibrium value as shown,

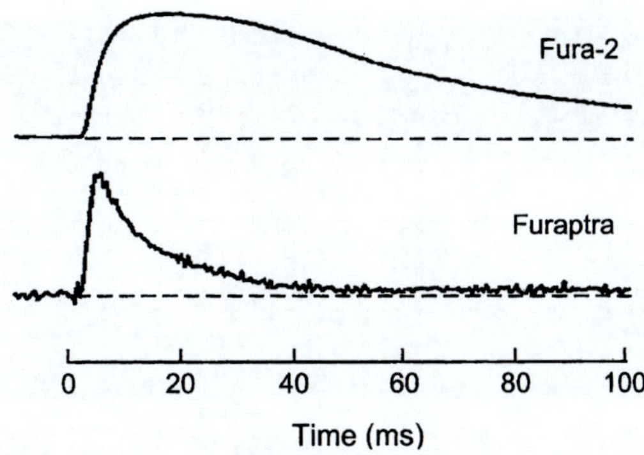

Fig. 3.14. Response to a brief Ca^{2+} influx event, scaled to the same peak amplitude, monitored in muscle cells with high affinity Ca^{2+}-indicator fura-2 and low affinity indicator furaptra (modified from Fig. 1.9 of (44)).

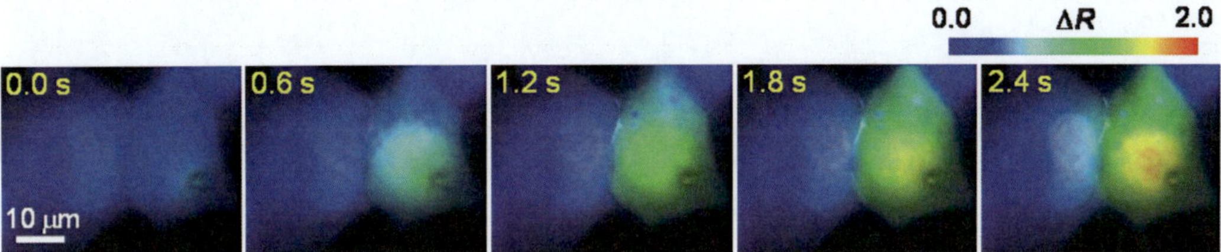

Fig. 3.15. At time $t = 0.0$s a glass micropipette starts to deliver IP_3 into the cytoplasm of the right cell under whole-cell patch clamp conditions. While IP_3 diffuses along its concentration gradient, Ca^{2+} is mobilised from intracellular stores of both cells, with different time courses. Pseudo colours encode the pixel-by-pixel fura-2 ratio change $\Delta R = R - R_0$, where R_0 indicates the pre-stimulus ratio, according to the colour-scale bar shown above. For further details, see (45).

for example, by the sequence of images in Fig. 3.15 collected from a pair of gap-junction coupled HeLa cells loaded with fura-2.

The key question to be asked (and answered!) whenever images of this sort are displayed is: how do the pseudo colours relate to the actual local $[Ca^{2+}]$? Obviously, the matter is not limited to wide-field ratiometric images (Fig. 3.15), but encompasses also the encoding of fluorescence emission from single wavelength dyes, e.g., in confocal line scan experiments (Fig. 3.16).

Despite the countless abuses present in the literature, the very presence of spatial and temporal $[Ca^{2+}]$ gradients which endow the rainbow-coloured pictures shown above with their (potentially) artistically attractive character implies that equilibrium formulae, e.g., (3.11), (3.12), and (3.17), simply *cannot* be used to quantify $[Ca^{2+}]$ under such non-equilibrium conditions. So, what needs to be done if one is to step out of the *pretty pictures* trap using optically measurable quantities to actually estimate $[Ca^{2+}]$ across time and space?

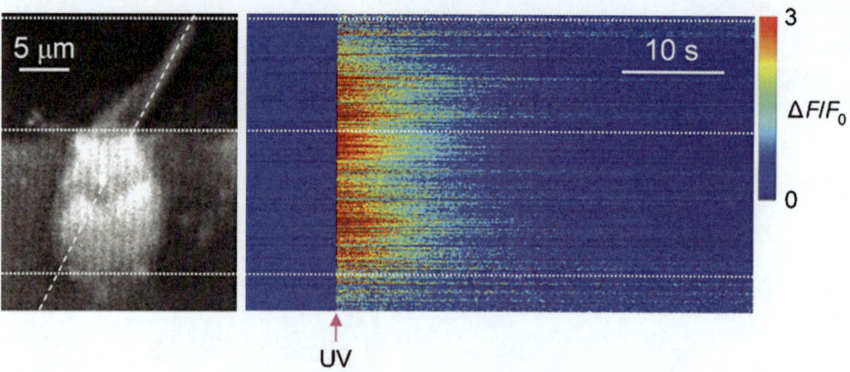

Fig. 3.16. *Left*: confocal image of baseline Fluo-4 fluorescence of a hair cell in an organotypic culture of mouse utricular macula. The diagonal dashed line represents the scan line during subsequent data acquisition. The hair bundle comprises the top and middle horizontal dotted lines (the latter crosses the cell cuticular plate). The cell soma comprises middle and lower lines. Right: in this line-scan image, ordinates are pixel positions along the scan line, abscissa is time and fluorescence transients, $\Delta F/F_0$, evoked by a 4-ns UV pulse (arrow), are color-coded according to the color scale-bar at right. Signals from pixels below the lowermost horizontal dotted line arise from adjacent cells. For further details, see (46).

5. Ca^{2+} Dynamics

Temporary and localized fluxes of Ca^{2+} ions entering the cytosol are responsible for carrying information to cellular targets in their micro-environment (*Ca^{2+} microdomains*), creating a dynamic situation where Ca^{2+} concentration is rapidly lowered by a number of intracellular mechanisms (basic mechanisms are represented in Fig. 3.17).

Besides passive diffusion of Ca^{2+} away from its site of action and buffering by variety of Ca^{2+} binding proteins, control mechanisms include the action of Ca^{2+}–ATPases (PMCA and SERCA pumps) as well as Na^{+}–Ca^{2+} exchangers (NCX) (17). Ca^{2+} signals are also amplified by Ca-induced Ca-release (CICR) mediated by ryanodine receptors (RyRs) and Ins(1,4,5)P_3 receptors (IP_3Rs) in the membrane of the endoplasmic/sarcoplasmic reticulum (ER/SR) (18,19). Other mechanisms, such as voltage- and or ligand-gated Ca^{2+}-permeable channels in the plasma membrane may be at play, serving specialised cell functions.

A mathematical description of Ca^{2+}-related mechanisms provides the only way to estimate the $[Ca^{2+}]_i$ when the system is far from the equilibrium. In this case, the law of mass action (3.1) for Ca^{2+} and one (representative) endogenous buffer C becomes:

$$\frac{\partial[\mathrm{Ca}^{2+}]_\mathrm{i}}{\partial t} = k_\mathrm{OFF}^\mathrm{C}[\mathrm{CaC}] - k_\mathrm{ON}^\mathrm{C}[\mathrm{Ca}^{2+}]_\mathrm{i}[\mathrm{C}] + J, \qquad (3.20)$$

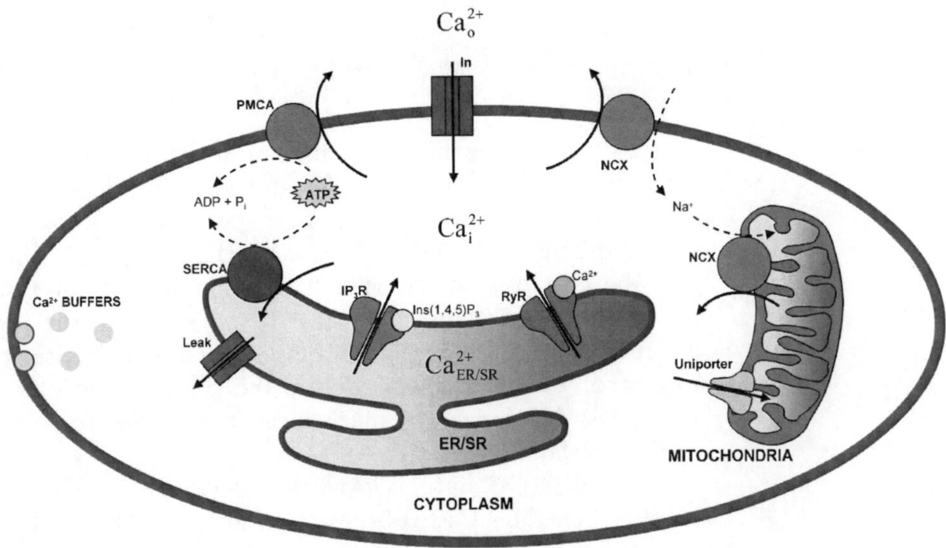

Fig. 3.17. Basic elements involved in Ca^{2+} signalling in eukaryotic cells.

where *t* is time, [C] and [CaC] are the concentrations of free buffer and Ca^{2+} bound to buffer, respectively. *J* (in M/ s^{-1}), the sum of the different Ca^{2+} fluxes entering/exiting the cytoplasm is a function of Ca^{2+} concentrations in various compartments and possibly other variables, e.g., IP_3 concentration. As for the latter, the opening of the IP_3Rs is modulated by IP_3, the binding of which increases the sensitivity of the channel to the surrounding $[Ca^{2+}]_i$. At high levels, cytoplasmic Ca^{2+} exerts a negative feedback action, decreasing the sensitivity to both IP_3 and Ca^{2+}: as $[Ca^{2+}]_i$ rises, the channel begins to close again. The contribution to *J* due to flux of Ca^{2+} out of the calcium store (ER/SR) into the cytosol can be described by the equation

$$J_{\mathrm{ER}} = \left[k_{\mathrm{LEAK}} + k_{\mathrm{IP_3R}}\right]\left(\left[\mathrm{Ca}^{2+}\right]_{\mathrm{ER/SR}} - \left[\mathrm{Ca}^{2+}\right]_{\mathrm{i}}\right), \qquad (3.21)$$

where k_{LEAK} (units of s^{-1}) describes Ca^{2+}-leakage through the ER/SR channels and k_{IP_3R} is a function of both $[Ca^{2+}]_i$ and $[IP_3]$ describing the IP_3 receptor kinetics. As mentioned above, other equations may have to be included in the description of *J* to account for the action of voltage- and/or ligand-gated Ca^{2+} permeable channels in the plasma membrane.

The increase of $[Ca^{2+}]_i$ due to Ca^{2+} influx is rapidly lowered by activation of SERCA and PMCA pumps, which accumulate Ca^{2+} in the ER/SR and outside the cell respectively. Pump action produces a negative Ca^{2+} flux (efflux) that can be included in *J* in the form

$$J_{\text{PUMP}} = -V_{\max} \frac{\left[\text{Ca}^{2+}\right]_{\text{i}}^{\text{m}}}{\left[\text{Ca}^{2+}\right]_{\text{i}}^{\text{m}} + K_{\text{M}}^{\text{m}}}, \tag{3.22}$$

where $V_{max} = n \cdot \nu$ is the pump maximum velocity (n is molar concentration of pump molecules and ν the maximal turnover rate in s^{-1}), K_M is the *Michaelis* constant and the exponent m equals two for the SERCA pumps and one for the PMCAs (2, 20–22). Since the PMCA is capable of effectively binding Ca^{2+} even when its concentrations is quite low, it is better suited to the task of restoring the resting $[Ca^{2+}]_i$. This task is vital when considerable amount of Ca^{2+} enters periodically the cell, e.g., in hair cells (Fig. 3.18) depolarized by mechanical stimuli (due to sound or acceleration).

In general, Ca^{2+} entry is localized and close to cellular targets and thus generates Ca^{2+} concentration gradients within the cell. The spatial and temporal description of diffusion processes is provided by Fick's second law (23):

$$\frac{\partial[\text{Ca}^{2+}]_{\text{i}}}{\partial t} \; D_{\text{Ca}} \cdot \Delta \, [\text{Ca}^{2+}]_{\text{i}}, \tag{3.23}$$

where

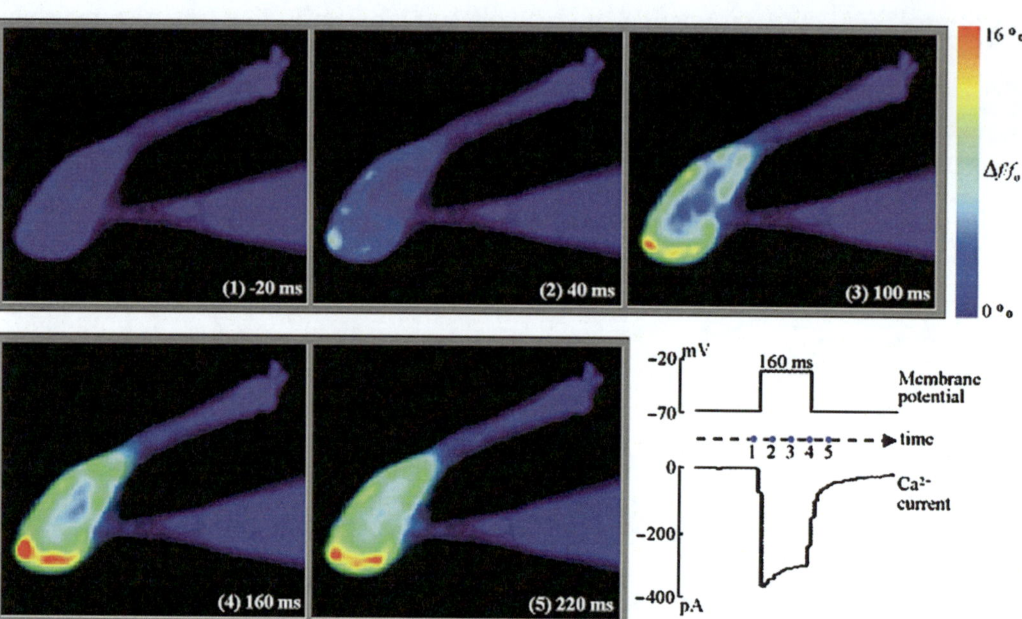

Fig. 3.18. Isolated hair cell from the frog crista ampullaris contacted by a patch pipette entering from the right and containing the membrane-impermeant Ca^{2+}-sensor Oregon Green 488 BAPTA-1. The hair cell was depolarized for 160 ms, from time zero, eliciting $[Ca^{2+}]_i$ transients near the active presynaptic zones (*hotspots*). Inward Ca^{2+} current through voltage-dependent Ca^{2+} channels ($Ca_V1.3$ type) was measured using a CsCl-based intracellular solution that selectively blocked K^+ currents (adapted from (47)).

$$\Delta = \frac{\partial^2}{\partial x^2} + \frac{\partial^2}{\partial y^2} + \frac{\partial^2}{\partial z^2}$$

is the Laplace operator and the dependence of $[Ca^{2+}]_i$ on spatial coordinates $\mathbf{r} = (x, y, z)$ and time t is understood. $D_{Ca} \cong 440 \mu m^2 s^{-1}$ is the diffusion constant of Ca^{2+} in an aqueous medium with the viscosity of the cytosol (3).

Including the contribution due to Ca^{2+} diffusion (3.23) in (3.20) and considering the presence of an exogenous buffer B and an endogenous buffer C, yields the system of reaction–diffusion equations:

$$\begin{cases} \frac{\partial [Ca^{2+}]_i}{\partial t} = k^C_{OFF}[CaC] - k^C_{ON}[Ca^{2+}]_i[C] + D_{Ca} \cdot \Delta[Ca^{2+}]_i + J \\ \frac{\partial [B]}{\partial t} = k^B_{OFF}[CaB] - k^B_{ON}[Ca^{2+}]_i[B] + D_B \cdot \Delta[B] \\ \frac{\partial [CaB]}{\partial t} = -k^B_{OFF}[CaB] + k^B_{ON}[Ca^{2+}]_i[B] + D_{CaB} \cdot \Delta[CaB] \\ \frac{\partial [C]}{\partial t} = k^C_{OFF}[CaC] - k^C_{ON}[Ca^{2+}]_i[C] + D_C \cdot \Delta[C] \\ \frac{\partial [CaC]}{\partial t} = -k^C_{OFF}[CaC] + k^C_{ON}[Ca^{2+}]_i[C] + D_{CaC} \cdot \Delta[CaC] \end{cases} \quad (3.24)$$

where the diffusion coefficients of free buffers and Ca^{2+}-bound to buffers are almost equal. In particular, D_C, D_{CaC} are between 70 and 90 $\mu m^2 s^{-1}$ for the three important Ca^{2+} buffers calbindin–D_{28K}, calretinin and parvalbumin (3) while D_B, D_{CaB} are generally larger, e.g., $D_B = 220\ \mu m^2 s^{-1}$ for Oregon Green 488 BAPTA-1.

Thus, to estimate $[Ca^{2+}]_i$ as a function of $\mathbf{r} = (x, y, z)$ and t, one needs to solve the above system of equations (3.24). Classical methods to numerically solve (systems of) partial differential equations (PDEs), subdivide the solution space (e.g. the cytoplasm of the cell) in several compartments (*voxels* or *shells*) (24–27). Discretization of the Laplace operator then leads to a system of ordinary differential equations (ODEs) for each voxel/shell of the 3D diffusion volume (Fig. 3.19), which take into account the contribution to J due to the adjacent compartments.

Numerical solution of very large set of simultaneous ODEs is typical in Systems Biology and represents an important constraint in the choice of the model parameters. In fact, CPU time consumption increases as the third power of the spatial resolution, e.g., a reaction–diffusion space of 1 μm^3 could be subdivided in 8,000 cubic voxels of side $L = 50$ nm or in 10^6 voxels of side $L = 10$ nm.

An alternative approach to simulate molecular diffusion is the Monte Carlo method. This type of simulations, which are essentially based upon the repetitive generation of random numbers, have been used to study reaction and diffusion processes in biological systems (28–34) including Ca^{2+} dynamics (35–40).

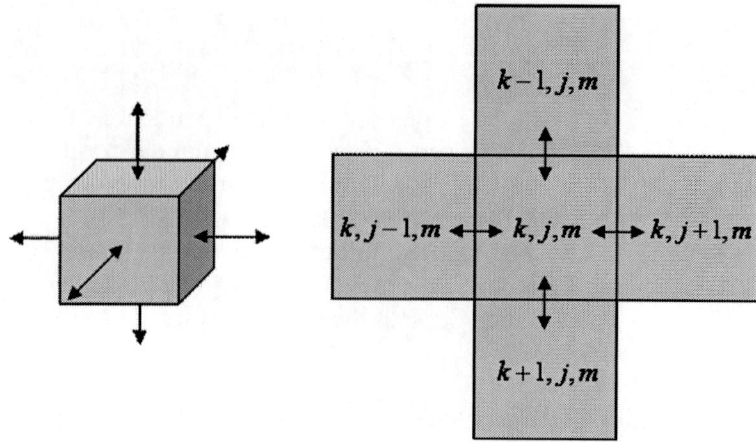

Fig. 3.19. Example of space discretization in cubic voxels. To solve (3.24) in the (*k*, *j*, *m*) indexed voxel we have to solve a system with further six equations corresponding to the orthogonal fluxes from the adjacent voxels.

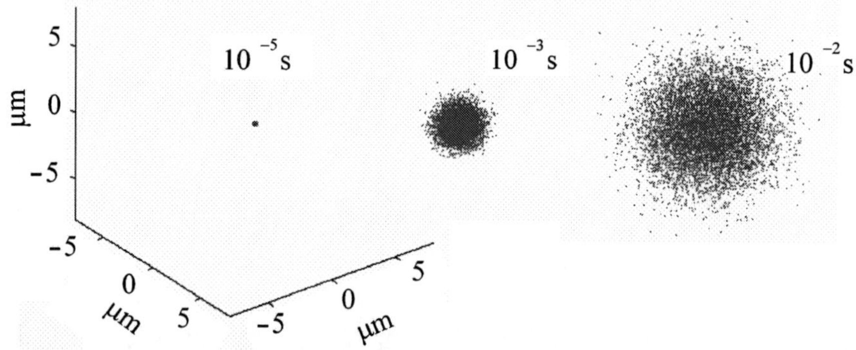

Fig. 3.20. Random walk simulation for 10^4 particles diffusing in an unlimited isotropic medium shown at three different times after the onset of diffusion from a point source.

The Monte Carlo algorithm can be exploited to approximate the Brownian motion of the real molecules (*random walk*) with that of a suitable number of *particles* (Fig. 3.20). One advantage of the method is also the possibility of simulating with great accuracy 3D diffusion in the presence of realistic boundaries (3).

The study of Ca^{2+} dynamics in living cells typically combines Ca^{2+}-sensitive fluorescent dyes, patch clamp and optical microscopy to produce images of the patterns of fluorescence of a Ca^{2+} indicator following various stimulation protocols. In the most favourable case, the only observable variable is a linear combination of the fluorescence emission of free dye and dye bound to Ca^{2+}. Thus, the most interesting unobservable variables such as $[Ca^{2+}]_i$ or [C] (Fig. 3.21) must be deduced only after optimization by trial and error of the model free parameters with a fitting procedure of the model output to the experimental data.

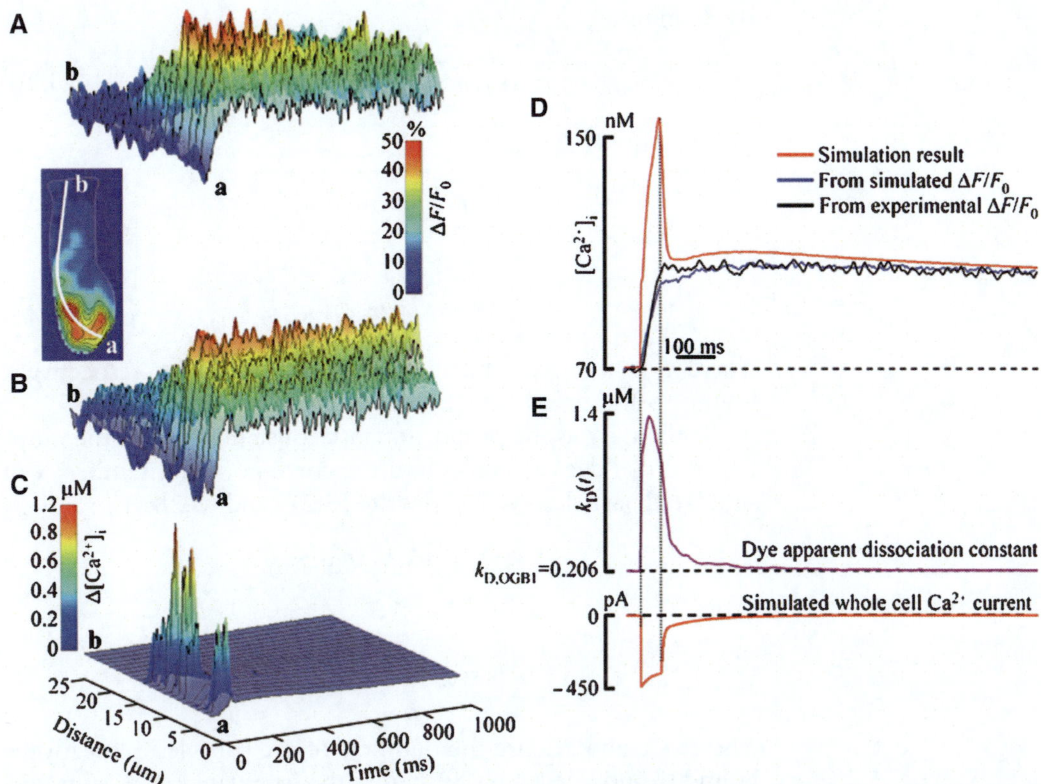

Fig. 3.21. Fluorescence signals appear as low pass filtered versions of the $[Ca^{2+}]_i$ in the presence of sustained Ca^{2+} influx. The model simulates the buffered diffusion of Ca^{2+} from presynaptic hotspots following 50-ms depolarization of the hair cell plasma membrane, reconstructed in three dimensions (3). (**A**) Pseudoline-scan representation of the experimental $\Delta F/F_0$ signal obtained with 50-μM Oregon Green 488 BAPTA-1; abscissas represent time and ordinates are distance along the *white line* (**a–b**) passing through a number of hotspots (*inset*). (**B**) Similar $\Delta F/F_0$ signals were obtained for the modelled cell containing 1.7 mM of a fast endogenous Ca^{2+} buffer. (**C**) $\Delta[Ca^{2+}]_i$ changes corresponding to the simulation in (**B**). (**D**) Time course of the simulated $[Ca^{2+}]_i$, integrated over the entire cell volume (*red trace*) and $[Ca^{2+}]_i$ values derived either from the simulated (*blue trace*) or from the experimental (*black trace*) $\Delta F/F_0$ whole cell signals, based on the law of mass action at equilibrium (3.11). (**E**) Average dissociation "constant" of the Oregon Green 488 BAPTA-1 dye (*top*) computed from (3.2) from the local simulated reactant concentrations accounts for the nonequilibrium of the system, perturbed by the whole cell Ca^{2+} current (*bottom*) due to cell depolarization. For further details of the model, see (3).

The original solution provided by the Monte Carlo model in (3) is the generation of a dye virtual fluorescence to be compared with that recorded during the experiments. At a given emission wavelength, the measured fluorescence signal F can be expressed by (3.7) as:

$$F = \bar{S}_f \cdot n_f + \bar{S}_b \cdot n_b, \tag{3.25}$$

where n_f is the number of free dye molecules and n_b is the number of dye molecules buffered to Ca^{2+}, $\bar{S}_f$ and $\bar{S}_b$ are proportional to S_f and S_b in (3.7), respectively.

By defining

$$\alpha = \bar{S}_b / \bar{S}_f = F_{max} / F_{min}, \tag{3.26}$$

we can write (3.25) as:

$$F = \bar{S}_f \cdot (\alpha n_b + n_f) \tag{3.27}$$

and

$$F_0 = \bar{S}_f \cdot (\alpha n_{b,0} + n_{f,0}) \tag{3.28}$$

where $n_{b,0}$ and $n_{f,0}$ are the numbers of dye molecules, respectively, bound to and free from Ca^{2+} before the Ca^{2+} stimulus.

Defining χ as the proportionality constant between the number n of real dye molecules and the corresponding number N of simulated particles, we can rewrite (3.27) and (3.28) as:

$$F = \chi \bar{S}_f \cdot (\alpha N_b + N_f)$$

and

$$F_0 = \chi \bar{S}_f \cdot (\alpha N_{b,0} + N_{f,0})$$

where $N_{b,0}$ and $N_{f,0}$ are the numbers of dye particles, respectively, bound to and free from Ca^{2+} before the (simulated) Ca^{2+} stimulus. The last two equations permit us to compute the useful quantity:

$$\frac{\Delta F}{F_0} = \frac{F - F_0}{F_0} = \frac{\alpha N_b + N_f - \alpha N_{b,0} - N_{f,0}}{\alpha N_{b,0} + N_{f,0}} \tag{3.29}$$

and thus to convert particle counts to simulated fluorescence signals. Note that the parameter α we used to generate the virtual $\Delta F/F_0$ signals is characteristic of the dye and is relatively insensitive to the experimental conditions, whereas the absolute values of F_{max} and F_{min} may vary from experiment to experiment. For example, Oregon Green 488 BAPTA-1 has $\alpha \cong 5$ (41) while for Fluo-4 $\alpha \cong 100$ (42).

It is also possible to include the effect of the poor axial resolution of wide field microscopy on $\Delta F/F_0$ by considering the relationship between the fluorescence intensity $F(z)$ due to a point source and the source distance, z, from the focal plane ($z=0$). In particular, the ratio $F(z)/F(0)$ can be estimated as shown in (43).

Acknowledgments

This work has been supported by grants to FM from Fondazione Cariparo (Progetti di Eccellenza 2006), MIUR PRIN Grant n. 2007BZ4RX3_003 and the European commission FP6 Integrated

Project EuroHear (LSHGCT20054512063) under the Sixth Research Frame Program of The European Union. We thank the Editor of this book, Alexej Verkhratsky, for discussions and constructive criticism.

References

1. Tsien RY (1980) New calcium indicators and buffers with high selectivity against magnesium and protons: design, synthesis, and properties of prototype structures. Biochemistry 19(11):2396–2404
2. Wu YC, Tucker T, Fettiplace R (1996) A theoretical study of calcium microdomains in turtle hair cells. Biophys J 71(5):2256–2275
3. Bortolozzi M, Lelli A, Mammano F (2008) Calcium microdomains at presynaptic active zones of vertebrate hair cells unmasked by stochastic deconvolution. Cell Calcium 44(2):158–168
4. Thomas D et al (2000) A comparison of fluorescent Ca2+ indicator properties and their use in measuring elementary and global Ca2+ signals. Cell Calcium 28(4):213–223
5. Hyrc KL, Bownik JM, Goldberg MP (2000) Ionic selectivity of low-affinity ratiometric calcium indicators: mag-Fura-2, Fura-2FF and BTC. Cell Calcium 27(2):75–86
6. Grynkiewicz G, Poenie M, Tsien RY (1985) A new generation of Ca2+ indicators with greatly improved fluorescence properties. J Biol Chem 260(6):3440–3450
7. Kao JPY (1994) Practical aspects of measuring [Ca2+] with fluorescent indicators. In: Nuccitelli R (ed) A practical guide to the study of calcium in living cells. Academic, San Diego, p 155–181
8. Lipp P, Niggli E (1993) Ratiometric confocal Ca(2+)-measurements with visible wavelength indicators in isolated cardiac myocytes. Cell Calcium 14(5):359–372
9. Miyawaki A et al (1997) Fluorescent indicators for Ca2+ based on green fluorescent proteins and calmodulin. Nature 388(6645):882–887
10. Truong K et al (2001) FRET-based in vivo Ca2+ imaging by a new calmodulin-GFP fusion molecule. Nat Struct Biol 8(12): 1069–1073
11. Palmer AE, Tsien RY (2006) Measuring calcium signaling using genetically targetable fluorescent indicators. Nat Protoc 1(3):1057–1065
12. Heim N et al (2007) Improved calcium imaging in transgenic mice expressing a troponin C-based biosensor. Nat Methods 4(2):127–129
13. Wallace DJ et al (2008) Single-spike detection in vitro and in vivo with a genetic Ca2+ sensor. Nat Methods 5(9):797–804
14. Hendel T et al (2008) Fluorescence changes of genetic calcium indicators and OGB-1 correlated with neural activity and calcium in vivo and in vitro. J Neurosci 28(29):7399–7411
15. Neher E (1995) The use of fura-2 for estimating Ca buffers and Ca fluxes. Neuropharmacology 34(11):1423–1442
16. Helmchen F, Imoto K, Sakmann B (1996) Ca2+ buffering and action potential-evoked Ca2+ signaling in dendrites of pyramidal neurons. Biophys J 70(2):1069–1081
17. Berridge MJ, Bootman MD, Roderick HL (2003) Calcium signalling: dynamics, homeostasis and remodelling. Nat Rev Mol Cell Biol 4(7):517–529
18. Verkhratsky A (2005) Physiology and pathophysiology of the calcium store in the endoplasmic reticulum of neurons. Physiol Rev 85(1):201–279
19. Rizzuto R, Pozzan T (2006) Microdomains of intracellular Ca2+: molecular determinants and functional consequences. Physiol Rev 86(1):369–408
20. Dumont RA et al (2001) Plasma membrane Ca2+-ATPase isoform 2a is the PMCA of hair bundles. J Neurosci 21(14):5066–5078
21. Goldbeter A, Dupont G, Berridge MJ (1990) Minimal model for signal-induced Ca2+ oscillations and for their frequency encoding through protein phosphorylation. Proc Natl Acad Sci USA 87(4):1461–1465
22. Lumpkin EA, Hudspeth AJ (1998) Regulation of free Ca2+ concentration in hair-cell stereocilia. J Neurosci 18(16):6300–6318
23. Crank J (1975) The mathematics of diffusion, 2nd edn. Oxford University Press, London, p 424
24. Canepari M, Mammano F (1999) Imaging neuronal calcium fluorescence at high spatiotemporal resolution. J Neurosci Methods 87(1):1–11
25. Roberts WM (1994) Localization of calcium signals by a mobile calcium buffer in frog saccular hair cells. J Neurosci 14(5 Pt 2): 3246–3262

26. Klingauf J, Neher E (1997) Modeling buffered Ca2+ diffusion near the membrane: implications for secretion in neuroendocrine cells. Biophys J 72(2 Pt 1):674–690
27. Nowycky MC, Pinter MJ (1993) Time courses of calcium and calcium-bound buffers following calcium influx in a model cell. Biophys J 64(1):77–91
28. Riley MR et al (1995) Monte Carlo simulation of diffusion and reaction in two-dimensional cell structures. Biophys J 68(5):1716–1726
29. Saxton MJ (1994) Anomalous diffusion due to obstacles: a Monte Carlo study. Biophys J 66(2 Pt 1):394–401
30. Saxton MJ (1996) Anomalous diffusion due to binding: a Monte Carlo study. Biophys J 70(3):1250–1262
31. Bartol TM Jr et al (1991) Monte Carlo simulation of miniature endplate current generation in the vertebrate neuromuscular junction. Biophys J 59(6):1290–1307
32. Kruk PJ, Korn H, Faber DS (1997) The effects of geometrical parameters on synaptic transmission: a Monte Carlo simulation study. Biophys J 73(6):2874–2890
33. Olveczky BP, Verkman AS (1998) Monte Carlo analysis of obstructed diffusion in three dimensions: application to molecular diffusion in organelles. Biophys J 74(5): 2722–2730
34. Stibitz GR (1969) Calculating diffusion in biological systems by random walks with special reference to gases diffusion in the lung. Respir Physiol 7(2):230–262
35. Gil A et al (2000) Monte carlo simulation of 3-D buffered Ca(2+) diffusion in neuroendocrine cells. Biophys J 78(1):13–33
36. Bennett MR, Farnell L, Gibson WG (2000) The probability of quantal secretion near a single calcium channel of an active zone. Biophys J 78(5):2201–2221
37. Segura J, Gil A, Soria B (2000) Modeling study of exocytosis in neuroendocrine cells: influence of the geometrical parameters. Biophys J 79(4):1771–1786
38. Coggan JS et al (2005) Evidence for ectopic neurotransmission at a neuronal synapse. Science 309(5733):446–451
39. He L et al (2006) Two modes of fusion pore opening revealed by cell-attached recordings at a synapse. Nature 444(7115):102–105
40. Stern MD, Cheng H (2004) Putting out the fire: what terminates calcium-induced calcium release in cardiac muscle? Cell Calcium 35(6):591–601
41. Maravall M et al (2000) Estimating intracellular calcium concentrations and buffering without wavelength ratioing. Biophys J 78(5):2655–2667
42. Woodruff ML et al (2002) Measurement of cytoplasmic calcium concentration in the rods of wild-type and transducin knock-out mice. J Physiol 542(Pt 3):843–854
43. Hiraoka Y, Sedat JW, Agard DA (1990) Determination of three-dimensional imaging properties of a light microscope system. Partial confocal behavior in epifluorescence microscopy. Biophys J 57(2):325–333
44. Poenie M (2006) Fluorescent calcium indicators based on BAPTA. In: Putney JW Jr (ed) Calcium signaling. CRC Taylor & Francis, Boca Raton, FL, p 1–50.
45. Beltramello M et al (2005) Impaired permeability to Ins(1, 4, 5)P3 in a mutant connexin underlies recessive hereditary deafness. Nat Cell Biol 7(1):63–69
46. Spiden SL et al (2008) The novel mouse mutation Oblivion inactivates the PMCA2 pump and causes progressive hearing loss. PLoS Genet 4(10):e1000238
47. Lelli A et al (2003) Presynaptic calcium stores modulate afferent release in vestibular hair cells. J Neurosci 23(17):6894–6903

Chapter 4

Bioluminescent Ca^{2+} Indicators

Laura Fedrizzi and Marisa Brini

Abstract

In the last two decades, the study of Ca^{2+} homeostasis in living cells received a great impulse by the explosive development of genetically encoded Ca^{2+}-indicators. The cloning of the Ca^{2+}-sensitive photoprotein aequorin and of the green fluorescent protein (GFP) from the jellyfish *Aequorea victoria* has been enormously advantageous for the biologists.

As polypeptides, aequorin and GFP allow their endogenous production in cell system as diverse as bacteria, yeast, slime moulds, plants and mammalian cells. Moreover, it is possible to specifically localize them within the cell by including defined targeting signals in the amino acid sequence.

These two proteins have been extensively engineerized to obtain several recombinant probes for different biological parameters, among which Ca^{2+} concentration reporters are probably the most relevant. In this review, we will not treat the GFP-based Ca^{2+} probes, but we will present the applications offered by aequorin in the study of intracellular Ca^{2+} homeostasis, discussing also the new generation of bioluminescent probes that couple the Ca^{2+} sensitivity of aequorin to GFP fluorescence emission. In these probes, aequorin Ca^{2+}-dependent photon emission delivers energy to the GFP acceptor in a bioluminescence resonance energy transfer (BRET): this process enhances the stability and the high signal-to noise ratio of the probes and permits real-time measurements of subcellular Ca^{2+} changes in single cell imaging experiments. Very recently, the development of transgenic animals expressing GFP–aequorin bi-functional probes has also permitted the video-imaging of Ca^{2+} concentrations changes in live animals.

Key words: Bioluminescence, Aequorin, GFP–aequorin, Ca^{2+} signalling, Organelles, In vivo imaging

1. Introduction

The field of the neuronal cell biology has undergone a rapid expansion in the recent last years. The development of new tools and elaborated technological systems for the investigation of neuronal cells has contributed extensively to this phenomenon. Ca^{2+} signalling represents one of the major areas of interest. The spatio-temporal specificity of Ca^{2+} transients is emerging as essential

A. Verkhratsky and Ole H. Petersen (eds.), *Calcium Measurement Methods*, Neuromethods, vol. 43,
DOI: 10.1007/978-1-60761-476-0_4, © Humana Press, a part of Springer Science+Business Media, LLC 2010

element to the correct delivery and the decoding of the Ca^{2+} messages. The possibility to monitor Ca^{2+} fluxes inside specific intracellular compartments or specialized regions within the cells contributes to define details in the understanding of the physiology of different cell types and tissues among which neuronal cells. The characterization of Ca^{2+} signalling during the development, the plasticity, and the functioning of the central nervous system starts also to contribute information on Ca^{2+} dysregulation implicated in many neurological disorders such as Alzheimer's, Parkinson's, Huntington's disease and migraine (1–4).

Numerous Ca^{2+} indicators became available during the years for the characterization of the Ca^{2+} signal. Actually, these probes are represented by the fluorescent dyes and the genetically encoded fluorescent or luminescent proteins. All of them show some advantages and disadvantages and the choice of one or the other depends on the biological problem that has to be investigated. In this chapter, we will discuss why we need to choose the bioluminescent proteins to monitor Ca^{2+} concentration. We will also provide a comprehensive overview about principles, applications, and the more recent developments in field of Ca^{2+} monitoring based on the detection of bioluminescence of the recombinant Ca^{2+} sensitive photoprotein aequorin.

2. Why Choosing Bioluminescence?

Ca^{2+} signals can be detected using different Ca^{2+} reporters, i.e. molecules sensitive to Ca^{2+} which undergo to physico-chemical changes after the binding with the ion. Usually, these variations are sufficiently prominent to be monitored and converted in Ca^{2+} concentration values. Significant developments in Ca^{2+} imaging, the suitability of different techniques and indicators have permitted the monitoring of local Ca^{2+} signals in different cell types and the acquisition of fine spatio-temporal details.

Fluorescent chemical probes as small organic molecules trappable in the cytoplasm and thanks to their high dynamic range, their easy use and calibration are wide employed (for details see Chap. 5 in this issue), but, in recent years, the molecular cloning and the subsequent engineering of aequorin and green fluorescent protein (GFP) from the bioluminescent jellyfish *Aequorea Victoria* have permitted a strong contribution to the development of new Ca^{2+} sensors particularly useful in the study of Ca^{2+} signalling in specific cell districts.

In the jellyfish *Aequorea victoria*, Ca^{2+} sensitive photoprotein aequorin is associated with the GFP to emit a bioluminescent green light signal (λ_{max} 510 nm) upon Ca^{2+} binding. The binding of Ca^{2+} to aequorin results in the peroxydation of an aromatic residue of the prosthetic group of aequorin, the coelenterazine. This reaction produces a blue light (λ_{max} 470 nm) (5) that, in turn, generates an intramolecular bio/chemioluminescence resonance energy transfer (BRET or CRET) to the acceptor GFP (6). These two proteins were initially utilized to develop independent probes (see also Chap. 5 in this issue), but recently they were combined to originate the so-called "bi-functional Ca^{2+} reporter gene" GFP–aequorin (7).

One of the big advancements derived from the developing of these genetically encoded probes is represented by the possibility to add target sequences to their sequences and thus localize them in a specific cell compartment where Ca^{2+} can be directly monitored. The Ca^{2+}-sensitive photoprotein aequorin has represented the first engineerized recombinant probe that has permitted to acquire important information on the organellar Ca^{2+} homeostasis. Successful expression has been obtained from different cell compartments, such as the endoplasmic reticulum, the Golgi apparatus, the nucleus, the mitochondrial matrix, the plasma membrane and the peroxisomes (for a review see Ref. (8)) and in different cell types such as epithelial cell lines (HeLa, CHO, HEK 293, etc.), primary cultures of neurons, skeletal and cardiac muscles, plant cells, yeast, etc.

In neuronal cells, and particularly in primary cultures of neuronal cells, unfortunately, the application of genetically encoded probes is limited by the difficulty to introduce them with the conventional transfection methods. However, in the last years, the development of different types of viral vectors has overcome this limitation and a number of paper describe their use in neurons as well (9–13).

GFP has been initially utilized for different cell biology applications, i.e. as reporter for gene expression, as transfection marker or marker for organelles, cell structures and specific proteins. In the last decade, GFP and its variants have undergone enormous molecular engineering to generate specific biochemical sensors and monitor several biologically relevant parameters, among which Ca^{2+} is the prominent one. GFP mutants were modified by adding a Ca^{2+}-sensor to their sequence thus generating the so-called cameleons, camgaroos, pericams, etc. (see Chap. 5 in this issue).

Although the above mentioned genetically encoded fluorescent indicators could represent a good choice for single cell analyses, the necessity of light excitation to reveal their fluorescence can cause some disadvantages like phototoxicity, photobleaching and autofluorescence. The recently developed bioluminescent aequorin-based reporters overcome these aspects since they do not require to be excited by artificial illumination: the light produced by the oxidation of the prosthetic group following Ca^{2+} binding is sufficient to trigger the emission of fluorescence through the BRET phenomena. Moreover, the production of transgenic animal models expressing these probes is improving the knowledge about the Ca^{2+}-regulated neuronal processes *in vivo* (14, 15).

2.1. The Ca^{2+}-Sensitive Photoprotein Aequorin

2.1.1. Aequorin History

Aequorin is a protein of 22 kDa containing four helix-loop-helix "EF-hand" domains, of which three can bind Ca^{2+}. The apoprotein binds an imidazopyrazine compound (the prosthetic group coelenterazine) and molecular oxygen to form a stable photoprotein complex (5). The resolution of its structure by X-ray crystallography has revealed that the four EF hand domains are arranged in pairs to form a globular moiety with a central hydrophobic core cavity that accommodates the ligand coelenterazine-2-hydroperoxide (16). Upon binding of Ca^{2+}, the photoprotein undergoes conformational changes leading to the oxidation of coelenterazine with the release of CO_2 and blue light.

Aequorin has been isolated, extracted, and purified from the jellyfish *Aequorea victoria* in 1962 (17). Purified apoaequorin can be regenerated into active aequorin in the absence of Ca^{2+} by incubation with coelenterazine, oxygen and a thiol agent. Thanks to its Ca^{2+} triggered light emission and the reliable calibration procedure, aequorin has been widely employed for studying intracellular Ca^{2+} during the 1960s and 1970s, before the development of synthetic Ca^{2+} fluorescent dyes (18–21). Since it is not cell permeable, aequorin has especially been used in giant cells, i.e. barnacle muscle fibres, oocytes, or cells from heart, liver and adrenal gland, where it was introduced by permeabilization or microinjection procedures.

In 1985, the cloning of the cDNA of aequorin (22) and the possibility to introduce it in expression vectors has re-discovered and expanded the use of aequorin for Ca^{2+} measurements in living cells. By these means, it is possible to express the recombinant protein in a wide variety of cell types (23) and to transduce by viral infection primary cell cultures of different embryological origin which are usually more resistant to the standard transfection protocols.

2.1.2. Characteristics of a Versatile Ca^{2+} Probe

The employment of aequorin in monitoring Ca^{2+} is particularly convenient for a number of advantages, among them the ease of use, the low cost and the simplicity of the equipment used for the measurements. Aequorin can be expressed in a variety of cell types and it efficiently folds into a fully functional protein. To reconstitute active aequorin, it is sufficient to incubate the synthetic prosthetic group (commercially available) in cell culture medium or in physiological saline buffer (see Sect. 4.3 for details). Recombinantly expressed aequorin is not exported or secreted, nor it is compartmentalized or sequestered within cells thus permitting to detect Ca^{2+} changes that occur over relatively long periods. Aequorin is not toxic and also does not perturb cell functions or embryo development (24).

As already mentioned, the enormous advantage of aequorin is that it can be employed to design probes directed to the desired specific cell localization by adding to its cDNA specific targeting sequences. Furthermore, aequorin can be more convenient in respect to chemical fluorescent dyes, also to measure cytosolic Ca^{2+} concentration in several conditions:

- *To have a real quantification of Ca^{2+} concentration values.* Aequorin has a very low buffering capacity when compared with the fluorescent dyes being its K_d around 10 μM. In addition, recombinant aequorin is usually expressed at concentration in the range of 0.1–1 μM, which is 2–3 orders of magnitude lower than that of trapped fluorescent dyes.
- *To monitor Ca^{2+} concentration in a subpopulation of cells which express a particular repertoire of proteins.* That is in some cases it is necessary to determine whether the expression of a specific protein of interest could alter the Ca^{2+} homeostasis. According to the most common transfection protocols, the co-transfected plasmids enter in the same subset of cells thus avoiding the necessity to select stable clones or identified the transfected cells by co-transfecting suitable markers and analyse them in imaging experiments.
- *To monitor cytosolic Ca^{2+} variations in the range of micromolar values that may occur for example in excitable cells such as neurons.* Native aequorin permits to reach concentrations at which most of the fluorescent indicators are saturated, being particularly suitable for measuring Ca^{2+} at concentration between 0.5 and 10 μM.

Furthermore, aequorin is insensitive to other divalent ions (with exception to Sr^{2+}) and to variations of the pH in the physiological range (6.6–7.4) thus representing an optimal probe to measure Ca^{2+} also when these parameters change.

Another important characteristic of recombinant aequorin is represented by the possibility to modify its Ca^{2+} sensitivity and thus to accurately measure different ranges of Ca^{2+} concentration. Unfortunately, it is not possible to increase its Ca^{2+} sensitivity thus this probe is not appropriate for accurate measurements of resting Ca^{2+} concentrations (i.e. in the range of 100–300 nM values), but different strategies can be adopted to reduce the Ca^{2+} sensitivity thus making aequorin suitable to measure Ca^{2+} in compartments where its concentration is high such as the intracellular stores and in the proximity of Ca^{2+} channels. Essentially, three modifications can be introduced: (1) the mutation of amino acid residues in the "EF-hand" domains to reduce the affinity of aequorin for Ca^{2+} (25, 26); (2) the use of surrogate cations which elicit a slower rate of photoprotein consumption (e.g. Sr^{2+}) (26, 27); (3) the employment of modified synthetic prosthetic groups which reduce the affinity of aequorin for Ca^{2+} (28). The combination of these approaches has permitted to expand the range of measurable Ca^{2+} concentration from 0.5–10 μM to mM values.

Aequorin is particularly suitable in cell population measurements since its quantum yield emission is low, and only sophisticated high sensitivity cameras detect the photon emission from single cell (29). Each aequorin molecule emits one photon and only a fraction of the total photoprotein pool emits light throughout the experiment: thus, in population studies (generally 10^3–10^4 cells) the light signals vary from 20–30 photons (at resting Ca^{2+} concentrations) to 10^5 photons per second. Since single cell imaging requires much higher emission of light for a high-quality temporal resolution of the image, this aspect represents a big limitation of the aequorin-based method. To overcome this limit, very recently, new bioluminescent Ca^{2+} probes have been developed and they will be discussed in the next section.

2.2. New Bioluminescent Ca^{2+} Probes

To overcome aequorin low light emission and to improve bioluminescence signal, new Ca^{2+} sensitive reporter genes have been developed by combining the fluorescent properties of GFP to the Ca^{2+} sensitivity of aequorin (7). In these reporters, the two proteins were fused: the aequorin moiety acts as Ca^{2+} sensor which delivers emission energy to the GFP acceptor in a BRET process similar to that observed in natural conditions in the jellyfish. Optimization of the energy transfer between the two molecules was obtained by the insertion of a covalent link which was designed on purpose after the evaluation of different ratio of green over blue light emission. Using the same approach adopted for the aequorin chimeras, several GFP–aequorins have been targeted to specific cell districts, including mitochondrial matrix, endoplasmic reticulum and to microdomains important in synaptic transmission such as synaptic vesicles and postsynaptic density. The stability and the high signal-to-noise ratio of these new reporters are important properties that enable real-time measurements of subcellular Ca^{2+} changes in single mammalian neurons (7). GFP fluorescence allows to visualize and choose the appropriate neuronal area on transfected cells. Ca^{2+} signal can be recorded using a modified epifluorescence microscope where the intensities of BRET activity were translated in pseudo-colour or using a luminometer according to a protocol similar to that used for the detection of aequorin alone (see below).

3. Methods

3.1. Available Recombinant Aequorins

Using appropriate regulatory elements and peptide signals, aequorin can be directed into virtually any specific location inside the cell. Several chimeras have been constructed by fusing the

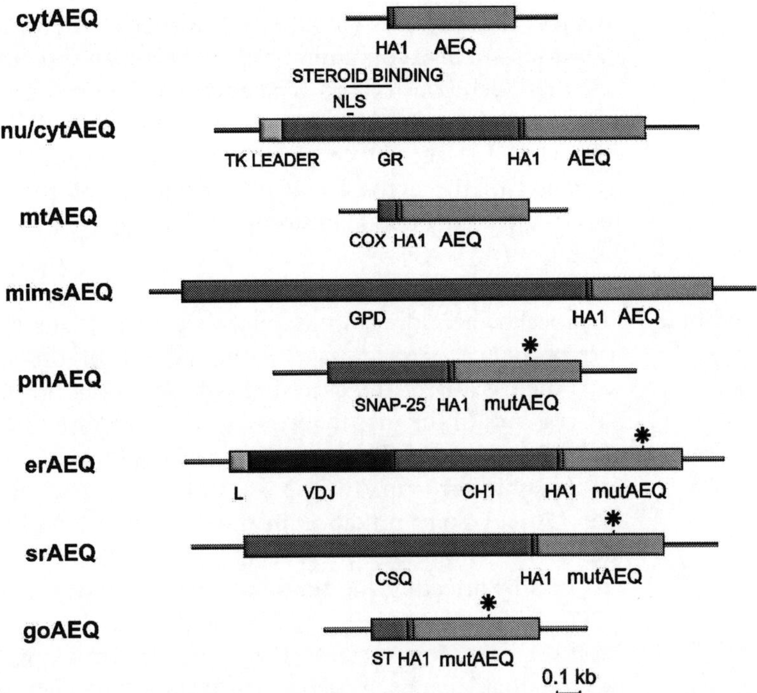

Fig. 4.1. Schematic representation of different HA1-targeted aequorin (AEQ) chimeras. The strategy for the specific targeting of the chimeras in the different cellular compartments is described in the text. *Asterisk* shows the position of the (Asp→119Ala) mutation in the second "EF-hand" Ca^{2+} binding site of aequorin. *mutAEQ* mutated aequorin; *cytAEQ* cytosolic aequorin; *nu/cytAEQ* nucleus/cytosol shuttling aequorin; *mtAEQ* mitochondrial aequorin; *mimsAEQ* mitochondrial intermembrane space aequorin; *pmAEQ* plasma membrane aequorin; *erAEQ* endoplasmic reticulum aequorin; *srAEQ* sarcoplasmic reticulum aequorin; *goAEQ* Golgi apparatus aequorin.

appropriate leader sequence to the aequorin cDNA. The only modification inserted to aequorin cDNA has been the addition of a short sequence coding for the hemagglutinin epitope (HA1), a strong tag of nine amino acids which has been helpful in the immunolocalization of the transfected recombinant proteins. Some of the chimeras are summarised in Fig. 4.1. Recently, commercial antibodies that directly recognize aequorin sequence also became available.

3.1.1. Cytosol

Wild-type aequorin is exclusively expressed in the cytosol and does not require any modification to detect Ca^{2+} in this region. The recombinant protein has only been modified by adding the sequence coding for the HA1 epitope tag (30).

3.1.2. Nucleus

Two different aequorin chimeras to monitor Ca^{2+} concentration in the nucleoplasm are available. The constitutive targeting of the aequorin to the nucleus has been obtained by fusing the sequence between amino acids 407–524 of the glucocorticoid receptor (GR). This sequence contains the DNA binding domain and the nuclear localization signal (31). Another chimera, similar to the previous but also containing the hormone binding domain of the GR has been constructed. The advantage of the latter probe consists in the fact that it acts as nucleus/cytosol shuttling Ca^{2+} reporter. In fact, in the absence of hormone, the protein is retained in the cytosol and it translocates to the nucleus only upon glucocorticoids addition (cell incubation with dexamethazone) (32).

3.1.3. Mitochondrial Matrix and Mitochondrial Intermembrane Space

To localize aequorin in mitochondrial matrix, the mitochondrial pre-sequence derived from subunit VIII of the human cytochrome c oxidase (the cleavable mitochondrial pre-sequence and six residues of the mature protein) has been fused in frame to the sequence encoding HA1-aequorin (33). A low Ca^{2+} affinity variant of this chimera (mutAEQ) has also been created by introducing the D119A point mutation in the second "EF-hand" domain of aequorin. The mutation decreases the affinity for Ca^{2+} of about 20-fold by affecting the cooperative binding of the photoprotein (26). These chimeras are available in different mammalian expression vectors. More recently, they were also developed in lentiviral vectors that have been particularly suitable to infect neuronal cells (13) and other primary cell cultures (Lim et al., in preparation).

Another aequorin chimera has been constructed to monitor Ca^{2+} changes in the close proximity of mitochondria. This chimera is localized in the mitochondrial intermembrane space and it has been obtained by fusing the cDNA encoding glycerol phosphate dehydrogenase, an integral protein of the inner mitochondrial membrane which protrudes in the intermembrane space, with aequorin N-terminal tail (34).

3.1.4. Plasma Membrane

The recombinant aequorin targeted to the sub-plasma membrane region has been constructed by fusing the cDNA encoding SNAP-25, a protein of the SNARE complex anchored to the cytosolic face of membranes via palmitoyl side chains in the middle of the molecule, to the low affinity aequorin (35).

3.1.5. Endoplasmic/ Sarcoplasmic Reticulum

The strategy adopted to target our aequorin chimera to the endoplasmic reticulum is not based on the addition of the KDEL retention sequence (see Ref. (25)), but on the fusion of a portion of the Igγ2b heavy chain. The translated polypeptide contains the leader sequence, the VDJ and CH1 domains fused at the N-terminal of the low affinity aequorin. The retention inside the lumen of the endoplasmic reticulum is due to the binding of the CH1 domain to the endoplasmic reticulum protein BiP, which in the absence of Ig light chain retains the Igγ2b heavy chain portion fused to aequorin (36).

Another chimera to selectively targeted aequorin to the sarcoplasmic reticulum has been constructed by fusing calsequestrin to the N-terminal of aequorin. This chimera selectively localizes to the terminal cisternae where calsequestrin is normally resident (27) and has permitted to monitor Ca^{2+} concentration heterogeneity inside the lumen of the sarcoplasmic reticulum (37).

3.1.6. Golgi Apparatus

A Golgi apparatus-targeted aequorin has been constructed by fusing in frame the cDNA encoding the transmembrane portion of sialyltransferase and HA1-low affinity aequorin. This protein is retained in the membrane of the *trans*-Golgi network (34).

3.2. Expressing Aequorin

To detect aequorin photons emission, at least 24 h before transfection, cells are seeded on coverslips of 13 mm of diameter (eventually covered by polylysine/gelatine/collagen, accordingly to the cell types to facilitate cell adhesion), and allowed to grown to 50% of confluence. All the most common transfection protocols are appropriate to introduce the aequorin chimeras in different cell types. They are applicable with success to a multitude of stable cell lines, i.e. to the commonly used immortalized cell models, whereas in primary cultures, the efficiency of transfection is usually low and the light emission yield is scarce thus impairing the quality of the signal. Neuronal primary cultures are in fact usually resistant to the common transfection protocol such as Ca^{2+}-phosphate and cationic lipid based methods. To overcome this limitation, several recombinant aequorins have been constructed into engineered plasmids of viral origin. Different viral vectors, among them adenoviral (11, 12) and herpes simplex virus type 1 (HSV-1) (9) vectors to deliver targeted aequorin into the endoplasmic reticulum of a number of different primary cells have been described. Recently, the availability of lentiviral vectors in our laboratory led us to insert the aequorin sequences into this delivery system. We have reported the detection of the mitochondrial targeted aequorin activity in precursors of striatal neurons (13), but we have also produced lentiviral vectors containing untargeted-aequorin, endoplasmic reticulum, and sub-plasma membrane targeted aequorin (Lim et al., in preparation). The targeting of all these chimeras is very selective to the desiderate compartment as documented by the images shown in Fig. 4.2.

3.3. Reconstitution of Active Protein

Coelenterazine is added to the culture medium of transfected cells allowing it to diffuse freely across the cell membrane, and once inside the cells it spontaneously binds to the polypeptide reconstituting the active probe (Fig. 4.3). Coelenterazine currently employed in the experiments is a synthetic hydrophobic prosthetic group, which is commercially available. Coelenterazine variants that reduce the rate of emission of luminescence by aequorin are also commercially available. Namely, coelenterazine n is largely employed in the measurement of cellular compartments endowed with high Ca^{2+} concentration, i.e. intracellular stores.

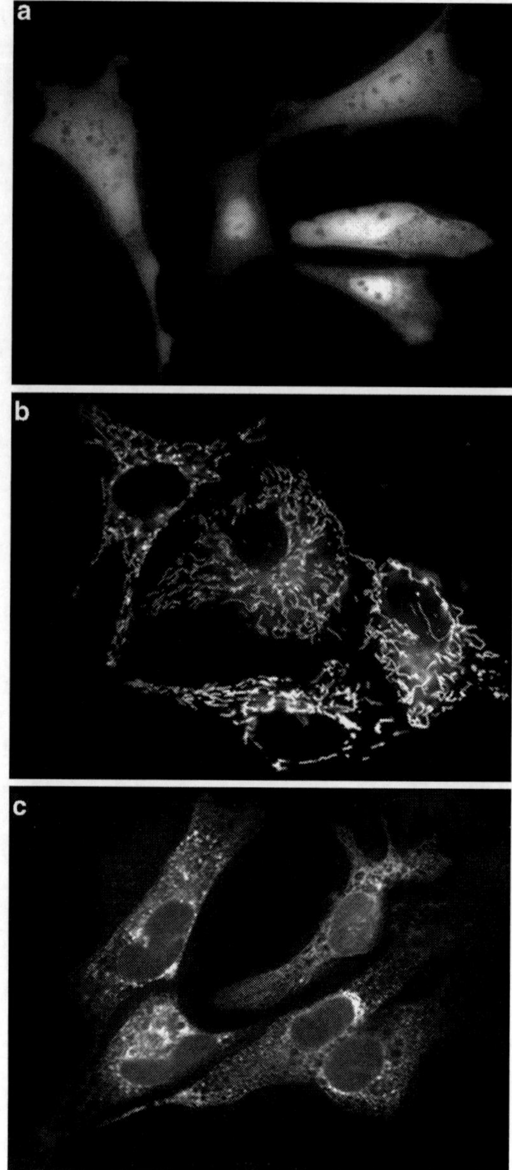

Fig. 4.2. Analysis of mouse clonal striatal cell lines expressing aequorin chimeras targeted to cytosol (**a**), mitochondria (**b**), and endoplasmic reticulum (**c**). Striatal cells were infected with lentiviral vectors encoding fusion RFP–, YFP–, and GFP–aequorin chimeras, respectively. Fluorescence emission was monitored to reveal the localization of the fusion proteins.

Coelenterazine is normally prepared as a stock solution in pure methanol at concentration of 0.5 mM. 50 μl aliquots of such stock solutions are kept at −80°C in the dark in glass tubes.

The reconstitution procedure is simple and it varies according to the chimera employed.

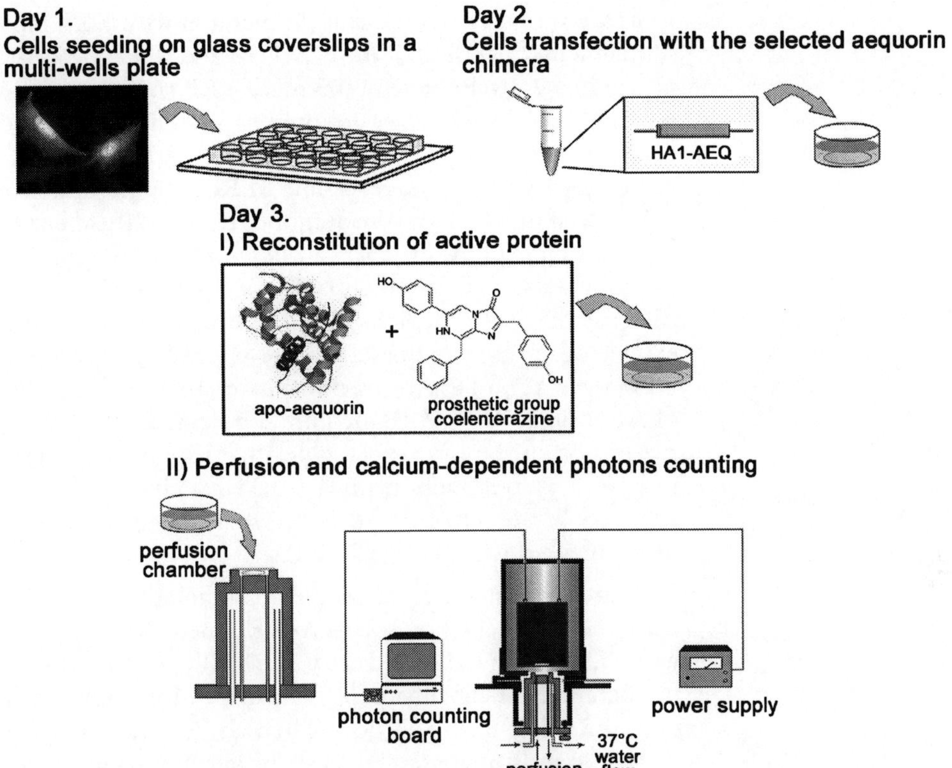

Fig. 4.3. Schematic representation of a typical experiment and of the aequorin measuring system. *Pmt* photomultiplier; *amp/discr* amplifier/discriminator.

3.3.1. Cytosol, Nucleus, and Mitochondria

To reconstitute the cytosol, nucleus, and mitochondria targeted aequorin, the coverslip with transfected cells is incubated with 5 μM coelenterazine wt for 2 h in DMEM supplemented with 1% FCS at 37°C in 5% CO_2 atmosphere. Then, the coverslip is directly transferred to the luminometer chamber (Fig. 4.3), where it is perfused with KRB saline solution (Krebs–Ringer modified buffer: 125 mM NaCl, 5 mM KCl, 1 mM Na_3PO4, 1 mM $MgSO_4$, 5.5 mM glucose, 20 mM HEPES, pH 7.4, 37°C). According to the experiment type, 1 mM $CaCl_2$ or 0.1 mM EGTA can be added to the KRB buffer.

3.3.2. Sarco/Endoplasmic Reticulum, Golgi Apparatus

To reconstitute the endo/sarcoplasmic reticulum and Golgi apparatus targeted aequorin, a drastic reduction of the luminal Ca^{2+} has to be made before adding coelenterazine since the high rate of aequorin consumption in high Ca^{2+} compartment would lead to the discharge of the active form before the measurements have been carried out. To this end, slightly different protocols can be applied:

1. Cells are washed from the culture medium with KRB medium supplemented with 0.6 mM EGTA and the coverslip was placed in a 24 wells plate in 0.3 ml of KRB supplemented with 0.6 mM EGTA, the Ca^{2+} ionophore ionomycin (5 μM) and coelenterazine n (5 μM). In the case of particularly sensitive cell types, e.g. neurons, reversible SERCA pump inhibitors as 2,5 ditert-butyl benzohydroquinone (tBHQ, 20 μM) or cyclopiazonic acid (CPA, 20 μM) can be employed in replacement of ionomycin. Cells are incubated for 1 h at 4°C in the dark. Alternatively, store depletion can be obtained by applying an extracellular agonist coupled to the generation of the second messenger 1,4,5 inositol trisphosphate ($InsP_3$) in the absence of extracellular Ca^{2+} (KRB medium supplemented with 0.6 mM EGTA) or, in the case of cell types that express the ryanodine receptor, by treatment with 10 mM caffeine. After agonist application, the cells are extensively washed with KRB medium supplemented with 0.6 mM EGTA.
2. Cells are washed as described above and incubated for 10 min with 0.5 ml of depletion solution (KRB medium supplemented with 0.6 mM EGTA and 20 μM tBHQ or CPA) at room temperature. The incubated depletion solution is then replaced with 0.3 ml of fresh solution. Then coelenterazine n (5 μM) is added and incubated for 1–2 h at room temperature in the dark.

In any case, after these incubations, cells are extensively washed with KRB supplemented with BSA 2% and 1 mM EGTA before being transferred to the luminometer chamber. It must be mentioned that even if mutated aequorin is used to measure Ca^{2+} concentration in the endo/sarcoplasmic reticulum lumen and in the Golgi compartment, it is necessary to further reduce the Ca^{2+} affinity of the photoprotein using a modified prosthetic group for the reconstitution, i.e. coelenterazine n (28).

3.3.3. Sub-plasma Membrane

To reconstitute plasma membrane aequorin, several procedures can be followed according to the efficiency of reconstitution reached in different cell types. In fact, the ideal situation will be the reconstitution of aequorin in the presence of extracellular Ca^{2+}, thus avoiding store depletions. Unfortunately, in the large majority of the cases, in these conditions, the high level of Ca^{2+} beneath the plasma membrane prevents efficient reconstitution, i.e. active aequorin is discharged as soon as it has been reconstituted. Thus, also in this case some manoeuvres need to be introduced: it is sufficient to reconstitute aequorin in KRB supplemented with 0.1 mM EGTA and 5 μM coelenterazine wt for 1 h at 37°C, then the coverslip is transferred to the luminometer chamber. In some cell types, the reconstitution can be obtained also in the presence of physiological concentration of extracellular Ca^{2+} (KRB supplemented with 1 mM $CaCl_2$ or in DMEM supplemented with 1% FCS and 5 μM coelenterazine). To measure Ca^{2+} concentration in this compartment, it is not necessary to use coelenterazine n.

3.4. Recording Setup and Calibration Procedure

The classical recording setup (Fig. 4.3) used to detect bioluminescence in cell population is composed by a low-noise photomultiplier which is kept in a dark box and positioned few millimetres away from the surface of the coverslip with the cells. Inside the box, coverslips are placed in a thermostatted chamber continuously perfused with physiological buffer using a peristaltic pump: agonists and drugs are added to the medium during perfusion of the cells. In the case of non-adherent cells, the perfusion chamber can be substituted with a chamber in which solutions can be injected. Following photons emission an amplifier discriminator generates pulses, which are captured by a Thorn EMI photon counting board connected to a compatible computer. The board allows storing of the experimental data for further analysis.

Aequorin shows a good ratio of the signal over background and allows a fast calibration of the signals using a simple algorithm that takes into account in physiological conditions of pH, temperature and ionic strength. The conversion of luminescence values into Ca^{2+} concentration values is based on the relationship between Ca^{2+} concentration (which in the 10^{-7}–10^{-5} M physiological range has a 2nd–3rd power relationship) and the fractional rate of aequorin consumption L/L_{max}, where L is the light recorded each second and L_{max} the total amount of light that can be generated by aequorin discharge of all available reconstitute molecules (18). A good estimation of L_{max} can be obtained from the total output recorded after exposure of cells to saturating Ca^{2+} concentration and detergents. Since the continuously consumption of aequorin during the experiment reduces the L_{max} value, the calibration requires to consider the ΔL_{max} between the begin and the end of the experiment. The model also contemplates the states of the Ca^{2+}-binding sites of aequorin and the association constant of the Ca^{2+}-activated sites (K_R) and that of the inactive sites (K_T). Considering to perform the experiment at standard conditions of temperature and perfusing cells in buffered physiological solutions, Ca^{2+} concentration can be calculated at each time point with the deduced algorithm.

Figure 4.4 shows typical Ca^{2+} transient traces obtained in neuroblastoma SH-SY5Y cells transfected with cytosolic, mitochondrial, plasma membrane and endoplasmic reticulum aequorin chimeras, respectively.

3.5. Available New Bioluminescent Reporters, GFP–Aequorin

New bioluminescent probes have been constructed by Brulet's group with the idea to create a "bi-functional reporter": GFP fluorescence acts as expression pattern probe, and aequorin moiety permits the measurements of Ca^{2+} changes at single cell-resolution. In fact the authors have reported that the fusion protein has a higher light emission

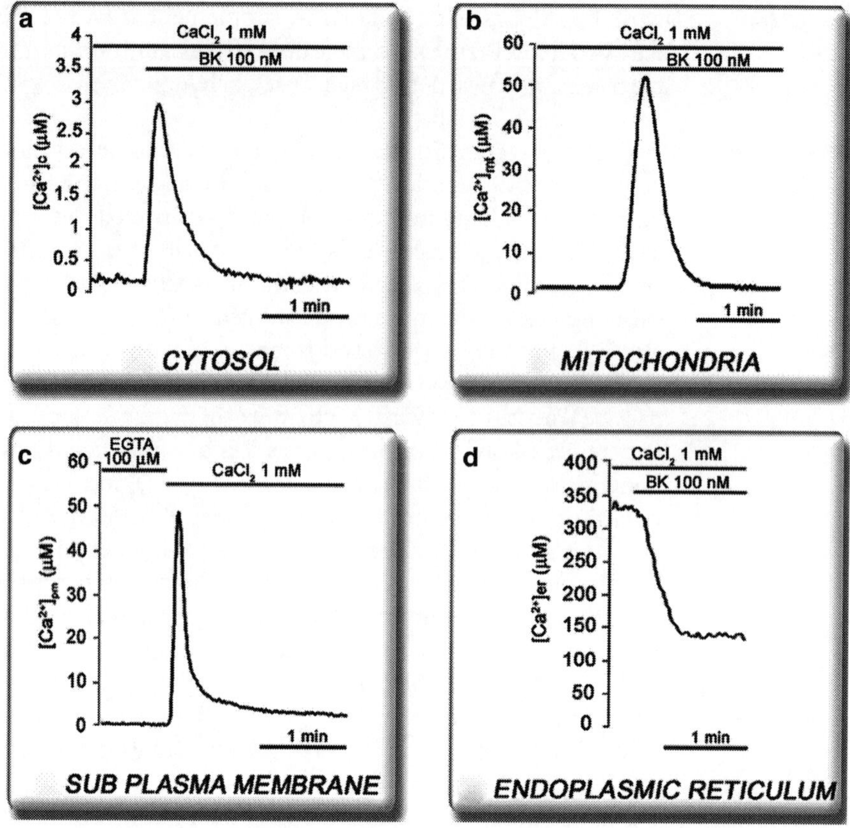

Fig. 4.4. Monitoring of cytosolic (**a**), mitochondria (**b**), sub-plasma membrane space (**c**) and endoplasmic reticulum (**d**) Ca^{2+} concentration in SH-SY5Y human neuroblastoma cells transfected with the selectively targeted aequorin chimeras. Where indicated cells were stimulated with bradykinin (BK), a G_q-coupled protein receptor agonist that induces the formation of inositol 1,4,5 trisphosphate.

when compared with aequorin chimera alone (7) thus permitting single-cell detection of photons with a cooled intensified charge-coupled device CCD camera.

3.5.1. Cytosol

GFP–aequorin fusion has been obtained by adding a 5-repeat flexible linker. The addition of linkers results in the raising of the ratio of green over blue light (500/470 nm). The enhanced GFP (EGFP) containing also F64L and S65T mutations in the chromophore that modify the excitation spectra and enhance fluorescence intensity was employed. An additional V163A mutation was introduced in the GFP sequence to enhance protein folding and increase fluorescence emission (7).

3.5.2. Mitochondrial Matrix and Endoplasmic Reticulum

The GFP–aequorin fusion chimeras have been targeted to the specific cell compartment utilizing the same strategies described in the section of aequorin chimeras (38). A single amino acid substitution D407A was introduced in the aequorin sequence to reduce the Ca^{2+} binding affinity of the photoprotein (39).

3.5.3. Synaptic Vesicles

Synaptotagmin I, a transmembrane protein of synaptic vesicles, has been fused in frame to GFP–aequorin cDNA to target the chimera to the cytosolic side of the synaptic vesicles (7). Also in this case, a D407A mutated aequorin was used to generate the chimera.

3.5.4. Postsynaptic Density

A chimera targeted to the postsynaptic density has been produced by adding the sequence coding the PSD-95 protein in frame to the D407A mutated GFP–aequorin (38). This probe is selectively localized in the dendritic synapses in close proximity of the NMDA receptor.

3.6. Introduction of GFP–Aequorin Based Probes in the Cells

These probes, similarly to aequorin probes, can be introduced into different cell types by transfection or infection procedures accordingly to the necessity and the cell type. They have been developed in expression mammalian vectors containing a cytomegalovirus promoter and the non-targeted GFP–aequorin has also been developed in a human E1-deleted adenovirus vector (38). Several cell types, including COS7, neuroblastoma cells and primary cortical neurons have been successfully transfected with these constructs and retinal organotypic slices from mouse have been infected with the adenoviral vector (38).

3.7. Reconstitution of GFP–Aequorin Probes Expressed in Cell Cultures and Detection of Photon Emission

For single cell bioluminescence studies, cells are grown on glass-bottomed dishes, transfected and kept in tyrodes buffer (129 mM NaCl, 5 mM KCl, 2 mM $MgCl_2$, 2 mM $CaCl_2$, 30 mM d-glucose, 25 mM Hepes, 10 μM glycine). The transfected GFP–aequorin chimeras are reconstituted with 5 μM coelenterazine for 45 min at 37°C or at room temperature, according to the protocols for each construct (38).

The photon emission can be collected either with a regular luminometer or with a combined bioluminescence/fluorescence detection system (see below).

The calibration of luminescence data can be achieved on the basis of a Ca^{2+} concentration response curve obtained in standard conditions (38) similarly to the calibration previously described for aequorin alone (18, 30).

3.8. Bioluminescence Imaging

Bioluminescence imaging has been performed by pioneering experiments (29) using mitochondria targeted aequorin. The expression of the protein has been optimized through intranuclear injection of the cDNA and the detection has been made by a charge-coupled device (CCD) camera fitted with a dual microchannel plate intensifier.

More recently, new systems to detect bioluminescence from single cell have been developed. One of them combines bioluminescence/fluorescence detection, thus permitting the simultaneous monitoring of BRET phenomenon and the GFP fluorescence.

The acquisition system is based on a highly sensitive photon cooled intensified CCD camera. The microscope is housed in a light tight dark box and the low level light emission is collected using an Imaging Photon Detector (IPD). The data acquisition software permits the conversion of single photon pulses into images and their superimposition to bright field or fluorescence images acquired with the CCD camera (38). This acquisition system can be applied to cell cultures but also to tissues slices, since the addition of GFP moiety to aequorin results in a longer wavelength light with higher tissue penetration depth in respect to aequorin alone.

Another very recent improvement in the detection of bioluminescence is represented by the development of a system based on the electron-multiplying charge-coupled-detector (EMCCD) imaging camera technology. This system can be mounted on either a bioluminescence or conventional microscope, and allows easy long-term monitoring of Ca^{2+} at the single cell level, obviating the need for expensive, fragile, and sophisticated equipment based on IPD (40).

4. Applications

The spatial resolution of the images acquired with the above mentioned bioluminescence detection systems depends essentially on light absorption and scattering in the samples and on the resolution of the camera. Even if recombinant fluorescence probes (i.e. cameleons) exhibit significantly better spatial and temporal resolution in respect to GFP–aequorin probes, the latter display characteristics that better fit with applications in animals *in vivo*. Among them, it must be mentioned that measurements can be carried out in a completely non-invasive way (no artificial illumination is required) both in anaesthetised or moving animal, thus ideally providing real physiological information. To an optimised real time Ca^{2+} dynamic visualization in the whole animal, GFP–aequorin probes have been improved by substituting GFP moiety with the yellow fluorescent protein Venus or the monomeric red fluorescent protein mRFP1 (41). These two new reporters display an increase in the detection of the bioluminescence activity in deep tissues in the whole animal. Venus-aequorin BRET is highly efficient and generates a red-shift in the peak emission of aequorin that improves photon transmission through tissues such as the skin and thoracic cage. mRFP1–aequorin fusion produces an additional emission peak above 600 nm thus obtaining an emission spectrum higher than the other reporters and permitting greater detection sensitivity in brain and heart tissues.

4.1. In Vivo Models

Real time visualization of Ca^{2+} dynamics in the whole animal represents an important goal to advance our understandings of the complexity of cellular functions and of the interplay of the different cell populations, especially at the neuronal level.

A lot of effort has thus put in developing whole animal bioluminescence imaging (BLI) (42, 43) and recently, the first *in vivo* GFP–aequorin-based measurement of mitochondrial Ca^{2+} has been described in *Drosophila* and in mice (14, 44). Whole brain *in vivo* imaging of transgenic *Drosophila* flies has permitted to detect the bioluminescence of Ca^{2+} signals from specific neuronal structures in response to different stimuli. The photon counting technique eliminates the necessity to pre-select exposure times and recording durations, therefore, Ca^{2+} signal can be monitored over a wide temporal range (from milliseconds to hours).

Transgenic mice have been produced with wild-type or targeted photoproteins in order to visualize different aspects of neuronal Ca^{2+} activity at different developmental stages. In one model, untargeted apoaequorin is ubiquitously expressed under the control of a strong promoter (15). In the other, mitochondrial-targeted aequorin is conditionally and ubiquitously expressed at all stages of the development (14). This model could be very useful to analyse the role of mitochondria in many physiological and pathological conditions (i.e. mitochondrial dysfunctions and Ca^{2+} homeostasis dysregulation are relevant in different neurodegenerative disorders). Interestingly, the authors have shown its applications to monitor whole body patterns of Ca^{2+} concentrations in freely moving mice. Video imaging of newborn mice has revealed mitochondrial Ca^{2+} fluxes in muscle during contraction/relaxation cycles, and mitochondrial variations in cephalic area related to the behaviour or kainic acid-induced seizure.

4.2. Visualization of Spatio-Temporal Ca^{2+} Dynamics in Tissue Slices

Acute tissue slices can be prepared from newborn transgenic animals or from wild type mice and then infected with viral approaches to express the desired GFP–aequorin chimera. These protocols, preserving the physiological connection between communicating cells, permit to reveal spatio-temporal Ca^{2+} dynamics in cell networks that are inevitable lost during cell dissociation. At the same time, they permit a relatively high resolution analysis, as documented in a study investigating Ca^{2+} dynamics in glial networks of dark-adapted mouse retina (45), that cannot still be reached by animal in toto analysis.

4.3. Reconstitution and Detection of GFP–Aequorin Probes in Tissue Slices and Animals

In the experiments on tissues isolated from GFP–aequorin transgenic mice, brain slices from 0 to 2 days old pups are cut using a vibrotome and incubated in a chamber containing an artificial cerebrospinal fluid and coelenterazine 5–10 μM for at least 1 h (14). In the case of dissociation of tissues from transgenic apoaequorin mice, cells are reconstituted in RPMI 1640 containing coelenterazione 10 μM at 37°C for 5 h (15).

For the experiments of *in vivo* imaging of the whole animal, coelenterazine is dissolved in ethanol (10 mM) and diluted in sodium phosphate buffer immediately before the injection in tail vein for adults or intra peritoneum for neonates (2–4 mg/kg mouse) (14). The maximum signal is obtained between 15 and 35 min after injection and, surprisingly and interestingly, Ca^{2+} signals could be studied in different tissues over several hours after the addition of coelenterazine via systemic injection. In fact, in living animals, aequorin seems to be not consumed immediately upon interaction with Ca^{2+}.

5. Conclusions

Bioluminescent Ca^{2+} probes are endowed with unique advantages. Among them, the possibility to measure Ca^{2+} in specific intracellular domains, with little background signal, no toxicity nor need of invasive loading and detecting procedures. Therefore, they represent an almost ideal Ca^{2+} indicator, being also suitable for Ca^{2+} variations monitoring in living animals. The development of BRET imaging methods coupled to higher sensitivity cameras promise to obtain important new data on the physiopathology of Ca^{2+} signalling in live organisms.

Acknowledgements

The authors are deeply indebted to past and present collaborators and thank the University of Padova (local funding and Ateneo Project 2008), the Telethon Foundation (Project GGP04169), the Italian Ministry of University and Research (PRIN 2003 and 2005), the Italian National Research Council (CNR, Agency 2000) for financial support.

References

1. Pietrobon D (2007) Familial hemiplegic migraine. Neurotherapeutics 4(2):274–284
2. Berridge MJ, Bootman MD, Roderick HL (2003) Calcium signalling: dynamics, homeostasis and remodelling. Nat Rev Mol Cell Biol 4(7):517–529
3. LaFerla FM (2002) Calcium dyshomeostasis and intracellular signalling in Alzheimer's disease. Nat Rev Neurosci 3(11):862–872
4. Rizzuto R, Pozzan T (2006) Microdomains of intracellular Ca2+: molecular determinants and functional consequences. Physiol Rev 86(1):369–408
5. Shimomura O, Johnson FH (1978) Peroxidized coelenterazine, the active group in the photoprotein aequorin. Proc Natl Acad Sci USA 75(6):2611–2615
6. Morin JG, Hastings JW (1971) Energy transfer in a bioluminescent system. J Cell Physiol 77(3):313–318
7. Baubet V, Le Mouellic H, Campbell AK et al (2000) Chimeric green fluorescent

protein-aequorin as bioluminescent Ca2+ reporters at the single-cell level. Proc Natl Acad Sci USA 97(13):7260–7265

8. Brini M (2008) Calcium-sensitive photoproteins. Methods 46(3):160–166
9. Alonso MT, Barrero MJ, Carnicero E et al (1998) Functional measurements of [Ca2+] in the endoplasmic reticulum using a herpes virus to deliver targeted aequorin. Cell Calcium 24(2):87–96
10. Bano D, Young KW, Guerin CJ et al (2005) Cleavage of the plasma membrane Na+/Ca2+ exchanger in excitotoxicity. Cell 120(2): 275–285
11. Kendall JM, Badminton MN, Sala-Newby GB et al (1996) Agonist-stimulated free calcium in subcellular compartments. Delivery of recombinant aequorin to organelles using a replication deficient adenovirus vector. Cell Calcium 19(2):133–142
12. Rembold CM, Kendall JM, Campbell AK (1997) Measurement of changes in sarcoplasmic reticulum [Ca2+] in rat tail artery with targeted apoaequorin delivered by an adenoviral vector. Cell Calcium 21(1):69–79
13. Lim D, Fedrizzi L, Tartari M et al (2008) Calcium homeostasis and mitochondrial dysfunction in striatal neurons of Huntington disease. J Biol Chem 283(9):5780–5789
14. Rogers KL, Picaud S, Roncali E et al (2007) Non-invasive *in vivo* imaging of calcium signaling in mice. PLoS One 2(10):e974
15. Yamano K, Mori K, Nakano R et al (2007) Identification of the functional expression of adenosine A3 receptor in pancreas using transgenic mice expressing jellyfish apoaequorin. Transgenic Res 16(4):429–435
16. Head JF, Inouye S, Teranishi K et al (2000) The crystal structure of the photoprotein aequorin at 2.3 A resolution. Nature 405(6784):372–376
17. Shimomura O, Johnson FH, Saiga Y (1962) Extraction, purification and properties of aequorin, a bioluminescent protein from the luminous hydromedusan, Aequorea. J Cell Comp Physiol 59:223–239
18. Allen DG, Blinks JR (1978) Calcium transients in aequorin-injected frog cardiac muscle. Nature 273(5663):509–513
19. Cobbold PH (1980) Cytoplasmic free calcium and amoeboid movement. Nature 285(5765): 441–446
20. Ridgway EB, Ashley CC (1967) Calcium transients in single muscle fibers. Biochem Biophys Res Commun 29(2):229–234
21. Ridgway EB, Gilkey JC, Jaffe LF (1977) Free calcium increases explosively in activating medaka eggs. Proc Natl Acad Sci USA 74(2): 623–627
22. Inouye S, Noguchi M, Sakaki Y et al (1985) Cloning and sequence analysis of cDNA for the luminescent protein aequorin. Proc Natl Acad Sci USA 82(10):3154–3158
23. Rizzuto R, Brini M, Pozzan T (1994) Targeting recombinant aequorin to specific intracellular organelles. Methods Cell Biol 40:339–358
24. Miller AL, Karplus E, Jaffe LF (1994) Imaging [Ca2+]i with aequorin using a photon imaging detector. Methods Cell Biol 40:305–338
25. Kendall JM, Dormer RL, Campbell AK (1992) Targeting aequorin to the endoplasmic reticulum of living cells. Biochem Biophys Res Commun 189(2):1008–1016
26. Montero M, Brini M, Marsault R et al (1995) Monitoring dynamic changes in free Ca2+ concentration in the endoplasmic reticulum of intact cells. EMBO J 14(22):5467–5475
27. Brini M, De Giorgi F, Murgia M et al (1997) Subcellular analysis of Ca2+ homeostasis in primary cultures of skeletal muscle myotubes. Mol Biol Cell 8(1):129–143
28. Barrero MJ, Montero M, Alvarez J (1997) Dynamics of [Ca2+] in the endoplasmic reticulum and cytoplasm of intact HeLa cells. A comparative study. J Biol Chem 272(44): 27694–27699
29. Rutter GA, Burnett P, Rizzuto R et al (1996) Subcellular imaging of intramitochondrial Ca2+ with recombinant targeted aequorin: significance for the regulation of pyruvate dehydrogenase activity. Proc Natl Acad Sci USA 93(11):5489–5494
30. Brini M, Marsault R, Bastianutto C et al (1995) Transfected aequorin in the measurement of cytosolic Ca2+ concentration ([Ca2+]c). A critical evaluation. J Biol Chem 270(17): 9896–9903
31. Brini M, Murgia M, Pasti L et al (1993) Nuclear Ca2+ concentration measured with specifically targeted recombinant aequorin. EMBO J 12(12):4813–4819
32. Brini M, Marsault R, Bastianutto C et al (1994) Nuclear targeting of aequorin. A new approach for measuring nuclear Ca2+ concentration in intact cells. Cell Calcium 16(4):259–268
33. Rizzuto R, Simpson AW, Brini M et al (1992) Rapid changes of mitochondrial Ca2+ revealed by specifically targeted recombinant aequorin. Nature 358(6384):325–327
34. Rizzuto R, Pinton P, Carrington W et al (1998) Close contacts with the endoplasmic reticulum as determinants of mitochondrial Ca2+ responses. Science 280(5370):1763–1766

35. Marsault R, Murgia M, Pozzan T et al (1997) Domains of high Ca2+ beneath the plasma membrane of living A7r5 cells. EMBO J 16(7):1575–1581
36. Sitia R, Meldolesi J (1992) Endoplasmic reticulum: a dynamic patchwork of specialized subregions. Mol Biol Cell 3(10):1067–1072
37. Robert V, De Giorgi F, Massimino ML et al (1998) Direct monitoring of the calcium concentration in the sarcoplasmic and endoplasmic reticulum of skeletal muscle myotubes. J Biol Chem 273(46):30372–30378
38. Rogers KL, Stinnakre J, Agulhon C et al (2005) Visualization of local Ca2+ dynamics with genetically encoded bioluminescent reporters. Eur J Neurosci 21(3):597–610
39. Kendall JM, Sala-Newby G, Ghalaut V et al (1992) Engineering the CA(2+)-activated photoprotein aequorin with reduced affinity for calcium. Biochem Biophys Res Commun 187(2):1091–1097
40. Rogers KL, Martin JR, Renaud O et al (2008) Electron-multiplying charge-coupled detector-based bioluminescence recording of single-cell Ca2+. J Biomed Opt 13(3):031211
41. Curie T, Rogers KL, Colasante C et al (2007) Red-shifted aequorin-based bioluminescent reporters for *in vivo* imaging of Ca2 signaling. Mol Imaging 6(1):30–42
42. Abraham U, Prior JL, Granados-Fuentes D et al (2005) Independent circadian oscillations of Period1 in specific brain areas *in vivo* and *in vitro*. J Neurosci 25(38):8620–8626
43. Hardy J, Francis KP, DeBoer M et al (2004) Extracellular replication of Listeria monocytogenes in the murine gall bladder. Science 303(5659):851–853
44. Martin JR, Rogers KL, Chagneau C et al (2007) In vivo bioluminescence imaging of Ca signalling in the brain of Drosophila. PLoS One 2(3):e275
45. Agulhon C, Platel JC, Kolomiets B et al (2007) Bioluminescent imaging of Ca2+ activity reveals spatiotemporal dynamics in glial networks of dark-adapted mouse retina. J Physiol 583(Pt 3):945–958

Chapter 5

Monitoring Calcium Levels With Genetically Encoded Indicators

Olga Garaschuk and Oliver Griesbeck

Abstract

Calcium indicators are widely used to monitor activity in living neuronal tissue because of the tight relation between action potential firing and increases in the intracellular calcium concentration. Here, we describe the use of genetically encoded calcium indicators (GECIs) of the latest generation for monitoring calcium levels in the mammalian brain. We discuss how to choose the sensor for a given experiment, how to introduce the sensor into the cells of interest and how to estimate the sensitivity of the sensor in situ and in vivo. Finally, we illustrate the application of these sensors for high resolution in vivo imaging of sensory-driven neuronal activity.

Key words: Genetically encoded calcium indicators, Two-photon imaging, In vivo, Calcium, Fluorescence resonance energy transfer

1. Introduction

Many physiological processes in living cells are tightly linked to changes in the intracellular free calcium concentration (1–3). Therefore, high resolution calcium imaging has become a fundamental technique in cell biology and physiology. In recent years, fluorescent calcium biosensors consisting exclusively of amino acids have complemented the wide spectrum of synthetic (small molecule) calcium indicator dyes. Genetically encoded calcium indicators (GECI) can be introduced into specific cells, tissues or even whole organisms by various gene transfer techniques established in molecular biology. This property is of great advantage in various experiments in which (1) synthetic dye loading via a patch pipette or via use of acetoxymethyl esters is not practicable or (2) cell-type specificity or subcellular localization of indicators is desired.

A. Verkhratsky and Ole H. Petersen (eds.), *Calcium Measurement Methods*, Neuromethods, vol. 43, DOI 10.1007/978-1-60761-476-0_5, © Humana Press, a part of Springer Science+Business Media, LLC 2010

Other potential advantages of GECI lie in the labeling of long-range neuronal projections and their applicability for repeated long term imaging of neuronal function (4–6).

Genetically encoded calcium indicators consist of a calcium-binding protein such as calmodulin or Troponin C and mutants of Green Fluorescent Protein (GFP) linked together in various configurations (7–11). In sensors employing fluorescence resonance energy transfer (FRET), the calcium binding protein is inserted between two mutants of GFP acting as donor and acceptor fluorophores. Typically, these are either Cyan Fluorescent Protein (CFP) and Yellow Fluorescent Protein (YFP) or improved or circularly permutated variants thereof. Binding of calcium ions results in a conformational change within the fusion protein, typically enhancing FRET efficiency from donor to acceptor protein (12–14). Conveniently, the changes in FRET efficacy can be read out in a ratiometric fashion and calibrated to measure free calcium concentrations. Currently, there are two major FRET-based calcium sensor families, using different calcium sensing proteins as FRET modulators. Cameleons and the Design-Series are based on the interaction of calmodulin with its binding peptide M13 or versions of these proteins with re-engineered interaction interfaces (13–16). Other FRET sensors are based on Troponin C – the calcium sensor in skeletal and cardiac muscle (5, 12, 17). Single fluorophore sensors combine the fluorescent properties of a green or yellow fluorescent protein variant with calmodulin or Calmodulin/M13 induced calcium binding (18–23). Calcium binding typically leads to intensity modulation of fluorescence emission. In case of ratiometric pericam, however, excitation ratioing was described (20). Many of these sensors and/or their improved versions perform rather well when transfected into various cultured cells. Moreover, they have been successfully used in neurons of transgenic organisms such as *C. elegans* (24–28), *Drosophila* (29–32) and recently also mouse (15, 33–35). In this chapter, we describe how to use GECIs for monitoring calcium levels in the mammalian brain. We provide some rationale for the choice of the sensor for a given experiment, discuss means to introduce the sensor into the cells of interest, show how to estimate the sensitivity of the censor in situ and in vivo and finally illustrate some potential applications.

2. Selecting a Suitable Ca^{2+} Indicator

The ideal calcium sensor protein will (1) bind exclusively calcium ions, (2) be sensitive to the smallest changes in the intracellular free Ca^{2+} concentration (in the range of several tens of nM), (3) have sufficient brightness to provide high signal-to-noise ratio,

(4) respond linearly in the physiologically-relevant range of intracellular Ca^{2+} concentrations (tens of nM – tens of μM), (5) have sufficiently fast binding and unbinding kinetics, (6) be highly photostable to ensure continuous recordings over at least several hours and (7) be targeted to specific cell types or cellular subcompartments. However, there is no sensor fulfilling all these requirements. Therefore, a number of points have to be considered on how to select the optimal GECI for a given experiment. As already mentioned, important criteria are calcium affinity, kinetics, brightness, photostability and mode of read-out. The decision on indicator's affinity will depend on which calcium concentrations are to be measured. For detecting subtle cytosolic changes in free calcium due to firing of one or few action potentials, high affinity indicators are required. For targeting to plasma membrane, synaptic sites or cellular organelles lower affinity sensors may be preferable.

Another important decision when choosing an indicator is whether a single wavelength or a FRET-based ratiometric sensor is to be used. From the theoretical point of view, single wavelength indicators do have an advantage over FRET sensors in terms of signal-to-noise ratio. This is because uncorrelated noise that is detected independently in both emission channels of a FRET sensor is enhanced by ratioing the data. In addition, splitting the emitted light of a FRET sensor into two channels invariably causes photon loss. This in turn often results in an increased intensity of excitation light, increased photobleaching and photodamage and may dramatically impede the performance of a dim fluorophore. Unfortunately, current single wavelength sensors show only a fraction of the brightness of FRET calcium sensors. The most successful single wavelength sensors are based upon a circular permutation (cp145–144) of GFP which possessed decreased brightness, thermotolerance and increased pH sensitivity (18). Although these features have been systematically improved, modern single wavelength sensors like G-CaMP2 still have only up to 10% of the brightness of FRET-based sensors such as D3cpv, TN-XL or TN-XXL (5, 8, 36). Thus, when using these indicators, recordings with single cell resolution are restricted to cells located close to the tissue surface or in tissue slices (37). On the other hand, although relatively slow kinetics is a common feature of all genetically encoded calcium indicators, the high affinity FRET sensors are considerably slower than the high affinity single fluorophore sensor G-CaMP2.

FRET indicators, on the other hand, provide more quantitative signals and are less susceptible to artifacts by motion, pH or non-linear bleaching. Correlated noise which affects both channels equally is quite effectively canceled out by ratioing. When working in vivo, motion artifacts caused by breathing and blood flow of an experimental animal are the most significant sources of noise. As these artifacts are reasonably taken care of by ratioing, a FRET sensor may be the indicator of choice when working with a preparation where motion artifacts occur.

Induction of phenotype by sensor expression may be another point to assess. In particular sensors, relying on wild type calmodulin may mimic calmodulin overexpression. For example, constitutive expression of G-CaMP2 in mouse heart results in cardiomegaly – a phenotype resembling overexpression of calmodulin (23). Effects of activating calmodulin-dependent signaling after expression of wild type calmodulin-based sensors may be more or less pronounced depending on the sensor and the cell of interest. Invertebrate systems seem to be more tolerant to wild type calmodulin-based sensors. Thus, when working in invertebrate preparations, G-CaMPs or Case 12 may be the sensor of choice, while the FRET-based sensors (Table 5.1) currently seem to be more suitable for deep tissue imaging in mammals and for preparations where motion artifacts are a problem.

3. Introducing a Probe into a Cell of Interest

Several gene transfer techniques are available for expressing GECIs in the cells of interest. Cell lines and dissociated primary neurons in culture can be acutely transfected with plasmid DNA using liposomes or calcium phosphate precipitates. Protocols for transfection of cultured cells are described in detail in Sambrook and Russell: Molecular Cloning (38). Particle-mediated biolistic gene transfer of plasmid DNA may be the method of choice for the transfection of slice cultures because it does not require any

Table 5.1
In vitro properties of last generation single fluorophore and FRET-based calcium biosensors

	Indicator	Binding moiety	Dynamic range ($\Delta F/F$, $\Delta R/R$ (%))	K_D	Hill Coeff.	Refs.
Single fluorophore indicators	Case 12	wtCaM/M13	1,200%	1.0 μM	–	(22)
	GCaMP2	wtCaM/M13	400%	0.15 μM	3.8	(23)
FRET-based indicators	YC3.60	wtCaM (E104Q)/ M13	560%	0.25 μM	1.7	(15)
	D3cpv	Redesigned CaM/ M13	410%	0.6 μM	–	(16)
	TN-XXL	Troponin C	230%	0.8 μM	1.5	(5)

laborious DNA cloning procedures (39). This technique, however, appears to have a low efficiency so that one should expect only sparse labeling of individual neurons scattered throughout the slice. In addition, with this technique, it is difficult to target specific anatomic subregions of the slice.

In utero electroporation of plasmid DNA (40, 41) is a fast method that can lead to expression of a sensor in the intact mouse brain without the need for long lasting procedures to generate transgenic mice. It is to some extent possible to direct expression to the major anatomical regions of the brain, but expression is more or less scattered and not entirely reproducible. Stuart Firestein and colleagues have recently extended the use of this technique to early postnatal age (42) and used it for labeling newborn olfactory interneurons. The postnatal electroporation works well in rats and mice and provides a long-lasting (up to 1 year) staining of cells with different GFP variants including GCaMP2.

Another efficient means to introduce foreign genes into postmitotic neurons are viral vectors (43–45). A number of different vectors are available. They differ in efficiency, preference for different cell types in the brain, insert capacity and potential toxicity. In particular the latest generations of lentiviral vectors and vectors based on adeno-associated virus (AAV) have proven to introduce genes into neurons with minimal adverse effects. Also this technique offers the possibility of long term expression of the construct (for at least several months, see (46, 47)). However, expertise in molecular biology is needed initially to generate vectors coding for the indicator of choice. Once high titer viral stocks have been produced, these can be aliquoted, stored and used over long periods of time providing reproducible results in dissociated neurons, slices and in vivo. Viral vectors can be used to transfect cells in many different species. However, the reproducibility of expression patterns remains a challenge. Using this technique, it is also difficult to achieve cell-type specificity of labeling, because only a very limited number of cell-type specific promoters are small enough to be incorporated into the viral vectors.

Cell-type specificity can be achieved when generating a transgenic organism stably incorporating the indicator gene in its genome. Using currently available genetic toolbox, it allows relatively precise targeting of a gene of interest to defined populations of cells and/or cellular subcompartments, maximal degree of reproducibility of the expression pattern from one animal to the other and the ease of reproduction of experimental animals. Reliable protocols for transgenesis have been established for the major model organisms such as *C. elegans*, *Drosophila*, zebrafish or mouse. However, this is also the most labor-, time-, and money-consuming technique. In addition, several general as well as GECI-specific caveats of the technique have to be considered. First, it seems that some GECIs behave differently when expressed transgenically as compared to acute transfection. For example, transgenic expression of YC3.60 in the mouse brain using the CMV-actin promoter was achieved via injection of purified indicator DNA into the pronuclei of mouse oocytes and screening of founder animals. Surprisingly, this procedure yields a non-functional protein (15), while acute delivery using AAV-based viral vectors leaves the sensor functional (48). Second, the use of some promoters gives rise to variability in expression strength and patterns. For example, the Thy 1.2 expression cassette (49, 50) that is used for neuron-specific expression of genes in the mouse brain, yields considerably different expression properties depending on site of integration into the genome. As a result, often many independent transgenic lines have to be generated initially to obtain the desired expression level in the cells of interest. Figure 5.1 illustrates this issue. It shows three out of 17 independent mouse lines that were generated using Thy 1.2 promoter-driven expression of TN-L15 (35). Among those PCR-positive lines, there were lines with differing expression strengths as well as slight differences in the expression pattern.

4. Expression Levels: How much Indicator Is Enough?

Over the last years, some effort in the field was dedicated to achieve the highest possible expression levels of GECI (see discussion on this issue in (6)). Indeed, high expression level of the indicator protein provides high amount of detected photons and thus an immediate increase in signal-to-noise ratio of the recordings (see below). Furthermore, it reduces the intensity of the excitation light and thus photobleaching and photodamage. Finally, tiny processes like dendritic spines or presynaptic boutons become visible only at high concentrations of the indicator.

The flip side of this coin, however, also has to be mentioned. Because GECIs are Ca^{2+} binding proteins, they inherently

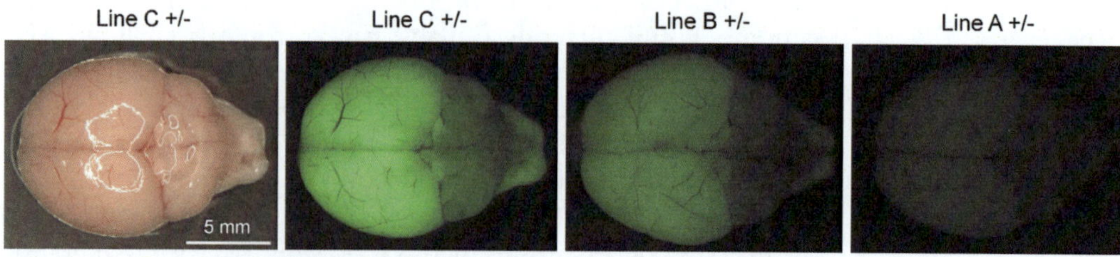

Fig. 5.1. Transgenic mice expressing the indicator TN-L15 using the Thy 1 promoter. Dissected brains from three different heterozygous lines (C, B and A) with expression levels from strong to weak are shown. All images were taken with a CCD camera under similar imaging conditions. Modified, with permission, from Ref. (35).

perturb intracellular Ca^{2+} signaling. This is especially the case for high affinity Ca^{2+} indicators such as GCaMP2 and YC3.60 (see Table 5.1). In this respect, the use of GECI faces similar problems as the use of small molecule Ca^{2+} indicators. As summarized by E. Neher (51) the "Ca-binding ratio" (κ_B) of the exogenously added Ca^{2+} chelator (e.g., GECI) has to be compared to "Ca-binding ratio" (κ_S) of endogenous Ca^{2+} buffers (see (51) for details). Changes in the intracellular free Ca^{2+} concentration can be measured only at very low indicator levels ($\kappa_B \ll \kappa_S$). At extremely high indicator concentrations ($\kappa_B \gg \kappa_S$) one can measure Ca^{2+} fluxes through the cell membrane. Intermediate or even worse unknown indicator concentrations significantly perturb the measured entity (intracellular free Ca^{2+} concentration) and can even influence behavior of cells and organisms (see (7, 8) and references therein).

It is relatively easy to estimate the concentration of added small molecule Ca^{2+} indicator (buffer) because it was either added via a patch pipette and the concentration in the pipette is known or it was introduced via a cell membrane in membrane-permeant form (e.g., Oregon green BAPTA-1 (OGB-1) AM). In the latter case, the brightness of the cell with unknown dye concentration can be compared to the brightness of the cell filled via a patch pipette (52). Similar technique could be used to determine the intracellular concentration of GECIs. For FRET-based indicators, we would suggest to fill patch pipettes with an intracellular solution containing purified recombinant indicator (preferable gel filtration grade) at various concentrations (for example, a series of four dilutions from 50 to 1 μM of recombinant protein). Patch and fill one of the cells with a pipette solution containing a given amount of added GECI, leave time for the indicator to diffuse into the cell and use two photon microscopy to compare intensities of such a "filled" wild-type cell with cells expressing the indicator after transfection or transgenic expression. Care should be taken to provide identical imaging conditions (the same setup, photomultiplier, excitation wavelength, etc.) and to directly excite acceptor protein (at excitation wavelengths longer than 960 nm (53)). Because direct excitation of the acceptor is insensitive to ambient calcium, this procedure will avoid artifacts caused by differences in the free intracellular calcium concentration in different cells.

This technique will allow estimating the concentration of fluorescent indicator protein. It should be kept in mind though that partially misfolded, proteolytically cleaved or folded, but non-fluorescent GECIs might still contribute to Ca^{2+} buffering. In addition, under certain circumstances, a significant fraction of fluorescent proteins can become nonresponsive, for example after transgenic expression of calmodulin-based sensors in the mouse brain (8, 34).

5. Experimental Setup

All modern imaging techniques (CCD camera-based imaging, confocal and two-photon microscopy) can be used for monitoring GECI fluorescence. In single photon excitation mode, GCaMP2 is excited at 488 nm whereas FRET-based indicators are excited at appr. 432 nm (using, for example, 420/30 nm band pass filter from Chroma technology Corp., Rockingham, VT USA). For two photon imaging, 900–910 mm excitation light is used to excite GCaMP2 (36, 54) whereas FRET indicators can be excited at wavelengths between 830 and 880 nm (5, 6, 29, 35). We noticed however, that the amount of basal CFP and YFP fluorescence changes when moving from shorter to longer excitation wavelengths. Therefore, in our laboratory we used 860 nm excitation light. This wavelength provides sufficient basal levels of fluorescence for both CFP and YFP and thus reduces noise in ratio images (35). To separate emission of CFP and YFP, we and others used a 515 nm beamsplitter (for example from Chroma). For FRET indicators employing CFP/YFP, 515 nm represents an isosbestic point at which indicator's fluorescence is calcium-independent. In addition to the beamsplitter bandpass emission filters (for example 480/30 nm for CFP and 535/30 nm for YFP, Chroma) are used to separate fluorescence of both proteins.

6. Examining Indicator's Properties in Living Cells

The properties of GECIs in living cells are expected to differ from their properties in vitro due to the differences in the ionic strength (55), biochemical milieu (56), intracellular Mg^{2+} concentration (8), etc. Moreover, several studies report changes in the dissociation constant (K_D) and dynamic range of the indicator protein in situ relative to in vitro (29, 34, 56, 57). Therefore, the ability of the indicator to signal changes in the intracellular free Ca^{2+} concentration has to be tested in living cells. If the final aim is to use the indicator for in vivo recordings of neuronal activity, this is best done in acute tissue slices. The basic properties of neurons in acute tissue slices are very similar to those recorded in vivo (58), but tissue slices provide more favorable conditions for imaging as well as for electrical recordings.

The quality of signal detection is generally described by the Signal-to-Noise Ratio (SNR). In the case of fluorescence measurements, the detection accuracy depends on the so-called shot noise (caused by the statistic nature of photon emission and detection; (59)). In a shot noise limited measurement, the SNR can be calculated according to the equation:

$$\mathrm{SNR} = \Delta F(R)/F(R)_0 \times \mathrm{N}^{1/2}$$

where F_0 or R_0 is the fluorescence or the fluorescence ratio at rest, ΔF or ΔR is the stimulus-evoked change in fluorescence or in the ratio and N is the number of detected photons (8, 60). Thus, the SNR of the recordings (which in a given experiment has to reach at least the value of 3) critically depends on brightness of the indicator (fluorescence-excitation cross section/quantum yield (61)), its affinity and dynamic range as well as the efficiency of the photon collection by the imaging system. Because in a real experiment all these parameters depend on many factors, the overall performance of the indicator has to be estimated experimentally. To do so, we and others simultaneously recorded electrical activity of a cell (action potential firing) and underlying changes in the GECI fluorescence. Figure 5.2 illustrates an experiment conducted using a Troponin-C based Ca^{2+} indicator CerTN-L15.

On a single trial basis, this indicator does not allow detecting single action potentials (Fig. 5.2c), but reliably reports bursts of APs (from doublets on; Fig. 5.2b). Another favorable feature of this indicator is the linearity of the response within physiological range of action potential firing (Fig. 5.2c). In similar experiments, GCaMP2 and D3cpv were shown to provide a single AP resolution at least in the subset of trials (6, 36). Although the responses of YC3.60 to a single action potential were not yet tested, other data (29, 48) suggest that also this indicator should be sensitive enough to resolve single action potentials (at least in situ; Table 5.2). However, these "calibration" experiments also revealed substantial drawbacks of some indicators. Thus, at resting

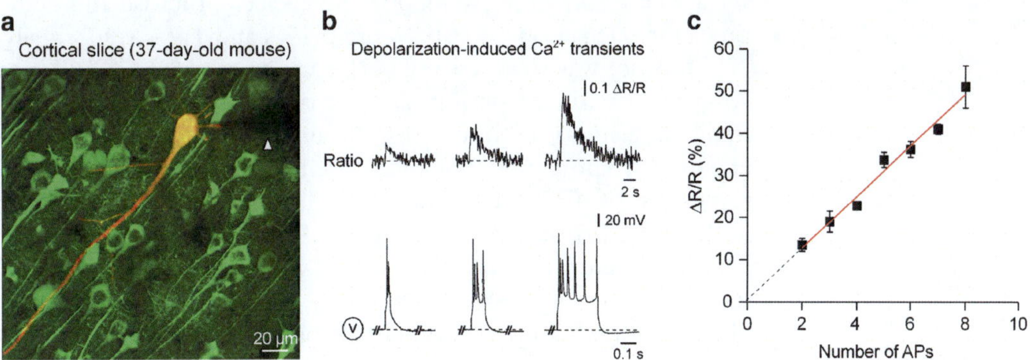

Fig. 5.2. In situ Ca^{2+} sensitivity of CerTN-L15. (**a**) An image of layer 2/3 neurons in a cortical slice. CerTN-L15 labeling is shown in *green*, a pyramidal neuron patched with an internal saline (composition: 140 K-gluconate, 12 KCl, 4 Mg-ATP, 0.4 Na-GTP and 10 HEPES (pH 7.3)) containing 100 µM Alexa Fluor 594 is shown in *yellow*. The shadow of the patch pipette is marked with an *arrowhead*. (**b**) Relative changes in the Citrine/Cerulean ratio (*upper*) and corresponding membrane depolarizations (*lower*, accompanied by 2, 4 and 7 action potentials, respectively) caused by three current injections (35, 100, and 300 ms long). (**c**) Relationship between the amplitude of the Ca^{2+} transients and the number of underlying action potentials Modified, with permission, from Ref. (35).

Table 5.2
Properties of last generation calcium indicators in brain slices

Indicator	Single AP resolution	Slope, $\Delta F/F$ ($\Delta R/R$) per 1 AP (%)	Behavior at physiological levels of activity	Refs.
CerTN-L15	No	6.4 2P excitation	Linear for 2–8 APs	(35)
GCaMP2	Yes, but in 25% of trials	6–13 2P excitation	Supralinear up to 40 APs	(36)
D3cpv	Yes	8.3 2P excitation	Sublinear, saturation already at 4AP	(6)
TN-XXL	Yes	0.9 (measured with a CCD camera)	Almost linear for 1–10 APs	(5)

conditions, GCaMP2 is too dim and has to be co-expressed with additional fluorescent indicator in order to visualize individual neurons (36), whereas D3cpv shows linear behavior only in a very narrow activity range (between 1 and 3 action potentials) and quickly saturates thereafter (6).

Figure 5.3 illustrates the in situ performance of the most recent Troponin-C-based GECI TN-XXL (5). Similar to its precursor CerTN-L15 (see Fig. 5.2), this indicator shows almost linear increase in the signal amplitude over a physiological range of neuronal activity (appr. up to 10 action potentials at 20 Hz) and a sublinear, saturating response at higher firing rates. Remarkably, however, this novel indicator reliably reports single action potentials with relatively good signal-to-noise ratio. In an experiment shown here (organotypic hippocampal slices), the rise and decay times for one action potential were 0.17 s and 1.9 s, respectively.

Another way to access the performance of a given GECI is to compare it with the performance of the one of commonly used small molecule Ca^{2+} indicators. In an elegant experiment Mao et al. (36) simultaneously monitored action potential-induced responses of GCaMP2 and a small molecule Ca^{2+} indicator X-Rhod-5F (loaded via a patch pipette) in cultured hippocampal pyramidal neurons. Whereas, X-Rhod-5F reliably detected even a single action potential (see their Fig. 5.4d) GCaMP2 did not. However, from 10 action potentials on GCaMP2 showed larger responses when compared with X-Rhod-5F because of the supralinearity of this GECI.

The same study showed an appr. 50% decrease in the amplitude of GCaMP2-mediated Ca^{2+} transients when changing from room (20–22°C) to "physiological" (in this study 34.5–35.5°C) temperature. This finding is in contrast with the performance of small molecule Ca^{2+} indicators, which show less than 10% difference

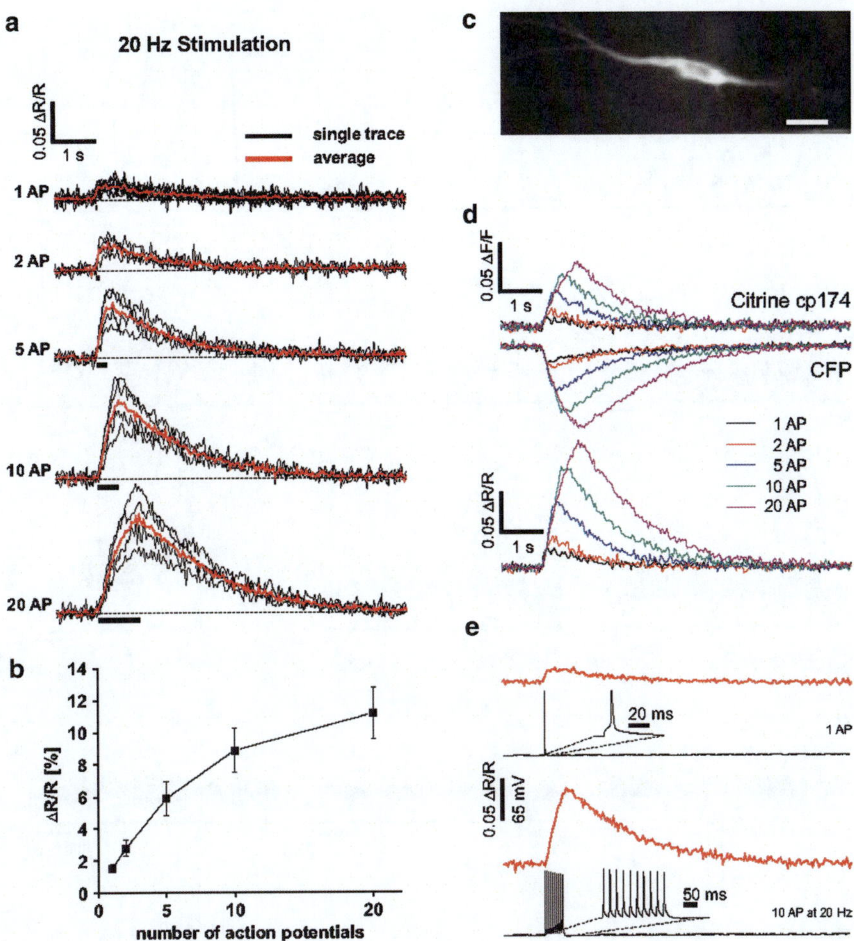

Fig. 5.3. Ca^{2+} sensitivity of TN-XXL. (**a**) TN-XXL-mediated Ca^{2+} transients recorded from a whole-cell patch clamped cultured hippocampal neuron (alike a cell shown in **c**) in response to 1, 2, 5, 10 and 20 action potentials (APs) evoked by injection of current pulses (0.2–0.4 nA for 10 ms/pulse) at 20 Hz. Each pulse elicited a single action potential. The black bar below each trace indicates the total time of the electrical stimulation. The *gray lines* show the $\Delta R/R$ of a single trial for each cell, the *red lines* – the average of traces recorded from four individual neurons. (**b**) Relationship between the amplitude of the Ca^{2+} transients and the number of underlying action potentials ($n=4$ cells, standard error bars). (**c**) A CA1 pyramidal neuron in a cultured hippocampal slice infected with a Semliki Forest virus coding for TN-XXL (24 h after infection). Scale bar, 5 μm. (**d**) A series of Ca^{2+} transients recorded from an individual CA1 cell. The *upper graph* displays the change in fluorescence intensity ($\Delta F/F$) for Citrine cp174 and CFP channels. The *lower graph* shows the ratio ($\Delta R/R$) calculated from background-subtracted $\Delta F/F$ traces. (**e**) Combined fluorometric Ca^{2+} measurements ($\Delta R/R$, first and third traces) and electrophysiological recordings (second and fourth traces) of neuronal responses to 1 AP and 10 AP at 20 Hz. Modified, with permission, from Ref. (5).

in the amplitude of the Ca^{2+} transient caused by a single action potential when changing the temperature from 24 to 34°C (62). This temperature-dependent reduction in the amplitude of GECI-mediated Ca^{2+} signals is most probably a reflection of unlucky combination of their slow Ca^{2+} binding kinetics and the smaller calcium entry (narrower action potentials) as well as its faster

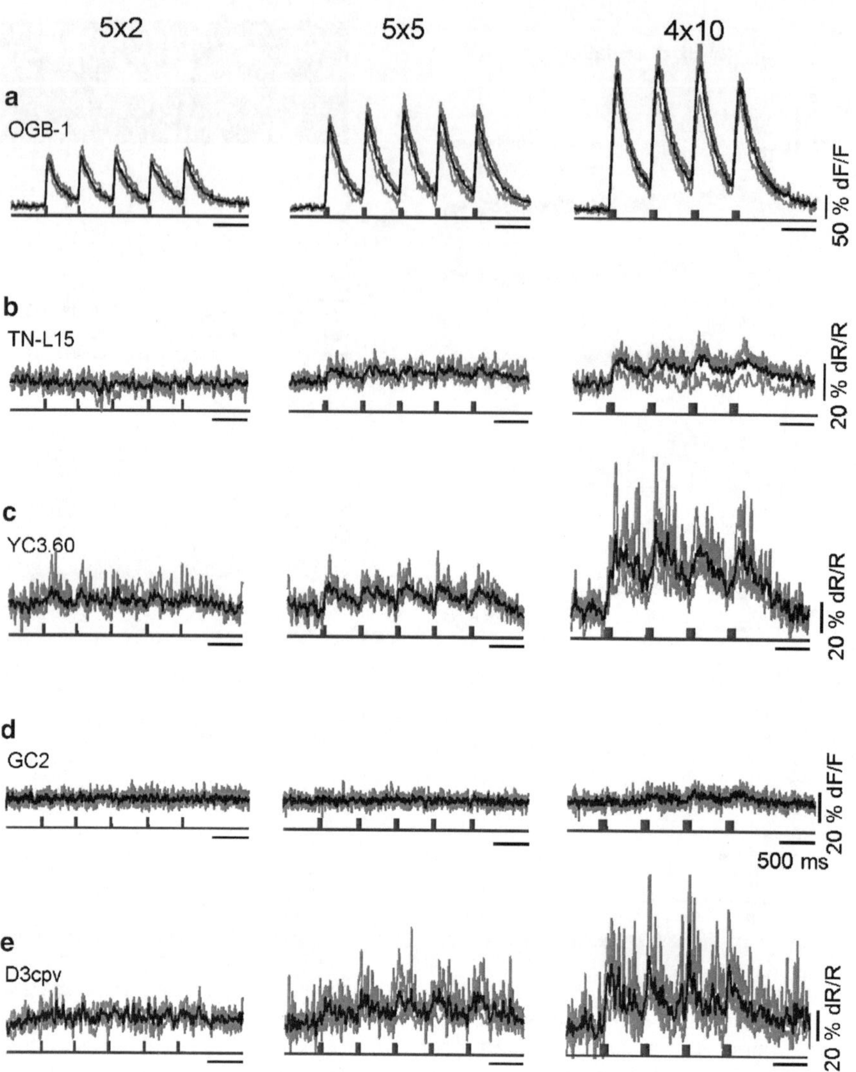

Fig. 5.4. In vivo comparison of the performance of GECIs vs. OGB-1. (**a–e**) In vivo two-photon measurements of Ca^{2+} transients in presynaptic boutons of *Drosophila* motoneurons. Motoneurons were either loaded with OGB-1 via a micropipette or genetically expressed one of the GECIs as indicated. Action potentials were induced by brief voltage pulses (5.5 V, 0.3 ms) applied extracellularly. AP volleys were spaced by 500 ms. Within a volley, APs were elicited at 100 Hz in packs of 2, 5, and 10 APs per volley (first, second, and third columns, respectively). Individual recording traces from four different boutons are shown in gray and their mean in *black*. Modified, with permission, from Ref. (29).

extrusion at physiological temperatures. Thus, several GECIs (TN-XXL, GCAMP2 and D3cpv) perform well when used in tissue slices at room temperature. At higher temperatures, however, small molecule Ca^{2+} indicators still offer a "golden standard" for GECI performance. It is therefore advisable to routinely perform dual calibration experiments, similar to those described in Mao et al. (36), to determine experimental conditions under which a given GECI can be used.

7. In Vivo Calcium Measurements

The main motivation to develop and to use GECIs, as well as their major theoretical advantage compared to small molecule Ca^{2+} indicators is the possibility to use these biosensors to relate activity of single identified neurons to animal's behavior in vivo. However, in vivo Ca^{2+} imaging using GECIs represents the most challenging type of experiment. Three different techniques were used so far to introduce GECIs into the living mouse brain: (1) transgenesis, (2) viral transfection and (3) in utero electroporation. Although some authors claim that virus-mediated gene transfer is crucial for a good in vivo performance of GECIs (6), the fact that similar GECI performance was encountered in acutely transfected and in utero electroporated neurons (5) seems to contradict this hypothesis.

It is indisputable, however, that the performance of GECIs worsens in vivo when compared with brain slices. For D3cpv, for example, the amplitudes of somatic Ca^{2+} transients caused by single action potentials decrease by a half (from 8.3% in slices to 3.5% in vivo, (6)). In line with this observation, even doublets of action potentials cause very modest changes in fluorescence of the majority of GECIs in vivo (29). The reason for this performance decrease is currently unclear, but may be caused at least in part by the mismatch between slow GECI kinetics and fast neuronal calcium handling (see above).

To calibrate the in vivo performance of a given GECI and to find out how well the particular indicator is suited for a given experiment, one again may want to compare its performance to that of a small molecule calcium indicator. Reiff and co-workers (29) did so by applying the same stimulation protocol to presynaptic boutons of *Drosophila* larvae motoneurons which were either microinjected with OGB-1 or were genetically-modified to express different GECIs.

In these two-photon-based in vivo recordings (Fig. 5.4) YC3.60 and D3cpv turned out to be best suited for reporting small Ca^{2+} changes. Good in vivo performance of YC3.60 was found also in the mouse cortex (48). D3cpv, however, again showed quite narrow dynamic range both in flies and in mice (6, 29), thus being more appropriate for the detection of neuronal activity (also sensory-driven, see Ref. (6)) than for the quantification of observed Ca^{2+} changes. TN-XXL was not yet characterized in vivo in much detail. It could be shown however that it is able to monitor sensory-driven neuronal activity in the mouse visual cortex (Fig. 5.5).

Interestingly, similar quality orientation tuning curves of individual neurons were measured in vivo using OGB-1 and TN-XXL. However, the amplitudes of Ca^{2+} transients evoked

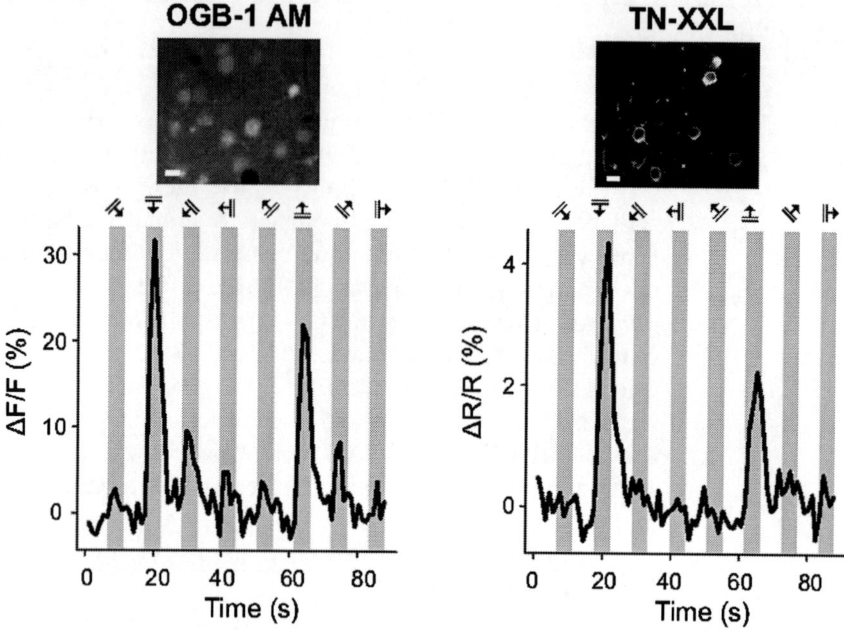

Fig. 5.5. Comparison of OGB-1- and TN-XXL-mediated Ca^{2+} signals in the mouse visual cortex. (**a**, **b**) *Upper:* images of layer 2/3 neurons loaded either with OGB-1 (*left*, multi cell bolus loading, (52)) or with TN-XXL (*right*, infection with the Semliki Forest virus). *Lower:* visually evoked calcium transients (averages of five trials) in an OGB-1 loaded neuron (*left*) and in a TN-XXL expressing neuron (*right*). Scale bars 10 μm. Modified, with permission, from Ref. (5).

by whole-field drifting grating stimuli were higher in OGB-1-stained neurons (Fig. 5.5). Also, the number of responding cells was almost three times higher in OGB-1 as compared to TN-XXL-stained preparations (5), suggesting that in many cells small light-evoked responses were detected only by OGB-1 but not by TN-XXL. It is worth to note that both D3cpv and TN-XXL remained functional in vivo for at least several weeks. This finding makes both indicators suitable for chronic in vivo imaging.

8. Conclusion

Genetically encoded calcium indicators improved dramatically over the last decade and still continue to improve at extremely high pace. By now, they remain functional in the mammalian brain in vivo are sensitive enough to report single action potential firing and are able to monitor sensory-driven neuronal activity. Whereas future improvements of indicator's design will mostly concentrate on speeding up its kinetics, many useful experiments can be done with already existing GECI. The "users," however, will need to quantify properties of these indicators under different experimental conditions, calibrate their responses to calcium and establish optimal expression levels in living cells.

Acknowledgments

Supported by Deutsche Forschungsgemeinschaft (GA654, SFB 596 and SP1172), and the Max-Planck Society.

References

1. Berridge MJ, Lipp P, Bootman MD (2000) The versatility and universality of calcium signalling. Nat Rev Mol Cell Biol 1:11–21
2. Berridge MJ, Bootman MD, Roderick HL (2003) Calcium signalling: dynamics, homeostasis and remodelling. Nat Rev Mol Cell Biol 4:517–529
3. Tsien RW, Tsien RY (1990) Calcium channels, stores, and oscillations. Annu Rev Cell Biol 6:715–760
4. Boldogkoi Z, Balint K, Awatramani GB, Balya D, Busskamp V, Viney TJ, Lagali PS, Duebel J, Pasti E, Tombacz D, Toth JS, Takacs IF, Scherf BG, Roska B (2009) Genetically timed, activity-sensor and rainbow transsynaptic viral tools. Nat Methods 6:127–130
5. Mank M, Santos AF, Direnberger S, Mrsic-Flogel TD, Hofer SB, Stein V, Hendel T, Reiff DF, Levelt C, Borst A, Bonhoeffer T, Hubener M, Griesbeck O (2008) A genetically encoded calcium indicator for chronic in vivo two-photon imaging. Nat Methods 5:805–811
6. Wallace DJ, Zum Alten Borgloh SM, Astori S, Yang Y, Bausen M, Kugler S, Palmer AE, Tsien RY, Sprengel R, Kerr JN, Denk W, Hasan MT (2008) Single-spike detection in vitro and in vivo with a genetic Ca^{2+} sensor. Nat Methods 5:797–804
7. Garaschuk O, Griesbeck O, Konnerth A (2007) Troponin C-based biosensors: a new family of genetically encoded indicators for in vivo calcium imaging in the nervous system. Cell Calcium 42:351–361
8. Hires SA, Tian L, Looger LL (2008) Reporting neural activity with genetically encoded calcium indicators. Brain Cell Biol 36:69–86
9. Mank M, Griesbeck O (2008) Genetically encoded calcium indicators. Chem Rev 108:1550–1564
10. VanEngelenburg SB, Palmer AE (2008) Fluorescent biosensors of protein function. Curr Opin Chem Biol 12:60–65
11. Zhang J, Campbell RE, Ting AY, Tsien RY (2002) Creating new fluorescent probes for cell biology. Nat Rev Mol Cell Biol 3:906–918
12. Heim N, Griesbeck O (2004) Genetically encoded indicators of cellular calcium dynamics based on troponin C and green fluorescent protein. J Biol Chem 279:14280–14286
13. Miyawaki A, Griesbeck O, Heim R, Tsien RY (1999) Dynamic and quantitative Ca^{2+} measurements using improved cameleons. Proc Natl Acad Sci USA 96:2135–2140
14. Miyawaki A, Llopis J, Heim R, McCaffery JM, Adams JA, Ikura M, Tsien RY (1997) Fluorescent indicators for Ca^{2+} based on green fluorescent proteins and calmodulin. Nature 388:882–887
15. Nagai T, Yamada S, Tominaga T, Ichikawa M, Miyawaki A (2004) Expanded dynamic range of fluorescent indicators for Ca^{2+} by circularly permuted yellow fluorescent proteins. Proc Natl Acad Sci USA 101:10554–10559
16. Palmer AE, Giacomello M, Kortemme T, Hires SA, Lev-Ram V, Baker D, Tsien RY (2006) Ca^{2+} indicators based on computationally redesigned calmodulin-peptide pairs. Chem Biol 13:521–530
17. Mank M, Reiff DF, Heim N, Friedrich MW, Borst A, Griesbeck O (2006) A FRET-based calcium biosensor with fast signal kinetics and high fluorescence change. Biophys J 90:1790–1796
18. Baird GS, Zacharias DA, Tsien RY (1999) Circular permutation and receptor insertion within green fluorescent proteins. Proc Natl Acad Sci USA 96:11241–11246
19. Griesbeck O, Baird GS, Campbell RE, Zacharias DA, Tsien RY (2001) Reducing the

environmental sensitivity of yellow fluorescent protein. Mechanism and applications. J Biol Chem 276:29188–29194

20. Nagai T, Sawano A, Park ES, Miyawaki A (2001) Circularly permuted green fluorescent proteins engineered to sense Ca^{2+}. Proc Natl Acad Sci USA 98:3197–3202
21. Nakai J, Ohkura M, Imoto K (2001) A high signal-to-noise Ca^{2+} probe composed of a single green fluorescent protein. Nat Biotechnol 19:137–141
22. Souslova EA, Belousov VV, Lock JG, Stromblad S, Kasparov S, Bolshakov AP, Pinelis VG, Labas YA, Lukyanov S, Mayr LM, Chudakov DM (2007) Single fluorescent protein-based Ca^{2+} sensors with increased dynamic range. BMC Biotechnol 7:37
23. Tallini YN, Ohkura M, Choi BR, Ji G, Imoto K, Doran R, Lee J, Plan P, Wilson J, Xin HB, Sanbe A, Gulick J, Mathai J, Robbins J, Salama G, Nakai J, Kotlikoff MI (2006) Imaging cellular signals in the heart in vivo: cardiac expression of the high-signal Ca^{2+} indicator GCaMP2. Proc Natl Acad Sci USA 103:4753–4758
24. Chalasani SH, Chronis N, Tsunozaki M, Gray JM, Ramot D, Goodman MB, Bargmann CI (2007) Dissecting a circuit for olfactory behaviour in Caenorhabditis elegans. Nature 450:63–70
25. Faumont S, Lockery SR (2006) The awake behaving worm: simultaneous imaging of neuronal activity and behavior in intact animals at millimeter scale. J Neurophysiol 95:1976–1981
26. Kerr R, Lev-Ram V, Baird G, Vincent P, Tsien RY, Schafer WR (2000) Optical imaging of calcium transients in neurons and pharyngeal muscle of C. elegans. Neuron 26:583–594
27. Kuhara A, Okumura M, Kimata T, Tanizawa Y, Takano R, Kimura KD, Inada H, Matsumoto K, Mori I (2008) Temperature sensing by an olfactory neuron in a circuit controlling behavior of C. elegans. Science 320:803–807
28. Suzuki H, Kerr R, Bianchi L, Frokjaer-Jensen C, Slone D, Xue J, Gerstbrein B, Driscoll M, Schafer WR (2003) In vivo imaging of C. elegans mechanosensory neurons demonstrates a specific role for the MEC-4 channel in the process of gentle touch sensation. Neuron 39:1005–1017
29. Hendel T, Mank M, Schnell B, Griesbeck O, Borst A, Reiff DF (2008) Fluorescence changes of genetic calcium indicators and OGB-1 correlated with neural activity and calcium in vivo and in vitro. J Neurosci 28:7399–7411
30. Liu L, Yermolaieva O, Johnson WA, Abboud FM, Welsh MJ (2003) Identification and function of thermosensory neurons in Drosophila larvae. Nat Neurosci 6:267–273
31. Marella S, Fischler W, Kong P, Asgarian S, Rueckert E, Scott K (2006) Imaging taste responses in the fly brain reveals a functional map of taste category and behavior. Neuron 49:285–295
32. Wang JW, Wong AM, Flores J, Vosshall LB, Axel R (2003) Two-photon calcium imaging reveals an odor-evoked map of activity in the fly brain. Cell 112:271–282
33. Diez-Garcia J, Matsushita S, Mutoh H, Nakai J, Ohkura M, Yokoyama J, Dimitrov D, Knopfel T (2005) Activation of cerebellar parallel fibers monitored in transgenic mice expressing a fluorescent Ca^{2+} indicator protein. Eur J Neurosci 22:627–635
34. Hasan MT, Friedrich RW, Euler T, Larkum ME, Giese G, Both M, Duebel J, Waters J, Bujard H, Griesbeck O, Tsien RY, Nagai T, Miyawaki A, Denk W (2004) Functional fluorescent Ca^{2+} indicator proteins in transgenic mice under TET control. PLoS Biol 2:763–775
35. Heim N, Garaschuk O, Friedrich MW, Mank M, Milos RI, Kovalchuk Y, Konnerth A, Griesbeck O (2007) Improved calcium imaging in transgenic mice expressing a troponin-C based biosensor. Nat Methods 4:127–129
36. Mao T, O'Connor DH, Scheuss V, Nakai J, Svoboda K (2008) Characterization and subcellular targeting of GCaMP-type genetically-encoded calcium indicators. PLoS One 3:e1796
37. He J, Ma L, Kim S, Nakai J, Yu CR (2008) Encoding gender and individual information in the mouse vomeronasal organ. Science 320:535–538
38. Sambrook J, Russell D (2001) Molecular cloning: a laboratory manual. CSHL Press, Cold Spring Harbor
39. Lo DC, McAllister AK, Katz LC (1994) Neuronal transfection in brain slices using particle-mediated gene transfer. Neuron 13:1263–1268
40. Tabata H, Nakajima K (2001) Efficient in utero gene transfer system to the developing mouse brain using electroporation: visualization of neuronal migration in the developing cortex. Neuroscience 103:865–872
41. Tabata H, Nakajima K (2008) Labeling embryonic mouse central nervous system cells by in utero electroporation. Dev Growth Differ 50:507–511
42. Chesler AT, Le Pichon CE, Brann JH, Araneda RC, Zou DJ, Firestein S (2008) Selective gene expression by postnatal electroporation during olfactory interneuron nurogenesis. PLoS One 3:e1517

43. Dittgen T, Nimmerjahn A, Komai S, Licznerski P, Waters J, Margrie TW, Helmchen F, Denk W, Brecht M, Osten P (2004) Lentivirus-based genetic manipulations of cortical neurons and their optical and electrophysiological monitoring in vivo. Proc Natl Acad Sci USA 101:18206–18211
44. Klein RL, Hamby ME, Gong Y, Hirko AC, Wang S, Hughes JA, King MA, Meyer EM (2002) Dose and promoter effects of adeno-associated viral vector for green fluorescent protein expression in the rat brain. Exp Neurol 176:66–74
45. Teschemacher AG, Wang S, Lonergan T, Duale H, Waki H, Paton JF, Kasparov S (2005) Targeting specific neuronal populations using adeno- and lentiviral vectors: applications for imaging and studies of cell function. Exp Physiol 90:61–69
46. Coura Rdos S, Nardi NB (2007) The state of the art of adeno-associated virus-based vectors in gene therapy. Virol J 4:99
47. Wong LF, Goodhead L, Prat C, Mitrophanous KA, Kingsman SM, Mazarakis ND (2006) Lentivirus-mediated gene transfer to the central nervous system: therapeutic and research applications. Hum Gene Ther 17:1–9
48. Kuchibhotla KV, Goldman ST, Lattarulo CR, Wu HY, Hyman BT, Bacskai BJ (2008) Abeta plaques lead to aberrant regulation of calcium homeostasis in vivo resulting in structural and functional disruption of neuronal networks. Neuron 59:214–225
49. Caroni P (1997) Overexpression of growth-associated proteins in the neurons of adult transgenic mice. J Neurosci Methods 71:3–9
50. Feng G, Mellor RH, Bernstein M, Keller-Peck C, Nguyen QT, Wallace M, Nerbonne JM, Lichtman JW, Sanes JR (2000) Imaging neuronal subsets in transgenic mice expressing multiple spectral variants of GFP. Neuron 28:41–51
51. Neher E (1995) The use of fura-2 for estimating Ca buffers and Ca fluxes. Neuropharmacology 34:1423–1442
52. Stosiek C, Garaschuk O, Holthoff K, Konnerth A (2003) In vivo two-photon calcium imaging of neuronal networks. Proc Natl Acad Sci USA 100:7319–7324
53. Zipfel WR, Williams RM, Webb WW (2003) Nonlinear magic: multiphoton microscopy in the biosciences. Nat Biotechnol 21:1369–1377
54. Diez-Garcia J, Akemann W, Knopfel T (2007) In vivo calcium imaging from genetically specified target cells in mouse cerebellum. Neuroimage 34:859–869
55. Hires SA, Zhu Y, Tsien RY (2008) Optical measurement of synaptic glutamate spillover and reuptake by linker optimized glutamate-sensitive fluorescent reporters. Proc Natl Acad Sci USA 105:4411–4416
56. Pologruto TA, Yasuda R, Svoboda K (2004) Monitoring neural activity and [Ca2+] with genetically encoded Ca^{2+} indicators. J Neurosci 24:9572–9579
57. Reiff DF, Ihring A, Guerrero G, Isacoff EY, Joesch M, Nakai J, Borst A (2005) In vivo performance of genetically encoded indicators of neural activity in flies. J Neurosci 25:4766–4778
58. Waters J, Larkum M, Sakmann B, Helmchen F (2003) Supralinear Ca^{2+} influx into dendritic tufts of layer 2/3 neocortical pyramidal neurons in vitro and in vivo. J Neurosci 23:8558–8567
59. Homma R, Baker BJ, Jin L, Garaschuk O, Konnerth A, Cohen LB, Bleau CX, Canepari M, Djurisic M, Zecevic D (2009) Wide-field and two-photon imaging of brain activity with voltage- and calcium-sensitive dyes. Methods Mol Biol 489:43–79
60. Yasuda R, Nimchinsky EA, Scheuss V, Pologruto TA, Oertner TG, Sabatini BL, Svoboda K (2004) Imaging calcium concentration dynamics in small neuronal compartments. Sci STKE 2004:pl5
61. Xu C (2000) Two-photon cross sections of indicators. In: Yuste R, Lanni F, Konnerth A (eds) Imaging: a laboratory manual. Cold Spring Harbor Press, Cold Spring Harbor, pp 19.11–19.19
62. Markram H, Helm PJ, Sakmann B (1995) Dendritic calcium transients evoked by single back-propagating action potentials in rat neocortical pyramidal neurons. J Physiol 485:1–20

Chapter 6

Intracellular Calcium-Sensitive Microelectrodes

Roger C. Thomas

Abstract

Ca^{2+}-sensitive microelectrodes are time-consuming to make and require large robust cells. But, they do not add to buffering and do not require expensive equipment. I describe how to make and use the electrodes and briefly consider the leakage problem.

Key words: Microelectrode, Leakage, Buffering

1. Introduction

Calcium-sensitive microelectrodes (CaSMs) are difficult to make and use, but offer several advantages over the much more widely-used optical methods. In spite of this, only a very few researchers are currently publishing data obtained with CaSMs. The advantages are that CaSMs are easy to calibrate, inexpensive, require only basic electrophysiological apparatus with a high-impedance amplifier, have a very wide sensitivity range, and do not add to intracellular Ca^{2+} buffering. The two sensors used, ETH 1001 (1) and ETH 129 (2), were developed by the late Willy Simon in Switzerland, and are available from Fluka now part of Sigma. 50 mg of sensor is enough for many thousands of CaSMs although you need solvents and additives to make the cocktail. It took some time to discover that optimal performance of these sensors in microelectrodes was only obtained with a PVC-gelled cocktail (3).

The first sensor used in a PVC-gelled cocktail was ETH 1001, but I prefer ETH 129 which has a superior selectivity in favour of calcium. A few years ago, I tried for some months to make effective CaSMs with minimal or no added PVC, but I failed. For some reason, CaSMs only work well if the cocktail contains

A. Verkhratsky and Ole H. Petersen (eds.), *Calcium Measurement Methods*, Neuromethods, vol. 43, DOI 10.1007/978-1-60761-476-0_6, © Humana Press, a part of Springer Science+Business Media, LLC 2010

enough PVC to make it solid once the solvent has evaporated. The PVC also allows pressure to be applied to the back of the CaSM, which I have found greatly improves its performance. I do not know why this works, but the pressure slowly leads to the extrusion of a small amount of cocktail which forms a visible blob. Perhaps, this increases the sensor surface area and reduces the tip resistance. I have not pursued this further.

The rare use of CaSMs is largely because of their many disadvantages. These include their slow responses, 1–2 μm tips requiring large and robust cells and tendency to create a significant leak at the point of insertion. All "sharp" microelectrodes probably create a Ca^{2+} leak at the point of insertion; sadly, the CaSMs are ideally placed to detect it. With care and patience, the leak can be reduced to levels which I hope are insignificant. I have recently discovered that CaSMs filled with my cocktail record intracellular barium levels about as well as calcium (4).

For those few who might be tempted to use CaSMs inside cells, and the larger number who might like to use them outside cells, I here describe the way I make them, evolved over about 15 years of trial and error.

2. Methods

2.1. Glass Tubing Types

The Ca^{2+}-sensing cocktail has a very high electrical resistance, and seems to be somewhat short-circuited even by borosilicate glass. In the past, I have used aluminosilicate glass since it has a suitably high resistivity and can be pulled with conventional electrode pullers. But, I now prefer quartz glass as the CaSMs made with it last longer and are easier to make with sharper tips (Fig. 6.1). Sadly, such glass can only be pulled with an expensive laser-powered puller such as the Sutter P-2000. I use tubing of 1 mm external diameter and 0.7 mm internal diameter.

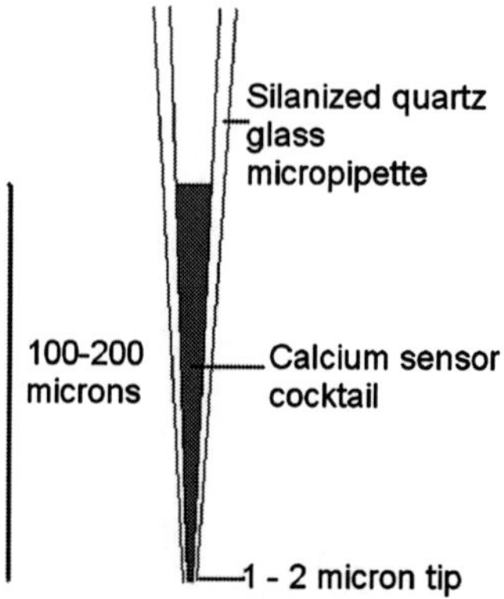

Fig. 6.1. Diagram of the tip of a Ca^{2+}-sensitive microelectrode.

2.2. Pulling the Micropipettes

If the glass is not supplied fire-polished, as the quartz supplied by Sutter now seems to be, the ends must be heated in a high-temperature flame (oxy-acetylene for example) to make them smooth and easy to handle. 75 mm lengths then need to be pulled at a high heat and pull to produce gently-tapering tips – typically with an outside diameter of 8 μm at 0.1 mm from the tip. The settings I use on the P-2000 are Heat 940, Filament 4, Velocity 45, Delay 120, and Pull 220. I normally pull about ten micropipettes before the next step.

2.3. Enlarging the Micropipette Tips

As pulled, the micropipettes typically have very small tips, less than 100 nm diameter. Such a tip filled with Ca^{2+} cocktail has too high an electrical resistance to function well. It is also impossible to suck up a suitable column of cocktail through it. The tips must therefore be enlarged by bevelling or simple breakage on a glass rod. Bevelling should produce a functionally sharper tip and be more controllable. While bevelling certainly works well, I have found in practise that simply lowering the tips onto a glass rod with a micromanipulator while observing with a microscope works perfectly well. I aim at an outside tip diameter of 1.5 ± 0.5 μm.

2.4. Silanizing the Micropipettes

All glass is naturally hydrophilic, so that aqueous solutions tend to displace and short-circuit the sensor cocktails used in ion-sensitive microelectrodes. There are many subtly different techniques which have been used to render the surface hydrophobic. Various silanes have been used in solution or as vapour to treat the glass, followed by baking to fix or cure the glass–silane combination. For many years, I have used a modification of the treatment developed by Munoz and Coles (5, 6), the pioneers of quartz ion-sensitive microelectrodes.

After pulling a batch of micropipettes and breaking their tips, I put them on a small nickel tray essentially the broken end of a spatula. Then, I leave them in air for about 20 min to adsorb a little water. I presume that is what happens, but all I really know is that silanization works better with a period in air beforehand. I then put the tray into a custom-made mini-vacuum oven as shown in Fig. 6.2. The oven is a 15 cm length of borosilicate tubing connected at one end to a vacuum pump and with the other end closed with the rubber piston from a 20 ml disposable syringe. I make sure that the micropipette tips are well within the heating coil, which is made of 12 turns of 22 swg Kanthal A resistance wire wound round about 3.5 cm of the glass tube. Having capped the tube/oven, I switch on the vacuum pump evacuate for about 30 s and then close the tube to the pump and switch it off.

I then inject into the oven about 3 μl of Bis (dimethylamino) dimethylsilane (Fluka 14755, Sigma) and switch on a heating current of about 3.6 A, which generates a temperature of about 400°C. I leave the current running for about 10 min, then switch

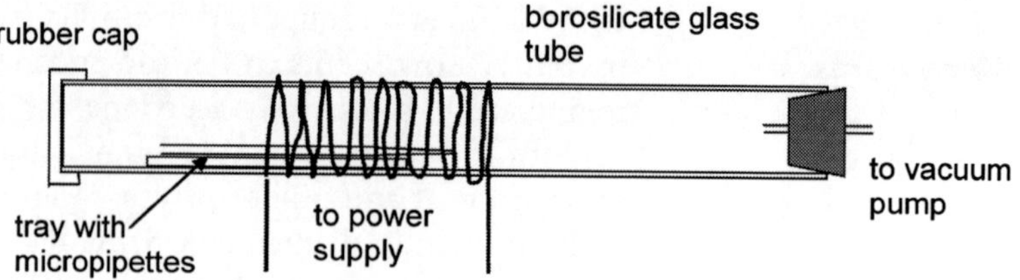

Fig. 6.2. Mini vacuum oven for silanizing micropipettes. For details see text.

it off and re-evacuate the oven. I leave it to cool under vacuum before removing the micropipettes to a storage dish.

2.5. Filling the Calcium-Sensitive Microelectrodes

The filled CaSMs are good for up to 3 days, so I normally fill 3–6 at one time depending on how pessimistic I feel about their likely performance and how hard I plan to work. First, I backfill the micropipettes with 1 mM $CaCl_2$ using a drawn-down 1 ml disposable syringe. Then, I mount each micropipette vertically in a microforge setup and connect it via silicone rubber tubing to a 10 ml syringe a quarter filled with air. Observing the tip through a horizontally-mounted microscope, I apply air pressure using the syringe to force the air out of the tip. The time it takes to push out the air is an indication of the tip size. If it takes more than about 10 s, the resulting microelectrode is likely to have too sharp a tip. With a well-silanized micropipette, the backfill solution will withdraw a few microns from the tip when the air pressure is released. Applying some pressure to stop this withdrawal, I then dip the pipette tip into a small aliquot of the Ca cocktail held in a J-tube (7) and suck up a column of 0.15–0.3 mm of the cocktail. The filled microelectrodes may be left in air until used. To some extent, I have found that the electrode lifetime is proportional to the column length while shorter columns give a stable response sooner.

My cocktail is very similar to that described by the Swiss inventors (2) and has the following composition in mg ml^{-1}: Ca ionophore ETH 129–12, Sodium tetrakis[3,5-bis(trifluoromethyl) phenyl]borate – 6, 2-nitrophenyloctylether – 200, high molecular weight PVC – 34, and tetrahydrofuran – 748. It remains usable for several years although losing tetrahydrofuran slowly. When the cocktail loses fluidity, more tetrahydrofuran can be added.

2.6. Testing the Electrodes and Applying Pressure

Once I have set up an experiment with all electrodes mounted over the experimental preparation. I test the CaSM by switching the flowing solution from normal to one with no added calcium and with 1 mM EGTA to reduce the free $[Ca^{2+}]$ below about 10 nM. The response of a potentially useful CaSM will be a voltage change of at least 120 mV. Then, I apply air pressure of about 1 bar to the back of the CaSM. This should increase the voltage response to at least 150 mV between normal solution with a few mM $CaCl_2$ and the almost Ca-free EGTA solution. If not either the tip is too sharp, the glass not well-silanized or there is an electrical problem. Insulation must be clean and the amplifier input impedance high. Note that at a minimum a second intracellular microelectrode is needed to measure the membrane potential which must be subtracted from that of the CaSM (all intracellular ion-sensitive microelectrodes record the sum of the voltage difference generated by the ion difference between cytoplasm and the bathing solution and the membrane potential.)

2.7. Experimental Protocol

Since CaSMs are necessarily rather blunt, it is advisable to insert them into the cell first. Inevitably, there will be some movement of the cell as the CaSM is pushed in. With snail neurones, I often have to move the electrode down by over 0.1 or even 0.2 mm before the tip will enter the cell. I show an example of this in Fig. 1 of a recent paper (8). An earlier experiment from 1993, in which I recorded both Fura 2 fluorescence and the potential from a CaSM is shown in Fig. 6.3 here. It is clear that the overall [Ca^{2+}] in the cell as measured by Fura 2 was essentially constant between

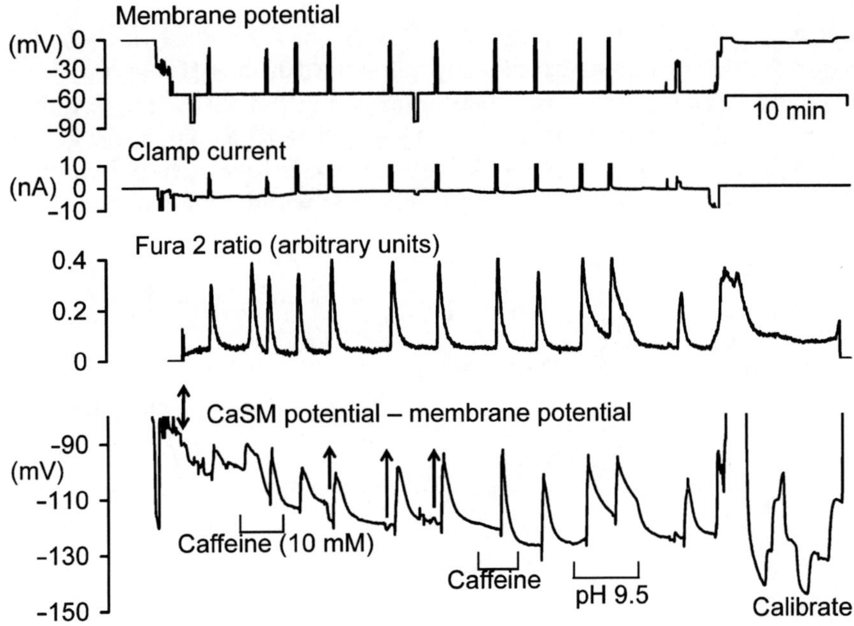

Fig. 6.3. Experiment on a voltage-clamped snail neurone comparing two ways of measuring intracellular [Ca^{2+}]. From the top, the four traces show the membrane potential recorded with a conventional KCl-filled microelectrode, the clamp current passed through a CsCl-filled microelectrode, the ratio of the fluorescence (excited at 340 and 380 nm) from Fura 2 injected into the cell at the beginning of the recording, and the potential from the CaSM from which has been subtracted the membrane potential. For most of the experiment, the membrane potential was clamped at –50 mV. At intervals, the clamp potential was changed to test the subtraction or open Ca channels. Caffeine was applied to empty the Ca stores and the Ringer pH was increased to 9.5 to inhibit the plasma membrane Ca-ATPase. At the end, the electrodes were withdrawn from the cell and the bath perfused with calibration solutions of pCa 8, 7 and 6.

depolarisations or applications or caffeine. But, the CaSM potential recorded from the same cell shows values indicating that at first the recorded [Ca^{2+}] at the electrode tip is much higher than is seen towards the end of the experiment. The arrows show where I moved the CaSM up or down in a mostly vain attempt to persuade the CaSM to seal in better.

I conclude that for at least the first part of the experiment, the CaSM record was distorted by leakage of Ca^{2+} ions from outside the cell as diagrammed in Fig. 6.4. Thus, experiments with CaSMs need to allow time for the CaSM to seal into the cell. I have also found that really deep penetrations help to reduce the problem.

2.8. The Leakage Problem

I suspect that all conventional microelectrodes that cross the cell membrane create a leakage pathway for ions to enter or leave down their gradients. A leakage for Na^{+} ions was clearly shown for crab muscle by Taylor and Thomas in 1984 (9). Patch-clamp electrodes are widely believed to create almost no leaks. I have myself made pH-sensitive patch electrodes (10)), but failed completely to make Ca^{2+}-sensitive patch electrodes.

The only real solution seems to be patience and hope that the ISM seals in before the extra load created by the leak kills the cell.

2.9. Calibration Problems

In my hands, CaSMs give a linear response of about 28 mV per decade from 7 mM to below 100 nM [Ca^{2+}]. Calibration solutions are available commercially, but I have not used them preferring to make my own using BAPTA as buffer. Details of the way I have calibrated these electrodes have been published (11). The main problem remains that on withdrawal from the cell the CaSM

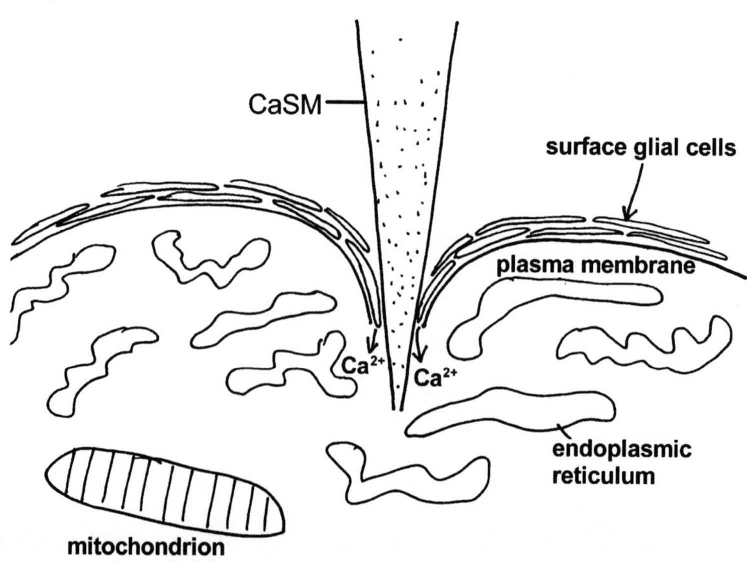

Fig. 6.4. Diagram showing Ca^{2+} ions leaking into a cell around the point of insertion.

response tended to change, and that calibration immediately before an experiment required high-potassium solutions which damaged the preparation in the bath. I have not in fact calibrated any CaSMs in the last 10 years since I am not concerned with the precise value of the [Ca^{2+}] at the tip only with physiologically-relevant changes.

3. Conclusions

It seems that in the last few years, CaSMs have been rarely used except extracellularly or in the large neurones of molluscs. Perhaps, technical advances or increased efforts will eventually allow their use on other cells, but it seems more likely that the idea of poking sharp glass needles into living cells will become less and less popular.

Acknowledgements

This work has been supported by the MRC and the Wellcome Trust, to which I am very grateful.

References

1. Oehme M, Kessler M, Simon W (1976) Neutral carrier Ca^{2+} microelectrode. Chimia 30:204–206
2. Ammann D, Bührer T, Schefer U, Müller M, Simon W (1987) Intracellular neutral carrier-based Ca^{2+} microelectrode with subnanomolar detection limit. Pflugers Arch 409: 223–228
3. Rink TJ, Tsien R (1980) Calcium-selective micro-electrodes with bevelled. sub-micron tips containing poly(vinylchloride)-gelled neutral-ligand sensor. J Physiol (Lond) 308:5P–6P
4. Thomas RC (2009) The plasma membrane calcium ATPase (PMCA) of neurones is electroneutral and exchanges 2 H^+ for each Ca^{2+} or Ba^{2+} ion extruded. J Physiol (Lond) 587 (Pt 2):315–327
5. Munoz JL, Coles JA (1987) Quartz micropipettes for intracellular voltage microelectrodes and ion-selective microelectrodes. J Neurosci Methods 22:57–64
6. Munoz JL, Deyhimi F, Coles JA (1983) Silanization of glass in the making of ion-sensitive microelectrodes. J Neurosci Methods 8:231–247
7. Thomas RC (2001) Electrophysiological measurements using Ca2+-sensitive microelectrodes. In: Petersen OH (ed) Measuring calcium and calmodulin inside and outside cells. Springer, Berlin, pp 91–102
8. Thomas RC, Postma M (2006) Dynamic and static calcium gradients inside large snail (*Helix aspersa*) neurones detected with calcium-sensitive microelectrodes. Cell Calcium 41:365–378
9. Taylor PS, Thomas RC (1984) The effect of leakage on microelectrode measurements of intracellular sodium activity in crab muscle fibres. J Physiol (Lond) 352:539–550
10. Thomas RC, Pagnotta SE, Nistri A (2003) Whole-cell recording of intracellular pH with silanized and oiled patch-type single or double-barreled microelectrodes. Pflugers Arch 447:259–265
11. Kennedy HJ, Thomas RC (1996) Effects of injecting calcium-buffer solution on $[Ca^{2+}]_i$ in voltage-clamped snail neurons. Biophys J 70:2120–2130

Chapter 7

Ca^{2+} Caging and Uncaging

Shin Hye Kim and Myoung Kyu Park

Abstract

Neuronal Ca^{2+} signals occur in a very complex way. Direct imaging of Ca^{2+} changes in the soma, dendrites, and even single spines on fast time-scales greatly helps us understand the generation mechanism of diverse Ca^{2+} signaling events. However, Ca^{2+} imaging itself does not give information about the causal relationships between specific Ca^{2+} signals and specific functions. With the rapid improvements of new caged compounds, application or usage of uncaging/caging techniques has been expanded widely in biological research. Using caged compounds, the Ca^{2+} concentration in the target area can be either increased or lowered on a given time scale to various degrees. The most important advantage of the uncaging/caging technique is to control the intensity, duration, and area of light with high precision at a single cell level, providing very accurate control of biomolecules within a limited space at a given time. This chapter discusses how to use caged Ca^{2+} compounds in studying neuronal functions and Ca^{2+} signaling.

Key words: Ca^{2+} signals, Ca^{2+} uncaging, Ca^{2+} caging, Local uncaging, Caged Ca^{2+} compounds, Light sensitive probes, Neurons, Confocal microscope

1. Introduction

Intracellular Ca^{2+} plays various roles in neuronal functions; neurotransmitter release at synaptic terminals, regulation of ion channels and excitability, induction of synaptic plasticity, modulation of neurite outgrowth, remodeling of dendritic spines, determining cell death and survival, and even pathologic processes such as neurodegeneration and necrosis (1–3). This versatility of Ca^{2+} signaling in neurons is purely based on the diversity of Ca^{2+} signals (1). In the time domain, Ca^{2+} signals operate over a wide spectrum, ranging from milliseconds to hours. In the spatial domain, Ca^{2+} events vary from submicron regions to the whole neuronal compartment. Since neuronal Ca^{2+} signals are characterized

A. Verkhratsky and Ole H. Petersen (eds.), *Calcium Measurement Methods,* Neuromethods, vol. 43
DOI: 10.1007/978-1-60761-476-0_7 © Humana Press, a part of Springer Science+Business Media, LLC 2010

by very complex patterns that are linked to the specific function, information about temporal and spatial patterns is essential for understating the role of Ca^{2+} signals in neuronal functions. Direct imaging of Ca^{2+} changes in the soma, dendrites, and even single spines on fast time-scales greatly helps us understand the generation mechanism of diverse Ca^{2+} signaling events. However, Ca^{2+} imaging itself does not give information about the causal relationships between specific Ca^{2+} signals and specific functions. In order to correlate a specific Ca^{2+} event to its neuronal function, sophisticated manipulation of Ca^{2+} signals and simultaneous observation of neuronal functions such as electrical activities or vesicle release is essential. To evoke Ca^{2+} events artificially in neuronal cells during experiments, neurons can be stimulated in various ways. Neurotransmitters, neurohormones, and chemicals have been applied by various methods, such as a simple perfusion system or a pressure injection apparatus. Neurons can also be stimulated electrically by patch clamp electrodes or focal or field stimulation wires or plates. Although all these stimuli can cause appropriate Ca^{2+} events according to the stimulus type, the spatial and temporal resolution of such approaches is limited.

Recent advances in molecular probe development, microscopy, and laser instrumentation now allow us to directly manipulate Ca^{2+} signals with unprecedented spatial and temporal resolution through the use of light within a single neuron. Using light-sensitive probes known as caged compounds, we can either increase or lower the Ca^{2+} concentration in the target area on a given time scale to various degrees. With the combination of Ca^{2+} imaging (4), patch clamping (5), and amperometry (6), this "Ca^{2+} uncaging and caging technology" opens a new way to study neuronal functions (7).

With the rapid improvements of new caged compounds, application or usage of uncaging/caging techniques has been expanded widely in biological research. Therefore, many suppliers of confocal microscopes now tend to consider optimization of hardware and software for uncaging experiments. Since caged ATP was first used in biology 30 years ago, several kinds of caged compounds have been developed (8). Regarding Ca^{2+} signaling, several steps of advances in chemicals and imaging machines enable us to use this valuable tool with high efficiency and versatility. The most important advantage of the uncaging/caging technique is that we can control the intensity, duration, and area of light with high precision at a single cell level (4, 7), providing very accurate control of biomolecules within a limited space at a given time. In addition, because light can pass easily through the plasma membrane, uncaging/caging can be performed without mechanical intervention or disruption of the plasma membrane. Therefore, we can expect that caged compounds will provide a great opportunity for studying neuronal functions, including Ca^{2+} signaling. This chapter discusses how to use caged Ca^{2+} compounds in studying neuronal functions and Ca^{2+} signaling.

2. Methods

2.1. Principles of Caged Compounds

Caged compounds are light-sensitive probes that contain functionally active biomolecules in an inactive form and release them upon exposure to UV (300–400 nm) light, thereby generating a sudden concentration jump of specific biomolecules or acting as a rapid functional switch. Basically, caging technology is based on the idea that a molecule of interest can be rendered functionally inert by forming covalent bonds with photoremovable protecting groups (9). However, the inorganic Ca^{2+} ion cannot form covalent bonds to caging groups. Therefore, instead of Ca^{2+} itself, high-affinity Ca^{2+} chelators such as EDTA, EGTA, and BAPTA are linked to photoreactive groups (8). These modified forms usually have more or less similar affinities for Ca^{2+} as compared to the original forms of EDTA, EGTA, and BAPTA. However, due to the added photoreactive groups, UV exposure leads to cleavage of another nearby vital bond connecting with the functional groups that are essential for holding the Ca^{2+} ion, thereby resulting in a dramatic decrease in Ca^{2+} affinity (a several-fold increase in K_d values, (4)). Figure 7.1 illustrates these types of chemical changes during photolysis. For example, EGTA is a highly selective Ca^{2+} chelator. Tetraacetic acids are essential for holding the Ca^{2+} ion. Covalent modification of EGTA with the nitrophenyl (NP) group leads to instability of the covalent bond of the nearby amino ethyl ester group, which is critical for Ca^{2+} binding (Fig. 7.1, see NP-EGTA). Thus, UV illumination easily converts the high Ca^{2+} affinity NP-EGTA into low Ca^{2+} affinity cleaved fragments. EDTA- and BAPTA-derived caged compounds such as DM-nitrophen and DMNPE-4, act in the same way as shown in Fig. 7.1. Including the above caged Ca^{2+} compounds, many other Ca^{2+} caged compounds are now commercially available. Among them, NP-EGTA has been most widely used and selected caged compounds are listed in Table 7.1. While photolysis causes most caged Ca^{2+} compounds to become low affinity chemicals for Ca^{2+}, some of the compounds are reverse-converted into higher affinity chelators for Ca^{2+} upon illumination. Diazo-2 and DMNP-EDTA are examples of this and are thus used as caged Ca^{2+} chelators. UV irradiation cleaves Diazo-2 to become a functional Ca^{2+} chelator, BAPTA, that binds Ca^{2+} ion, thereby reducing the free Ca^{2+} concentration very quickly and potently. While the former is referred to as Ca^{2+} uncaging, the latter is referred to Ca^{2+} caging.

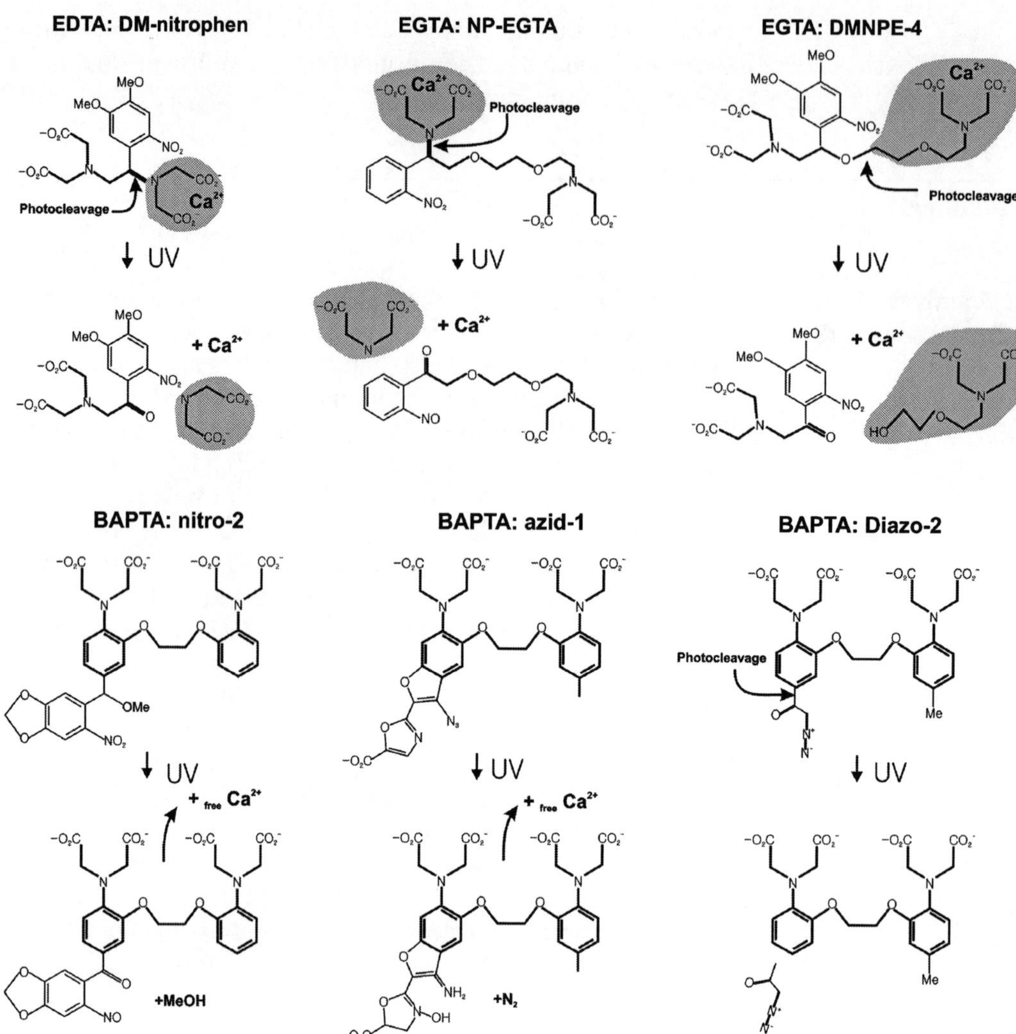

Fig. 7.1. Structures of caged Ca^{2+} compounds and their photolysis. Caged Ca^{2+} compounds are synthesized from EDTA, EGTA, and BAPTA that are presented as *thick lines*. Strong UV illumination caused these compounds to cleave into fragments having different K_d values for Ca^{2+}.

In order to liberate Ca^{2+} upon irradiation, these caged Ca^{2+} compounds should exist in the Ca^{2+} bound form. In this case, the dissociation constant (K_d value) is very important for estimating the quantity of caged compounds present which can be bound to Ca^{2+} ions. The dissociation constant between Ca^{2+} and caged compound is defined by the following equation:

$$K_d = \frac{[\text{free}\,Ca^{2+}]\times[\text{caged compound with no}\,Ca^{2+}\,\text{bound}]}{[\text{caged compound with}\,Ca^{2+}\,\text{bound}]}$$

The K_d value of caged Ca^{2+} compounds corresponds to the concentration of Ca^{2+} at which one-half of the compounds exists

Table 7.1
General properties of selected caged Ca^{2+} compounds

Caged compound	Ca^{2+} K_d in caged form (nM)	Ca^{2+} K_d in uncaged form (nM)	ε (M^{-1} cm^{-1})	ϕ	Uncaging index (ε φ)	Commercially available form
DM-nitrophen	5	3×10^9	4,300	0.18	774	Salt/AM
NP-EGTA	80	1×10^9	975	0.23	224	Salt/AM
NDBF-EGTA	100	2×10^9	18,400	0.7	12,880	Salt
DMNPE-4	48	2×10^9	5,120	0.09	461	AM
nitr-5	145	63×10^4	5,500	0.012	66	Salt/AM
azid-1	230	12×10^7	33,000	1	33,000	Salt
Diazo-2	22×10^2	73	22,800	0.03	1,596	Salt

DM-nitrophen 1-(2-nitro-4,5-dimethoxyphenyl)-*N*,*N*,*N'*,*N'*-tetrakis[(oxycarbonyl)methyl]-1,2-ethanediamine; *NP-EGTA* *o*-nitrophenyl-EGTA; *NDBF-EGTA* nitrodibenzofuran-EGTA; *DMNPE-4* dimethoxynitrophenyl-EGTA-4

as a Ca^{2+} bound form, i.e., the concentration of free Ca^{2+} at which the concentration of the caged Ca^{2+} compound with the bound Ca^{2+} equals the concentration of the compounds with no bound Ca^{2+}. The smaller the K_d value is, the more tightly bound is the Ca^{2+}. Therefore, since the free Ca^{2+} concentration in the cytosol is between 50 and 100 nM, selection of caged Ca^{2+} compounds having a lower K_d value is better for liberating more Ca^{2+} ions. In Table 7.1, the widely used caged Ca^{2+} compounds are summarized together with their basic properties. When looking at the K_d values before photolysis, DM-nitrophen has the lowest K_d value (~7 nM) at physiological conditions, indicating that most of the DM-nitrophen within the cytosol exists as the bound Ca^{2+} form. Thus, illumination can efficiently liberate Ca^{2+}. Unfortunately, DM-nitrophen is synthesized from EDTA, which also has a high affinity for Mg^{2+} ions (10). Thus, at physiological levels of Mg^{2+} ions (millimolar range), much of DM-nitrophen exists as a bound form with Mg^{2+} rather than Ca^{2+}. Therefore, it requires lowering of the Mg^{2+} concentration to non-physiological levels for effective liberation of Ca^{2+}. However, since Mg^{2+} causes a substantial decrease in both open probability and conductance of ryanodine receptors, interpretation of experimental data should be done with caution (11).

In contrast, NP-EGTA has no serious problems with Mg^{2+} ion in physiological situations since EGTA is more specific for Ca^{2+} than Mg^{2+}. However, the disadvantage is that it has a higher K_d value (80 nM) for Ca^{2+} than DM-nitrophen (7 nM). At resting conditions in which the free Ca^{2+} concentration in the cytosol is assumed to be ~100 nM; ~45% of NP-EGTA exists in the Ca^{2+}-free form and acts as a mobile Ca^{2+} buffer, thus affecting the diffusion of free Ca^{2+}. This problem would be the same with the other caged Ca^{2+} compounds having higher K_d values for Ca^{2+} than the cytosolic Ca^{2+} concentration. Thus, the observed Ca^{2+} spots in fluorescence imaging would be wider than the actual Ca^{2+} events generated (12, 13).

2.2. Uncaging Index: Extinction Coefficient and Quantum Yield

Caged Ca^{2+} compounds must release Ca^{2+} efficiently and quickly in response to illumination. To measure the efficiency of caged compounds, two parameters are widely considered (extinction coefficient (ε) and quantum yield (ϕ)). Since the caged compounds absorb photons (light) and undergo chemical reactions, the above two parameters reflect these properties very well (4). The propensity for light absorption is expressed as the extinction coefficient (ε) and the ease of the reaction is expressed as the quantum yield (φ). The extinction coefficient is determined by the optical density, the concentration of caged probes (c), and the optical path length (l). The optical density of a solution is the log (I_0/I_t), in which I_0 and I_t are the initial and transmitted intensities of light, respectively.

$$\varepsilon = \log\left(\frac{I_0}{I_t}\right)cl$$

The quantum yield is the measure of how many excited state molecules become uncaged. Thus, the quantum yield is a measure of the efficiency with which absorbed light produces some effect (in this case Ca^{2+} liberation). The product (ε·ϕ) of the extinction coefficient and the quantum yield is defined as the uncaging index.

$$\text{uncaging index} = \varepsilon \cdot \phi$$

Higher values of the uncaging index are considered better caged compounds; the higher the uncaging index, the lower the amount of light needed to uncage. If one use a compound having a higher uncaging index, it could be safer in ordinary experiments using most commercially available photolysis instruments. On the home pages of the companies that supply caged compounds (i.e., Invitrogen-www.invitrogen.com, Calbiochem-www.merckbiosciences.com, and Sigma-Aldrich-www.sigmaaldrich.com), the above parameters are conveniently supplied. If the light source is weak, optics are used having less UV transmittance, thick samples, such as brain slices, are used or if one wants a very rapid and large increase in the Ca^{2+} concentration, selection of caged compounds having a higher uncaging index could be beneficial. In this regard, some recently developed caged compounds, such as NDBF-EGTA or azid-1, would be a good choice (4).

2.3. Loading of Caged Compounds

Caged Ca^{2+} compounds were developed by the addition of photoreactive groups such as the nitrophenyl (NP) group and the 4,5-dimethoxy-2-nitrophenyl (DMNP) group, to the existing Ca^{2+} buffers (4). All these forms are designed to maximally interfere with the structures that are essential for Ca^{2+} binding. Despite attachment of the caged moiety, the permeability of the cell membrane for the caged Ca^{2+} compound still depends on the nature of the original compound. Most caged Ca^{2+} compounds cannot permeate the plasma membrane. Therefore, loading of the caged Ca^{2+} compounds into the cells requires specific techniques. For convenience, some of the compounds are commercially available as acetoxymethyl (AM)-forms which can pass through the plasma membrane, but some are not available commercially. Covering the charged parts of the caged compounds with AM ester bonds makes them membrane permeable. Once the caged compounds have entered the cells, endogenous esterases cleave the AM ester bonds generating the original caged Ca^{2+} form inside the cells, similar to what happens to AM forms of Ca^{2+} dyes. However in this case, it should be noted that some of the caged compounds may exist within the intracellular compartments such as the endoplasmic reticulum, Golgi complex, vesicles, and mitochondria (14). In Table 7.1, the commercially available forms such as the salt or AM forms are listed. As shown in Table 7.1, most caged compounds are in the salt form. In this case, intracellular loading requires specific procedures such as patch clamping, microinjection, ATP-induced permeabilization, electroporation, hypoosmotic shock, influx pinocytic cell-loading reagent, or ballistic microprojectile delivery (15). The most widely used are the patch clamp pipettes or injection systems. In the case of using a patch pipette, it is advantageous to know the accurate concentrations of the caged Ca^{2+} compounds and the dyes used. The injection methods need some skills and the success rate often depends on cell types, the operator's skills, and experimental conditions.

2.4. Handling Caged Compounds

Caged Ca^{2+} compounds should be handled carefully. Commercially available caged compounds are sold in small quantities (1–10 mg aliquots). Depending on the form (salt or AM), water or DMSO could be used to make a high concentration stock solution (at least 100 mM). These aliquots should be stored at –80°C until needed. Solutions of caged compounds are stable when frozen for many years (4). Exposure to white light should be kept to a minimum. Although caged compounds are not extremely sensitive to room light, it is best to avoid exposure for a long time. If one uses caged glutamate, it should be handled more carefully than other caged compounds since a small amount of photolysed products produce and increase the concentration of free glutamate, which affects neuronal excitability to some degree. The high sensitivity of glutamate receptors, especially metabotropic glutamate receptors, could be easily excited by traces of uncaged glutamate compounds. However, in the case of caged Ca^{2+} compounds it is not serious since the fraction of Ca^{2+} released from already damaged caged compounds would be negligible when compared with the free Ca^{2+} concentration in the cytosol. However, accurate calculation or estimation of Ca^{2+} liberation by photolysis requires the exact known amount of intact caged compounds since the appropriate amount of free Ca^{2+} should be added into a patch pipette or injection solution, in order to bind the compound with Ca^{2+}, as well as to maintain a free Ca^{2+} concentration in the cell. To calculate how much free Ca^{2+} is added to a solution, computer programs such as MaxChelator (http://www.stanford.edu/~cpatton/maxc.html) and CalC (http://web.njit.edu/~matveev/calc.html), are widely used.

2.5. Uncaging Procedures

To perform photolysis experiments with flash lamps, laser photolysis instruments or confocal microscopes both software and hardware systems should be organized properly. In the case of flash lamps, local photolysis especially at the single cell level in which one designs to liberate Ca^{2+} within a small area of a cell is relatively limited (16). Depending on the system, slits for UV light onto a small area of the visual field may be controlled. However, local illumination at the single cell level is usually very difficult. In contrast, most commercially available confocal microscopes that are equipped with a continuous UV laser support local photolysis at the single cell level. With the multiphoton confocal microscope, femtoliter photolysis is possible (17).

In a confocal microscope, the local uncaging procedure at the single cell level is described in Fig. 7.2. At first, appropriate amounts of caged compounds and Ca^{2+} dye should be loaded into the cells. Cultured or isolated neurons are incubated in a standard Na-HEPES solution containing 2–10 μM fluo-4 AM (Ca^{2+} dye) and 2–10 μM NP-EGTA AM (caged Ca^{2+} compound) for 20–40 min at room temperature. Depending on the systems, other Ca^{2+} dyes such as fluo-3, Oregon Green BAPTA, rhod-2, or fura-red are possible. Since caged Ca^{2+} compounds require UV light (300–400 nm), other longer wavelength lights should be reserved as excitation sources for Ca^{2+} dyes. Samples should be kept in the dark by wrapping in foil. After loading the caged Ca^{2+} compounds and dye, the cells were washed with normal physiological Na-HEPES solution. Then, these cells were mounted on the microscope for the experiments. If one uses a patch pipette, the internal solution containing 100–200 μM fluo-4 and 100–1000 μM NP-EGTA as a salt form are usually used with the addition of 80–300 μM Ca^{2+}. For exocytosis, a higher concentration of caged compounds may be required. In brain slices, it is customary to use a patch pipette (5, 18). The added free Ca^{2+} can be roughly estimated using the K_d value, the estimated free Ca^{2+} concentration, the concentrations of caged Ca^{2+} compounds, and the Ca^{2+} buffers. But, the exact calculation requires a complex solution considering all K_d values for all the Ca^{2+} buffers. Many free software programs (e.g., MaxChelator and CalC) are helpful in calculating the free Ca^{2+} concentration and the amount of Ca^{2+} to be added. After successful loading of the caged Ca^{2+} compound and dye using the imaging software that controls hardware, both of the areas to be imaged (Ca^{2+} measurement) and uncaged (photolysis) as a part of a cell or in the entirety are selected (drawing of the region of interest (ROI) and the region of uncaging (ROU)) respectively. Depending on the company's software and hardware, there would be a big difference in versatility. For example, the recovery time after photolysis, the independence between drawings of the ROI and the ROU, and remote, nearby, or onsite photolysis (or bleaching) during spot or line scanning are commonly required in physiological experiments, but some systems or versions are not fully supportive. In some flash lamp systems or laser photolysis instruments, slits for uncaging light may be

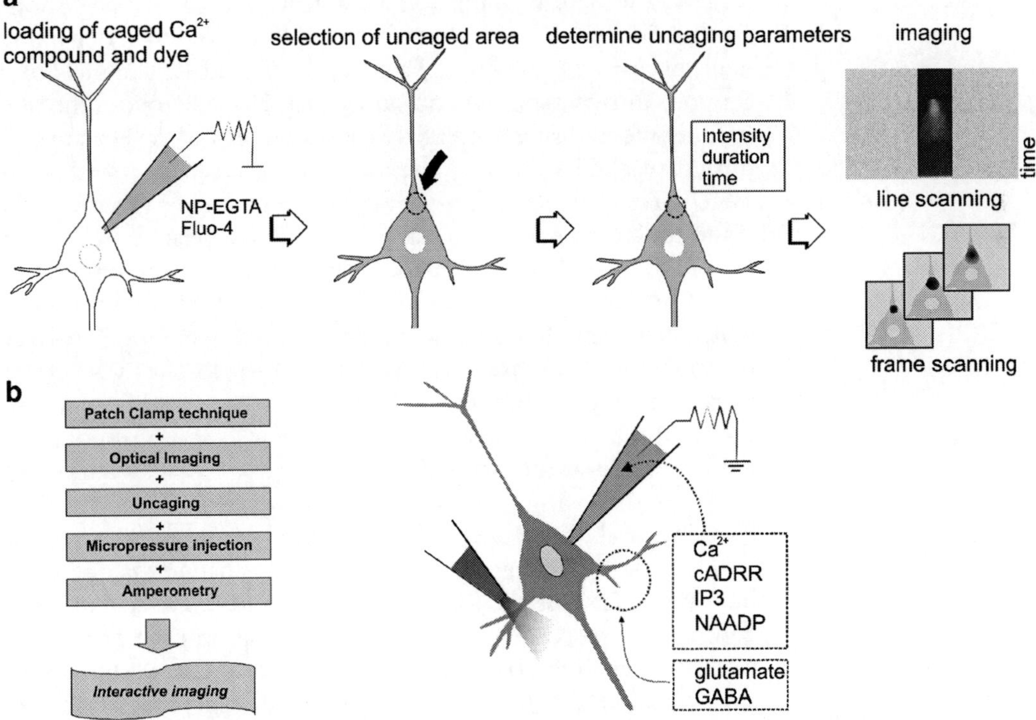

Fig. 7.2. Uncaging procedure and its applications to functional studies of neurons. (**a**) Cytosolic Ca^{2+} uncaging requires several steps; loading of appropriate amount of caged Ca^{2+} compounds and dye, mounting on the imaging machine, selection of the region of interest and the region of uncaging areas, adjusting uncaging parameters, and suitable scanning or imaging. (**b**) Uncaging broadens its application in combination with other techniques. Local control of Ca^{2+} and neurotransmitters and the following observation of electrical signals and chemical signals in live neurons are key methods to understand neuronal functions.

manually or electromechanically adjusted. Finally, researchers should determine the appropriate intensity, duration, and exposure time for uncaging (UV illumination), and perform uncaging function while taking a series of images from the ROI. For fast signals, line or spot scanning is preferred in confocal microscopy, but some commercial confocal systems do not support line scanning under the fine control of uncaging areas separately. Most systems support frame scanning with the editing function of the ROU independently.

Figures 7.3 and 7.4 show typical examples showing global or local Ca^{2+} uncaging experiments in the midbrain dopamine neurons. In the experiment shown in Fig. 7.3, isolated midbrain dopamine neurons were loaded with fluo-4-AM and NP-EGTA-AM, and then the neuron loaded with NP-EGTA was exposed to a low intensity of UV laser continuously (<1% of total power) and spontaneous action potentials were recorded at the same time by a cell-attached patch pipette. Upon irradiation (marked as an arrow), the cytosolic Ca^{2+} level was increasing gradually and slowly through the whole area of a neuron. In this case, by exposing a whole cell to low intensity of UV laser continuously, we are able to generate the Ca^{2+} ramp and are able to find cessation of spontaneous firing at a certain level of cytosolic Ca^{2+}. Midbrain dopamine neurons are well-known to have Ca^{2+}-activated K^+ channels that suppress intrinsic spontaneous firing (19, 20). Therefore, the rate of spontaneous firing decreased as the cytosolic Ca^{2+} level increased until it reached a threshold at which firing was suppressed completely. This protocol is very helpful to detect such a threshold for some functions related to Ca^{2+} levels. In contrast, as shown in Fig. 7.3B, it is also possible to control step-like changes in cytosolic Ca^{2+} levels by step-like increases in UV light transmittance, which can be controlled in a confocal microscope. Upon changes in irradiation intensity (transmittance), cytosolic Ca^{2+} level increased very quickly through the whole area of the neuron. Each letter in the fluorescence image in Fig. 7.3B corresponds to the Ca^{2+} levels at the lower graph.

Figure 7.4 shows typical two ways to raise local dendritic Ca^{2+} levels by local photolysis. Either increasing exposure time for UV light (A) or increasing UV intensity (B) can change the degree of increase in Ca^{2+} in the local area of a dendrite. On a Zeiss 510 confocal microscope, by increasing the iteration time of the UV laser, we were able to control the liberation of Ca^{2+} within a very small area of the dendrite to a different degree as shown in Fig. 7.4A. The iteration times of the UV laser in a, a′, b, and c were 1, 1, 3, and 9, respectively, and the relative increases in Ca^{2+} pulses in the dendrite are shown in the lower panel. One iteration time on the uncaging area was 97 ms. Repetitive exposures of the same duration of the UV light gave exactly the same shape of Ca^{2+} transient as shown in a and a′, in which the same pulse was applied

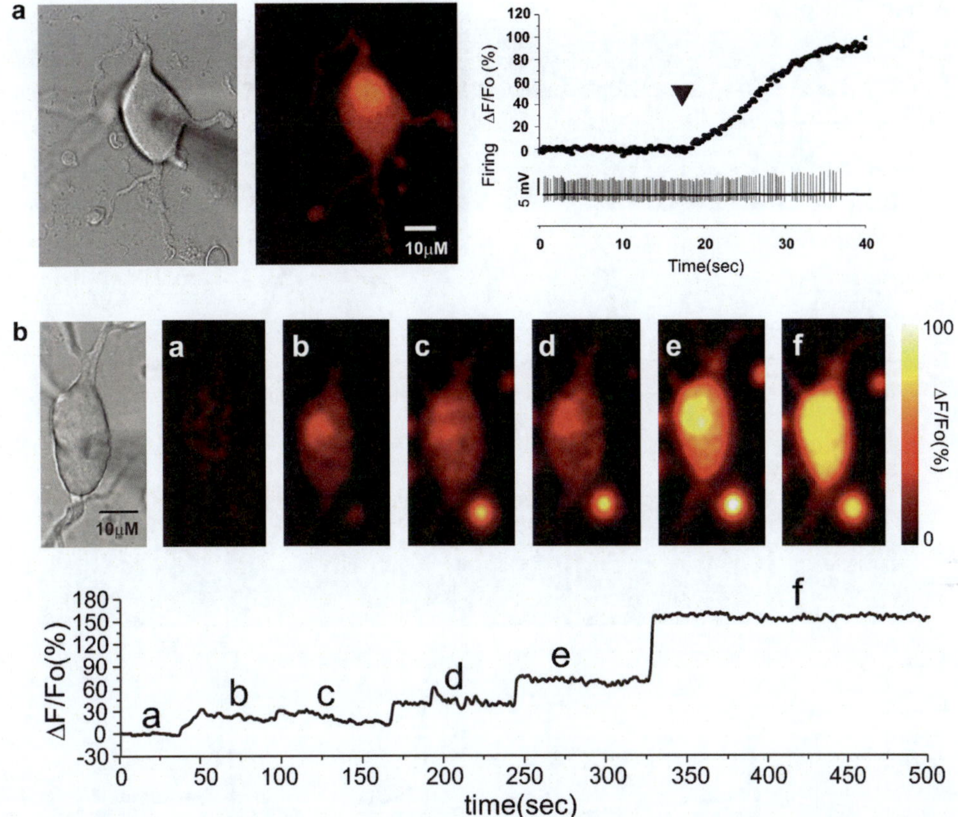

Fig. 7.3. Ramp or step-like global uncaging of caged Ca^{2+} compound, NP-EGTA. Acutely isolated dopamine neurons were loaded with 10 μM NP-EGTA AM and 3 μM fluo-4 AM. Recording was performed with a Zeiss 510 confocal microscope. Spontaneous firing was recorded by the cell-attached current clamp mode. (**A**) Continuous global illumination of a neuron at low levels of 351 and 364 nm lines led to a slow, but continuous rise in cytosolic Ca^{2+} levels (Ca^{2+} ramp). Consequently, the spontaneous firing rate decreased and finally disappeared. (**B**) Step-like continuous exposure to different intensities of 364 nm UV light led to step-like increases in cytosolic Ca^{2+} level in an acutely isolated midbrain dopamine neuron. The transmittance of UV light was controlled by Zeiss software and hardware, (**a**) 0% transmittance, (**b**) 0.05%, (**c**) 1.05%, (**d**) 3.05%, (**e**) 5.05%, (**f**) 7.05%, respectively.

repetitively. The duration of exposures were increased by increasing the iteration times (b and c are 3 and 9 times longer than a, respectively). As the duration of photolysis increased, the widths and amplitudes of the Ca^{2+} transient become longer and higher.

By contrast, in Fig. 7.4B, a small area of the left dendrite was selected as a region of uncaging and short pulses of UV light were applied with different UV transmittances, repetitively. The first uncaging was performed at 30% transmittance and subsequent uncaging on the same area was followed with 50, 70, and 100% transmittance. As shown in the lower panel, the Ca^{2+} levels in the uncaged area of the dendrite increased to different degrees depending on the uncaging intensities of the UV light.

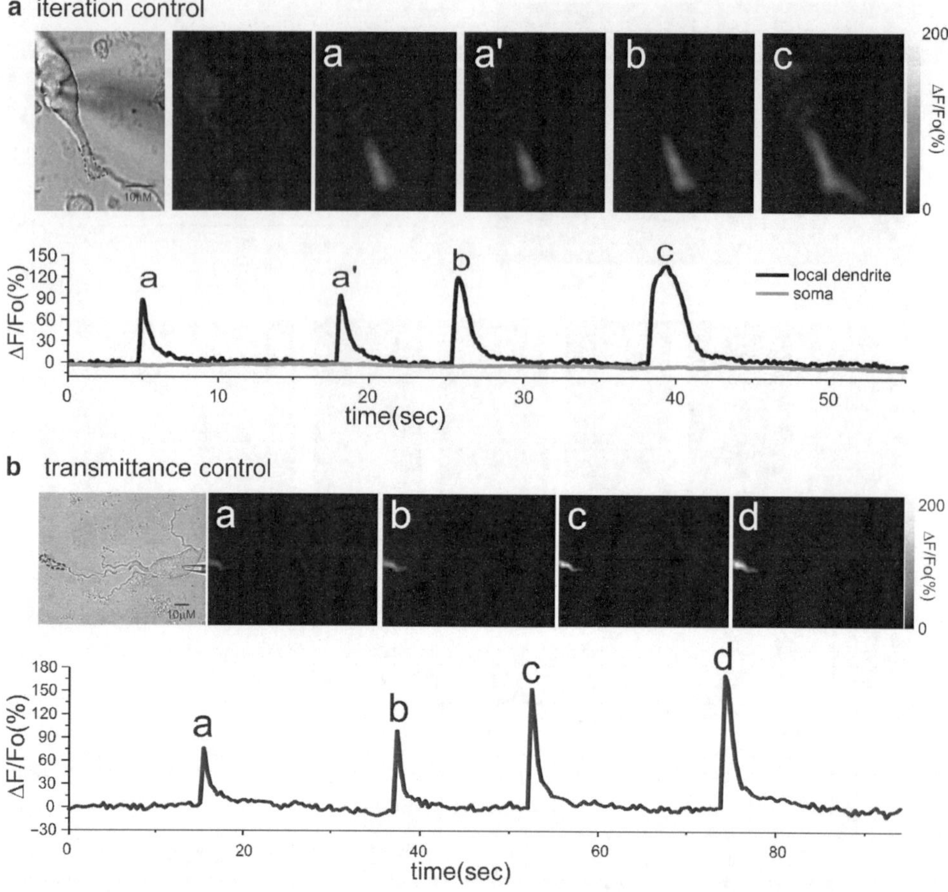

Fig. 7.4. Local uncaging in dopamine neurons. The neurons were loaded with 10 μM NP-EGTA AM and 3 μM fluo-4 AM. Local uncaging was performed serially by either increasing exposure time to UV light (**A**) or UV transmittance (**B**). (**A**) On the small area of the dendrite (*dotted circle*), a 364 nm UV laser was locally exposed for 78 ms to each scan iteration. The local dendritic Ca^{2+} concentration increased by irradiation of 1 iteration (**a**) and the second exposure to the same duration (a′) gave rise to the same amplitude of Ca^{2+} spike. When the iteration numbers (**b**: two times, **c**: nine times) were increased, the amplitudes and widths of the Ca^{2+} spikes increased. Output percentage of exposure with a 364 nm UV laser was 20%. (**B**) After exposure of a small area of the dendrite to a 364 nm UV laser, the local dendritic Ca^{2+} concentration was increased. When the 364 nm UV laser transmittance was increased, the amplitudes in local dendritic Ca^{2+} spikes increased with increased UV transmittances ((**a**) 30% of output, (**b**) 50% of output, (**c**) 70% of output, (**d**) 100% of output). The time of exposure to the 364 nm line is 31 ms.

However, in this case, the exposure time of the UV laser was all the same, and therefore the widths of each Ca^{2+} spike were not much different by comparison with those in Fig. 7.4A. Nevertheless, the amplitudes of each Ca^{2+} spike increased dramatically. As shown in the above examples, the uncaging experiments could be performed with fine control of uncaging levels by adjusting the illumination areas, intensities, exposure times, and frequencies.

In some cases, Ca^{2+} uncaging may be performed within the intracellular organelles such as mitochondria and endoplasmic reticulum (ER). Since caged Ca^{2+} compounds are supplied as the

AM-form and nearly all intracellular organelles have esterase activity, it is possible to load caged Ca^{2+} compounds into the intracellular organelles. In this case, however, removal of caged compounds from the cytosolic compartment was necessary using either a patch pipette or permeabilization to uncage within the organelles with the exception of the surrounding space of cytoplasm (14). However, it was also impossible to load caged compounds into specific organelles exclusively. For example, although one may want to uncage in the luminal space of the ER, caged compounds within the mitochondria and other membrane bound organelles were also uncaged. Thus, it is only meaningful if one is able to separate these signals using compartment-specific dyes or anatomical positions within the cell. In the pancreatic acinar cells, uncaging within the luminal space of the ER was performed successfully (21). In this experiment, NP-EGTA was not sufficient to increase Ca^{2+} within the ER since the ER lumen has a very high free Ca^{2+} concentration. Thus, it required a complete emptying of the ER Ca^{2+} store to a minimal level. In this case, it was possible to increase Ca^{2+} within the ER lumen with NP-EGTA. Distinction of fluorescence of the ER from mitochondria was based on the distinct localization of mitochondria in this specific cell (22).

3. Experimental Setup

3.1. Excitation Sources

Photolysis requires a light source that is amenable to emit in the UV-light range (300–400 nm). Light sources for this wavelength interval include a laser, mercury-lamp, flash lamp, or even a monochromator (16). It has to be determined whether only a part of the field of view needs to be illuminated (local uncaging) to control the release of the caged compound spatially, or whether the whole field of view needs to be illuminated (global uncaging) for rapid and quantitative release of the caged compound. Uncaging on a very small area such as an area covering a single synapse or single spine is often referred to as focal uncaging. Since light intensity is a key factor for photolysis, the light source determines the limits for temporal and spatial resolution. Depending on the plans of experiments, different spatial and time resolutions are required. Laser photolysis instruments and confocal microscopes equipped with a UV laser can support focal or local uncaging/caging at the single cell level (submicron level). But, they are much more expensive than flash lamps. If there is no need for fine local uncaging, cheaper flash lamps are the choice. With flash lamps global, uncaging/caging can be easily performed under the fluorescence microscope. In brain slices, two photon confocal microscopy efficiently supports focal or local uncaging since it can minimize an uncaging spot in the out-of-focus plane and achieve deep penetration into tissues by using a longer wavelength light. Recently, there are many commercially available flash or laser photolysis instruments that can be easily attached to fluorescence microscopes through an epifluorescence port.

3.1.1. Flash Lamps

While a laser light source generates coherent light, flash lamps produce non-coherent light. Thus, there is a limit to the light intensity that can be achieved by focusing a non-coherent light source. In the absence of light scattering, while the entire output of a laser can be focused onto an ~0.5 μm spot diameter, the output of a non-coherent flash lamp light can be maximally focused on an ~100 μm spot in diameter (16). If temporal and spatial resolution is not important, conventional arc lamps (mercury arc lamps) can also be used for global uncaging. Since the light intensity of a conventional arc lamps is very weak, the speed of photolysis is relatively slow. Therefore, selection of caged Ca^{2+} probes having a high uncaging index and the use of a higher concentration of caged compounds may be beneficial.

To increase the light intensity, commercially available flash lamps are widely used. Many companies offer a full-package flash photolysis instrument that includes a light source, controller, condenser, light delivery tube, and adaptor for a microscope. For example, Till Photonics (Germany) provides a photolysis unit (UV-flash). In this case, with a given fiber diameter of 1.25 mm and a numerical aperture of the fiber of 0.25, the light at the exit face plate of the fiber may be concentrated to a 240 μm spot with a NA = 1.3 objective. A full shot yields $>10^{15}$ photons between 340 and 390 nm in the back focal plane of the microscope. Depending on the objective used, this corresponds to 2×10^{23} photons/m^2 (Fluar 40x, 1.3 oil) or 5×10^{21} photons/m^2 (Achroplan W 63x, 0.9W) in the specimen plane. The Rapp OptoElectronic flash unit has similar properties when they use the same type of flash lamps. However, flash lamps have many limitations. The discharge of high energy from flash lamps generates large electrical noises. Thus, patch clamp measurements and amperometry may be influenced by the discharge at the very initial point of the flashing. In addition, it is also very difficult to manipulate pulse durations or pulse repetition intervals at a fast time scale. Nevertheless, flash lamps are the best choice for experiments that require large areas of illumination at a low cost. In this case, one should be careful whether illumination is homogeneous in the visual field since the center of the visual field is most strongly illuminated. In order to check whether photolysis is homogeneous or not, calibration or validation using a thin-layered caged fluorescein film may be helpful (Fig. 7.5b, see Sect. 7.3.2). To validate whether optimum light is reproducibly generated, the UV light intensity can be measured by a UV-sensitive photodiode.

3.1.2. Photolysis Laser Instrument

The major advantage of a laser in photolysis is to use coherent light. Therefore, the entire output of a laser can be focused onto a diffraction-limited spot <0.5 μm in diameter, under ideal conditions (16). Since the light intensity determines the speed and yield of photolysis, a photolysis laser is the instrument of choice for experiments requiring high temporal and spatial resolution (if financial resources are sufficient). Nevertheless, if a laser light source is not a built-in UV laser like a confocal microscope, it can be attached into the epifluorescence port through an adaptor. In this case, spatial resolution would exceed the diffraction-limited spot by several-fold (16). The total system includes a continuous or pulsed UV light source with fiber optic coupling, a mechanical shutter, a filter slider for photolysis, and a controller (supplied by many companies, such as Applied Scientific Instrumentation, Inc). In this case, the spot size depends on the diameter of the optic fibers. UV light should be correctly positioned on the region of uncaging. Since this system uses parallel light pathways (epifluorescence port) apart from confocal scanning light pathways, it is not easy to focus on the submicron regions, but it is rather used to illuminate a region about several micrometers in diameter. Especially in the brain slices, severe light scattering within the tissue and focus mismatch between imaging and uncaging lights are another problem for focused photolysis. Thus, there could be some

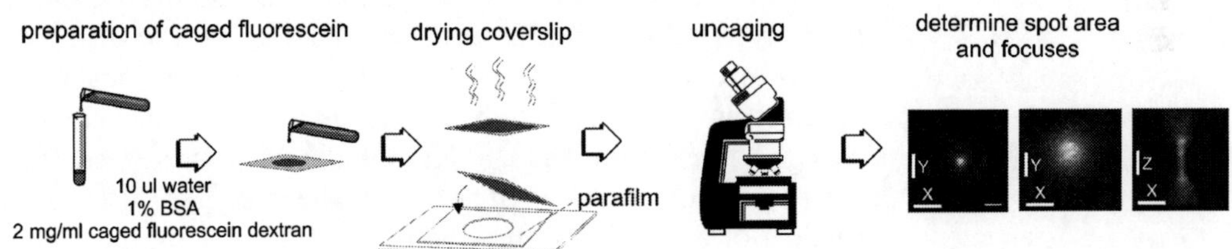

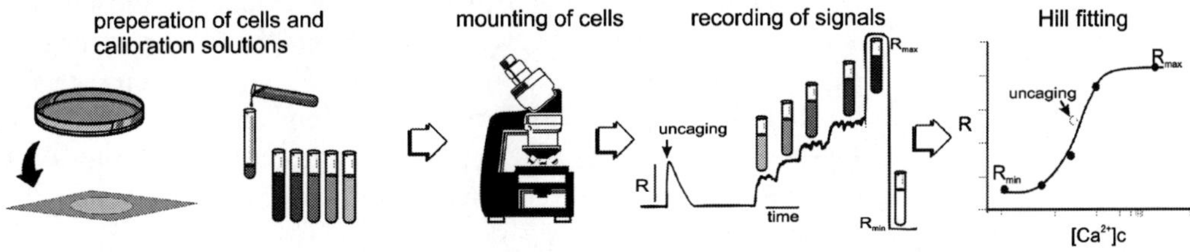

Fig. 7.5. Validation of uncaging spots and calibration of Ca^{2+} levels during uncaging. (**a**) Validation of uncaging spot or area with fluorescein film. After preparing an immobile caged fluorescein film, uncaging spots can be imaged by *X–Y* or *Z*-scanning mode. (**b**) Calibration of Ca^{2+} transient during photolysis. After making calibration solutions containing known variable concentrations of free Ca^{2+}, fluorescence changes can be measured under the ionomycin-treated Ca^{2+}-permeabilized cells.

limitations in this simple system for the experiments requiring highly focused photolysis at the submicron level.

In contrast, commercial confocal microscopes attached with continuous UV lasers or quasi-continuous UV lasers use both uncaging and imaging lights through the same acousto-optical tunable filter (AOTF) scanning module, thus it is possible to conveniently control uncaging rate, size, and repetition rates. However, in this circumstance, the UV laser uses the same light scanner so there is a little delay between switching of photolysis and imaging scanning. To protect sensitive photomultiplier detectors, imaging is impossible during photolysis. However, some recent confocal microscopes (i.e., Zeiss 700 duo scanning module) use a dual beam combiner handling two independent scanner groups into one thus allowing great flexibility for photolysis. By delivering uncaging light and imaging light through the same beam combiner on the same pixel at the same time, it is possible to perform real time imaging and uncaging without a time difference.

For uncaging in the confocal microscope, a continuous argon ion laser was most widely used as a UV laser several years ago. The argon ion laser can generate a light beam at 351 and 364 nm. Thus, using these wavelength lines, focal, local, and global uncaging/caging has been supported. One advantage of this continuous wave laser is to easily vary the duration of illumination using an electronic shutter; the disadvantages are the requirement of complex electrical facilities (208 V and 60 A) and a large water cooling system due to its low efficiency. This cooling system also generates some mechanical noise and may require a separate space to prevent circulation noises. Recently, a quasi-continuous solid state laser such as Nd:YVO_4, has been widely used. Since Nd:YVO_4 lasers generate 355 nm at 100 KHz, Nd:YVO_4 lasers operate similar to a continuous laser. In addition, they use regular electricity and do not require a water-cooling system. Solid-sate 405 nm violet lasers may be an alternative for photolysis. Thus, these lasers have become the first choice for uncaging experiments. For the rapid kinetic studies of biochemical interactions, pulsed lasers such as the frequency-tripled Nd:Yag laser, the frequency doubled ruby, and nitrogen lasers may be used. The high intensity would be helpful to overcome the limitation of photolysis speed. However, due to its low repetition rates, it is not suitable for experiments requiring continuous control of interpulse intervals or durations (16).

3.2. Delivery of Uncaging Light

For work on cultured or isolated cells, inverted microscopes are usually preferable as shown in Figs. 7.3 and 7.4, but experiments using brain slices are mostly performed under an upright microscope. For effective photolysis, there are not many problems in cultured and isolated cells since high numerical aperture (NA) objective lenses are readily available under conditions of no severe light scattering. Therefore, for photolysis, one only needs to check whether objective lenses are suitable and UV light is correctly delivered into the objective lens through an optic fiber. Ordinary objective lenses usually have low UV transmittance thus for effective photolysis, it should be determined whether the objective lens has a high UV transmittance. Since the UV transmittance is variable depending on the suppliers and the UV transmitted objective lens is also expensive, it is recommended to check the performance or have a demonstration before purchase. Since the experiments with brain slices requires an upright microscope and a water immersion lens and more importantly, it usually demands a combination with a patch clamp system, there are more points to be considered. For handling of patch pipettes and thick tissues, a long working-distance objective lens is required together with a long wavelength infrared (IR) light transmittance. Thus for uncaging, this system requires a specific lens having a wide range of transmittance from UV to IR light. Light scattering in thick brain tissues also adds more complexity (5, 23, 24). Thus, due to the variability of light transmittance of the objective lens and the chromatic aberration, it is recommended to check whether the objective lens would be suitable for the planned experiments.

Optic fibers are also fragile and the tips can be easily contaminated by dust, thus careful management is required. The diameter of an optic fiber and its type determines the illuminated spot size (16, 25). Focusing assemblies are available commercially (e.g., OptoElectronic, Hermany; Oz Optics, Canada). Finally, photolysis and imaging signals should be synchronized. For this, a transistor–transistor (TTL) pulse can be used to send information about shutters or time on flash. Many types of imaging software and hardware support a TTL pulse input and output through computer parallel ports or their own instruments.

For work using UV flashes or lasers, one should be careful not to be exposed to intense UV light pulses. Therefore, one should always wear goggles to protect the eyes if the UV Flash is fired or to align laser lines. Turning off the UV flash will always ignite a single flash by the final discharge of the capacitor banks.

3.3. Performance Validation of Uncaging Spots

Uncaging requires UV light that is invisible. Therefore, it is necessary to validate how the focused light leads to even photolysis of caged compounds in the visual field. Flash photolysis usually supports the strobe light function so that one can easily notice where the light is focused. In the strobe mode, only very weak photons are discharged per flash. In the case of the Till UV-flash, the condenser of a photolysis unit is equipped with a special short pass filter which blocks >390 nm and transmits below. When >580 nm, this filter starts to transmit again allowing the illuminated spot to be watched with naked eyes for alignment and determination of the uncaging area.

Nevertheless, the UV flash illuminates a relatively a wide area which may contain many cells. Thus, to check how the UV flash evenly evokes photolysis within the illumination area, it may be helpful to use the thin immobile layer of a caged fluorescein film (18, 26). Figure 7.5a briefly describes how to prepare a thin immobile layer of caged fluorescein and how to calibrate an uncaging spot. In the case of focal or local photolysis using a confocal microscope, the same procedure can be applied. At first, to make a thin film coated with caged fluorescein, a 10 μm water solution of 1% bovine serum albumin is mixed with 2 mg/ml of caged fluorescein dextran and then evenly spread on the coverslip. For >1 h, this coverslip is dried in air. This immobile film can be used as a test sample for uncaging performance and calibration (26). Thereafter, in a glass slide firmly attached to a layer of Parafilm, a large hole is made at the center in order to make an observation field for the dried coverslip. Using this thin film, *X–Y* scanning during photolysis gives information about whether a spot size or area is homogeneous. *Z*-scanning gives information about how uncaging light in a given condition affects photolysis above or below the focal plane. Simply, an uncaging procedure can be performed at the solution layer containing fluorescein and Ca^{2+} dye (mobile caged dye). In this case, by comparison with an uncaging profile at the immobile layer, it is possible to estimate how the diffusion of caged compounds affects spot size or working area of the caged molecule. In some cases, a glass pipette or bath filled with or containing caged solutions could be used for testing. Glycerol may be added to slow diffusion without affecting uncaging efficiency (26).

In the case of Ca^{2+} uncaging, if one wants to know how much the free Ca^{2+} concentration at the uncaging area increases, it is necessary to perform a calibration. Figure 7.5b describes how to estimate the increase in free Ca^{2+} concentration during the uncaging procedure. First, prepare a series of solutions having different concentrations of free Ca^{2+} ion. The free Ca^{2+} concentration in the series of solutions that were prepared by addition of different amounts of Ca^{2+} can be determined by using the Ca^{2+} calibration kit solutions (solutions having a series of known free Ca^{2+} concentrations, e.g., Invitrogen). Among them, a Ca^{2+}-free solution (5 mM EGTA or BAPTA only) and a 10 mM Ca^{2+} solution (no EGTA or BAPTA) should be included in order to obtain minimum and maximum fluorescence values. If a series of solutions having different Ca^{2+} concentrations are ready, ionomycin is added (5–10 μM). Then, with the normal Na-HEPES solution, perform uncaging experiments using cells loaded with caged compounds and dye.

After one or several uncagings (local or global), apply 5–10 μM ionomycin to make the plasma membrane permeable to Ca^{2+} and apply one of the above solutions containing different free Ca^{2+} concentrations in the presence of ionomycin. Take images until cytosolic Ca^{2+} levels increase to become stable, and then apply the next solution step-by-step. Finally, a Ca^{2+}-free solution and a 10 mM Ca^{2+} solution to determine the minimum and maximum values of fluorescence intensity. By observing these series of fluorescence intensities, amplitudes of Ca^{2+} spikes evoked by uncaging can be easily converted into real Ca^{2+} concentrations. For more accurate estimation, the known values of fluorescence intensities can be fit by using the Hill equation ($F/F_0 = (Ca^{2+})^n/(K_d + (Ca^{2+})^n)$), in which F and F_0 are measured and minimal fluorescence intensities.

4. Combination of Uncaging with Other Techniques

Ca^{2+} signals in neurons occur over a very complex spatial and time domain. Specific Ca^{2+} transients and patterns confer specific function in neurons. Local uncaging offers manually-confined generation of specific Ca^{2+} transients and patterns in live neurons, thereby allowing to observe specific functions linked to Ca^{2+} signals such as exocytosis, shapes of action potentials, firing patterns, excitability, synaptic plasticity, spine morphology, neurite outgrowth, and many others. Thus, its best usage comes from a combination with other techniques. Since uncaging is usually performed in the fluorescence microscope, optical imaging and uncaging is the inborn combination. Patch clamping adds more versatility not only observing electrical activities but also controlling intracellular components and adding or removing signaling molecules. As shown in Fig. 7.2b, multiple combinations with the patch-clamp, injection system, and amperometry are more beneficial. Since the uncaging/caging technique finely controls key steps in biological signaling with relative ease, it could be expected that the uncaging technology will become more important and applied widely in biological research.

Acknowledgments

This work was supported by the Korea Research Foundation Grant funded by the Korean Government (KRF-2008-E00011). The authors thank Prof. Petersen for the kind help with the revision of this manuscript.

References

1. Augustine GJ, Santamaria F, Tanaka K (2003) Local calcium signaling in neurons. Neuron 40:331–348
2. Berridge MJ (1998) Neuronal calcium signaling. Neuron 21:13–26
3. Verkhratsky A (2005) Physiology and pathophysiology of the calcium store in the endoplasmic reticulum of neurons. Physiol Rev 85:201–279
4. Ellis-Davies GC (2007) Caged compounds: photorelease technology for control of cellular chemistry and physiology. Nat Methods 4:619–628
5. Augustine GJ (1994) Combining patch-clamp and optical methods in brain slices. J Neurosci Methods 54:163–169
6. Oberhauser AF, Robinson IM, Fernandez JM (1996) Simultaneous capacitance and amperometric measurements of exocytosis: a comparison. Biophys J 71:1131–1139
7. Thompson SM, Kao JP, Kramer RH et al (2005) Flashy science: controlling neural function with light. J Neurosci 25:10358–10365
8. Ellis-Davies GC (2008) Neurobiology with caged calcium. Chem Rev 108:1603–1613
9. Adams SR, Tsien RY (1993) Controlling cell chemistry with caged compounds. Annu Rev Physiol 55:755–784
10. Ellis-Davies GC (2006) DM-nitrophen AM is caged magnesium. Cell Calcium 39:471–473
11. Fill M, Copello JA (2002) Ryanodine receptor calcium release channels. Physiol Rev 82:893–922
12. Goldberg JH, Tamas G, Aronov D, Yuste R (2003) Calcium microdomains in aspiny dendrites. Neuron 40:807–821
13. Neher E, Augustine GJ (1992) Calcium gradients and buffers in bovine chromaffin cells. J Physiol 450:273–301
14. Park MK, Tepikin AV, Petersen OH (2002) What can we learn about cell signalling by combining optical imaging and patch clamp techniques? Pflugers Archiv 444:305–316
15. Takahashi A, Camacho P, Lechleiter JD, Herman B (1999) Measurement of intracellular calcium. Physiol Rev 79:1089–1125
16. Tang CM (2006) Photolysis of caged neurotransmitters: theory and procedures for light delivery. In: *Current protocols in neuroscience*. Wiley, New York
17. Pettit DL, Wang SS, Gee KR, Augustine GJ (1997) Chemical two-photon uncaging: a novel approach to mapping glutamate receptors. Neuron 19:465–471
18. Wang SS, Augustine GJ (1995) Confocal imaging and local photolysis of caged compounds: dual probes of synaptic function. Neuron 15:755–760
19. Kim SH, Choi YM, Jang JY, Chung S, Kang YK, Park MK (2007) Nonselective cation channels are essential for maintaining intracellular Ca^{2+} levels and spontaneous firing activity in the midbrain dopamine neurons. Pflugers Arch 455:309–321
20. Kitai ST, Shepard PD, Callaway JC, Scroggs R (1999) Afferent modulation of dopamine neuron firing patterns. Curr Opin Neurobiol 9:690–697
21. Park MK, Petersen OH, Tepikin AV (2000) The endoplasmic reticulum as one continuous Ca^{2+} pool: visualization of rapid Ca^{2+} movements and equilibration. EMBO J 19:5729–5739
22. Park MK, Ashby MC, Erdemli G, Petersen OH, Tepikin AV (2001) Perinuclear, perigranular and sub-plasmalemmal mitochondria have distinct functions in the regulation of cellular calcium transport. EMBO J 20:1863–1874
23. Korkotian E, Oron D, Silberberg Y, Segal M (2004) Confocal microscopic imaging of fast UV-laser photolysis of caged compounds. J Neurosci Methods 133:153–159
24. Yasuda R, Nimchinsky EA, Scheuss V et al (2004) Imaging calcium concentration dynamics in small neuronal compartments. Sci STKE 219:l5
25. Eberius C, Schild D (2001) Local photolysis using tapered quartz fibres. Pflugers Arch 443:323–330
26. Sarkisov DV, Wang SS (2006) Alignment and calibration of a focal neurotransmitter uncaging system. Nat Protoc 1:828–832

Chapter 8

Ca^{2+} Imaging of Intracellular Organelles: Endoplasmic Reticulum

Robert Blum, Ole H. Petersen, and Alexei Verkhratsky

Abstract

The endoplasmic reticulum (ER) is a complex and highly dynamic three-dimensional intracellular membranous system, which acts as a dynamic calcium store in the majority of eukaryotic cells. The special arrangement of intra-ER Ca^{2+} buffers, characterized by low affinity for Ca^{2+}, in combination with SERCA pump activity keeps intraluminal Ca^{2+} ($[Ca^{2+}]_L$) at ~0.1–0.8 mM (Cell Calcium 38:303–310, 2005), thus creating a steep electrochemical gradient aimed at the cytosol. Activation of ER Ca^{2+} channels results in Ca^{2+} release, which contributes to $[Ca^{2+}]_i$ elevation, whereas SERCA-dependent Ca^{2+} uptake assists termination of cytosolic Ca^{2+} signals. In addition, the continuous luminal space can act as a travelling route for free Ca^{2+} ions ("Ca^{2+} tunnels"), thus bypassing cytosolic Ca^{2+} buffers and preventing mitochondrial Ca^{2+} uptake or loss of Ca^{2+} over the plasma membrane. Furthermore, changes in $[Ca^{2+}]_L$ regulate ER-resident chaperones, responsible for postranslational protein processing. Thus, $[Ca^{2+}]_L$ integrates various signalling events and establishes a link between fast signalling, associated with the ER Ca^{2+}release/uptake, and long-lasting adaptive responses relying primarily on the regulation of protein synthesis. This paper overviews modern techniques for the imaging of $[Ca^{2+}]_L$ using synthetic fluorescent Ca^{2+} dyes. The methods for ER dye loading, with a particular emphasis on employment of ER targeted esterases (the Targeted-Esterase induced Dye loading, TED) to increase specific accumulation of the probes within the ER lumen are described in detail.

Key words: Calcium imaging, Calcium indicator, Esterase, Endoplasmic reticulum, Protein targeting, Neurons

1. Introduction: Endoplasmic Reticulum as a Dynamic Ca^{2+} Store

The endoplasmic reticulum (ER) is a complex and highly dynamic three-dimensional intracellular membranous system (1–4). Its integrity is critical for the accuracy of membrane flow within the ER between the ER-to-Golgi intermediate compartment and the Golgi apparatus, in both anterograde and retrograde directions.

A. Verkhratsky and Ole H. Petersen (eds.), *Calcium Measurement Methods*, Neuromethods, vol. 43,
DOI 10.1007/978-1-60761-476-0_8, © Humana Press, a part of Springer Science+Business Media, LLC 2010

The membrane constitution and spatial organization of the ER underlies continuous flow of intracellular membranes, proteins, lipids and other milieu components that bud off from and fuse with the ER due to intracellular trafficking (5). Rapid morphological remodelling of the ER tubular structures permits dynamic interactions with other organelles (6, 7). In neurones, the ER is present throughout the cell extending from the nuclear envelope to the dendrites and postsynaptic structures (e.g. spines) within axons and presynaptic terminals, as well as in growth cones (4, 8). The lumen of the ER of neurones and non-neuronal cells forms a continuous, aqueous space in which Ca^{2+} and small molecules, (e.g. fluorescent dyes) can readily diffuse (9–14). The functional properties of the ER are also extraordinarily heterogeneous depending on spatial distribution of various enzymatic cascades (4, 15–17).

The ER acts as a dynamic intracellular Ca^{2+} store (1, 2, 18, 19), the function, which is supported by a complement of endomembrane-resident Ca^{2+} channels ($InsP_3$ receptors, $InsP_3Rs$, and ryanodine receptors, RyRs) and Ca^{2+} pumps of sarco-(endo)-plasmic reticulum Ca^{2+} ATP-ase (SERCA) type. Conceptually, the ER produces and shapes cytosolic Ca^{2+} signals by acting as source and sink for Ca^{2+} ions (20, 21). The special arrangement of intra-ER Ca^{2+} buffers characterized by low affinity for Ca^{2+} in combination with SERCA pump activity keeps intraluminal Ca^{2+} ($[Ca^{2+}]_L$) at ~0.1–0.8 mM (2), thus creating a steep electrochemical gradient aimed at the cytosol. Activation of ER Ca^{2+} channels results in Ca^{2+} release, which contributes to $[Ca^{2+}]_i$ elevation, whereas SERCA-dependent Ca^{2+} uptake assists termination of cytosolic Ca^{2+} signals. In addition, the continuous luminal space can act as a travelling route for free Ca^{2+} ions ("Ca^{2+} tunnels"), thus bypassing cytosolic Ca^{2+} buffers and preventing mitochondrial Ca^{2+} uptake or loss of Ca^{2+} over the plasma membrane (6, 9, 11, 22, 23).

In all cell types, intracellular Ca^{2+} dynamics are most commonly assessed by monitoring concentration of free Ca^{2+} in the cytosol ($[Ca^{2+}]_i$). Although this approach is highly informative and simple in principle, it may limit the interpretation of physiologically relevant signalling processes associated with the ER (13, 24–27). Upon stimulation, the release of Ca^{2+} from the ER may be obscured by Ca^{2+} influx/extrusion over the plasma membrane. In neurones, Ca^{2+} signalling in synaptic transmission, after stimulation with hormones, transmitters, co-transmitters, or pharmaceuticals is extraordinarily complex. Ca^{2+} influx over the plasma membrane occurs through multiple, functionally different mechanisms; either via ionotropic receptors, voltage-gated calcium channels or transient receptor potential (TRP)-like channels, all of which can be regulated by ER Ca^{2+} release.

The relevance of ER-mediated Ca^{2+} signalling in neural cells became clear when new techniques allowed the monitoring of

Ca^{2+} dynamics in the ER, and biochemical or genetic approaches have revealed a causal link between ER-derived Ca^{2+} signalling and specific neuronal functions. Here, the ER not only acts as a Ca^{2+} buffer but also as a complex source of Ca^{2+} signals involved in neurotransmitter release, release of growth factors and neurotrophins, synaptic plasticity, growth cone guidance, activity-dependent gene expression, cell survival and many other physiological processes (3, 4, 6, 18, 28, 29).

2. General Principles of ER Ca^{2+} Homeostasis and Signalling

The handling of free Ca^{2+} ions in the ER lumen is unique and inherently different from that in the cytosol or in mitochondria. In sensory neurones of the rat, the resting $[Ca^{2+}]_L$ reportedly varies from ~100 to 500 µM (12, 13), whereas $[Ca^{2+}]_i$ is of the order of 50–100 nM. Thus, there is a high electrochemical driving force for Ca^{2+} from both the extracellular space (~2 mM Ca^{2+}) and from the ER lumen into the cytosol. Loss of Ca^{2+} from the ER lumen occurs both at rest (via incompletely identified leakage pathways – (30)), and following activation of signalling cascades involving opening of $InsP_3Rs$ and RyRs. In neurites, continuous loss of ER Ca^{2+} to the cytosol and also over the plasma membrane has been observed in Ca^{2+}-free Ringer solution (26). Following $[Ca^{2+}]_i$ elevation, Ca^{2+} pumps and Ca^{2+} transporters restore the low $[Ca^{2+}]_i$ either by removing Ca^{2+} over the plasma membrane or by pumping Ca^{2+} back into the ER against the electrochemical gradient. Both $InsP_3Rs$ and RyRs are operative in nerve cells. A representative example of $InsP_3$-induced Ca^{2+} release (IICR) is metabotropic glutamate signalling at the Purkinje cell synapse, where fast activation of mGluR-receptors and subsequent IICR is of particular importance for motor function (31, 32). The RyRs, responsible for Ca^{2+}-induced Ca^{2+} release (CICR) act as a link for fast communication between the ER and depolarizing Ca^{2+} entry during neuronal activation (16, 33–35). A hotly debated example of fast-versus-slow Ca^{2+} signalling in neurones concerns the neurotrophins. Here, a complex network of dynamically coupled receptor tyrosine kinases, IICR, indirect activation of postsynaptic NMDA receptors or the fast activation of voltage-gated Ca^{2+} channels accomplishes the physiological function (36). Cytosolic Ca^{2+}-binding proteins such as parvalbumin or calbindin act as high-affinity buffers that control the dynamic and spatial spread of Ca^{2+} signals.

Activation of Ca^{2+} release from the ER as well as the refilling of the ER Ca^{2+} store in neurones can occur in a time frame of milliseconds to seconds and can be restricted to small neuronal microdomains, like spines and their adjacent dendritic structures

(9, 37). Thus, the direct real-time ER Ca^{2+} imaging is imperative for in-depth analysis of Ca^{2+} signalling within the ER (26, 38, 39).

3. Methods for Ca^{2+} Imaging in the ER

In principle, three techniques have been used for direct measurements of ER free Ca^{2+} concentration ($[Ca^{2+}]_L$); all include the analysis of Ca^{2+}-dependent changes in fluorescent or bioluminescent signals (38–40). Two of these techniques are based on the selective targeting of Ca^{2+}-indicator proteins to the ER lumen (25, 40–42). The third, and most widely used method relies on hydrophobic, acetoxymethyl (AM)-derivatized fluorescent Ca^{2+} indicators that can cross both plasmalemma and endomembranes. In the ER lumen, endogenous esterase activity hydrolyses the ester groups and releases a hydrophilic, non-diffusible Ca^{2+} indicator, which, under certain conditions, becomes the reporter of ER free Ca^{2+} (ester-loading technique, Fig. 8.1a, see (24, 26, 39, 43)).

3.1. Genetically Engineered Protein-Indicators

The genetically engineered protein-indicators are based on aequorin or green fluorescent protein (GFP) and recognize free Ca^{2+} through Ca^{2+}-binding protein domains (25, 40–42, 44). In recent years, spectacular progress in optimizing protein-based indicators has been made especially for the imaging of cytosolic Ca^{2+} in neurones in vitro, in situ and in vivo (45, 46). Protein-based indicators can be targeted selectively to the nucleus, mitochondria, the plasma membrane and the ER. Nonetheless, as far as $[Ca^{2+}]_L$ measurements are concerned, the genetically encoded Ca^{2+} indicators have shortcomings. In particular, protein-based indicators offer a rather small dynamic range, which, in combination with gaps in the Ca^{2+} sensitivity, makes it difficult to analyse Ca^{2+} dynamics in the neuronal ER accurately (42, 44). Real-time imaging of $[Ca^{2+}]_L$ fluctuations requires rather sophisticated experimental setups and offline data analysis, yet excellent protocols developed by several groups have helped to circumvent the problems (42). The best currently available ER indicators are: the GFP-based Cameleon YC4.3 (biphasic K_D for $Ca^{2+} = 0.8\,\mu M$, $700\,\mu M$ – (41, 47)); the recently developed Cameleon split YC7.3ER (K_D for Ca^{2+} $130\,\mu M$ – (48)), and the redesigned Cameleon D1 (biphasic K_D for Ca^{2+} $0.8\,\mu M$, $60\,\mu M$ – (49)), which all have been used successfully to monitor Ca^{2+} dynamics in the ER of non-neuronal cells. Even though progress has been substantial, the development of a low-affinity protein-based indicator to analyse Ca^{2+} signalling in the ER of neurones remains a major challenge.

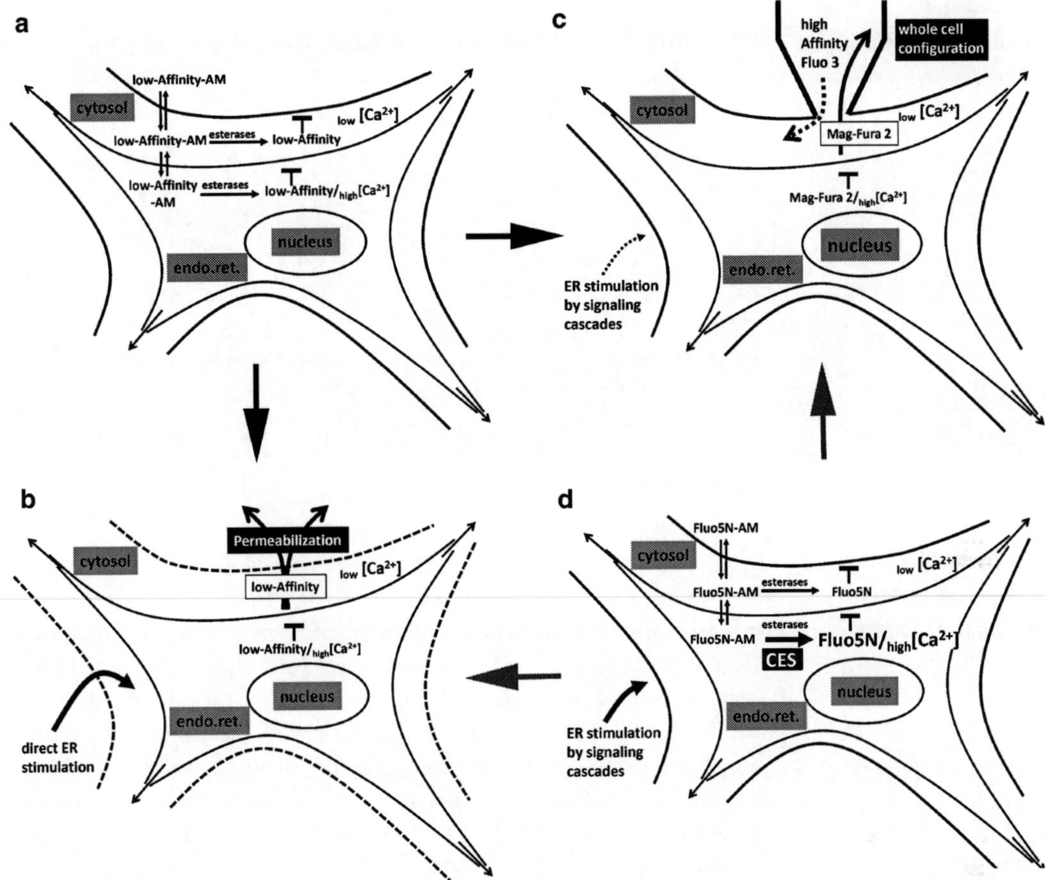

Fig. 8.1. Principles of intraluminal Ca^{2+} measurements. (**a**) *Esterase-based dye loading.* Low-affinity Ca^{2+} indicators such as Mag-Fura-2/AM, Mag-Fluo-4/AM or Fluo-5N/AM (see Table 8.1) are used for cell loading. Under normal conditions, AM-derivatives of the indicator dyes pass the biological membranes in a lipophilic, ion-insensitive state. For intraluminal imaging, the dye must also cross both the plasmalemma and the endomembrane. In the cytosol as well as in the ER lumen, endogenous esterases cleave of the AM-group and release a polar, Ca^{2+} sensitive fluorescent indicator. The low affinity for Ca^{2+} makes dyes useful under conditions of a high Ca^{2+} concentration in the ER. (**b**) *AM-ester loading in combination with plasmamembrane permeabilization.* A certain amount of the active dye remains in the cytosol because the AM-ester group is also cleaved by cytosolic esterases. This cytosolic portion of the Ca^{2+} indicator makes it difficult to distinguish cytosolic Ca^{2+} dynamics from ER signals. To remove the cytosolic indicator, the plasma membrane is permeabilized with low amounts of detergent in an artificial intracellular buffer. The method enables the direct stimulation of the endomembrane, thus bypassing signalling cascades. (**c**) *Dialysis of the cytosol under whole-cell configuration and simultaneous measurements of Ca^{2+} in the ER lumen and the cytosol.* For single cell analysis, cytosolic indicator can be dialyzed with a patch pipette. Combined with a simultaneous filling of the cytosol with a polar high-affinity Ca^{2+} indicator, this approach permits the simultaneous recording of cytosolic (Fluo-3, visible light) and ER derived signals (ratiometric Mag-Fura-2, UV-light) (**d**) *Targeted-esterase-induced dye loading.* A carboxylesterase (CES) is targeted to the lumen of the ER, thus providing a high esterase activity. This esterase activity releases the hydrophilic, Ca^{2+}-sensitive indicator dye Fluo-5N in the ER lumen by hydrolyzing the AM groups from the Ca^{2+}-insensitive compound Fluo-5N/AM. High concentrations of Fluo-5N are trapped in the ER lumen and form a highly fluorescent complex with Ca^{2+}. Combinations of different strategies are indicated by *arrows* in *black* (removal of cytosolic dye, without improvement of dye trapping) or *grey* (removal of cytosolic dye in combination with increased dye release in the ER lumen).

Table 8.1
Commonly used Ca^{2+} indicators for measurement of neuronal ER Ca^{2+} stores

Indicator	K_D for Ca^{2+}	Labelling strategy	Comments
Mag-Fura-2/AM (furaptra)	25–50 µM	Permeabilization (0.0001% saponin) Whole-cell dialysis TED-loading	Dorsal root ganglia; sensory neurones (in vitro) (12)(13, 39) Cortical neurones (in vitro) (removal of cytosolic dye needed) (26)
Mag-Fluo-4/AM	~20–25 µM	Direct loading	Dopamine neurones; *Substantia nigra* (in vitro) (9)
Fluo-5N/AM	~90 µM	TED-loading	Hippocampal and cortical neurones (in vitro) (26)

3.2. Synthetic Ca^{2+} Indicators and the Problem of Targeting

Synthetic dyes continue to be the most widely used Ca^{2+} indicators for imaging of Ca^{2+} dynamics in the ER (Table 8.1). These dyes offer advantages when compared with ER-targeted protein-based indicators. They provide a high photon emission rate, good Ca^{2+} specificity and an excellent signal-to-noise ratio (38, 39). In addition, synthetic Ca^{2+} probes have a wide range of Ca^{2+} affinities, very fast binding and dissociation ("on-off") kinetics and reasonably linear responses. Low-affinity Ca^{2+} indicators are available as both ratiometric (i.e. demonstrating spectral shift upon Ca^{2+} binding) as well as non-ratiometric, single wavelength indicators (i.e. showing an enhancement of fluorescence upon Ca^{2+} binding).

As a rule, the ER is loaded using cell incubation with the AM form of Ca^{2+} dyes (26, 39, 50, 51). However, the imaging of $[Ca^{2+}]_L$ is complicated by the fact that synthetic indicators do not have the specific ability to target the ER lumen, although ER loading may be improved by keeping the cells at 35–37°C (13, 52). Nonetheless, almost invariably some of the Ca^{2+} indicator will remain in the cytosol because the AM-ester group is cleaved by ubiquitous cytosolic esterases (Fig. 8.1a). The low-affinity Ca^{2+} indicators present in the cytosol do not disturb ER Ca^{2+} imaging in many types of non-excitable cells, because characteristic amplitudes of $[Ca^{2+}]_i$ transients (<5 µM) stay well below the detection threshold, whereas high levels of $[Ca^{2+}]_L$ are readily reported by the ER-trapped Ca^{2+} probe (10, 53). This is not the case, however, for excitable cells in which high-amplitude $[Ca^{2+}]_i$ transients can be detected by low-affinity dyes, making discrimination between cytosolic and ER Ca^{2+} signals difficult (13, 39).

Various protocols have been established to remove the unwanted dye from the cytosol. Initial strategies utilized the permeabilization of the plasma membrane with agents such as streptolysin O (19), digitonin (24) or saponin (12, 39) applied in an artificial intracellular buffer (permeabilization technique, Fig. 8.1b). Usage of the permeabilization technique in neurones is difficult to control. Consequently, the permeabilization reagents (39) may damage neurones rapidly and irreversibly. In addition, cytosolic components are washed out of the cell and intracellular signalling cascades are disturbed. Nonetheless, this technique has an advantage because various membrane-impermeable agents, for example, $InsP_3$, can be washed in and used to stimulate the ER directly (Fig. 8.1b).

3.3. Synthetic Ca^{2+} Dyes and Dialysis Under Whole-Cell Patch-Clamp Configuration

In experiments on both non-excitable and excitable cells, the whole-cell patch clamp technique has been employed to remove the cytosolic portion of the low-affinity Ca^{2+} probe (39, 51). In these experiments, the cells were loaded with membrane-permeable, low-affinity indicators (Fig. 8.1a), under conditions that favour ER-trapping, and subsequently the cytosolic portion of the indicator removed by dialysing the cell via a patch pipette during whole cell recording (Fig. 8.1c). Furthermore, using this approach, the high-affinity Ca^{2+} probe can be washed into the cytosol thus enabling simultaneous, independent imaging of free Ca^{2+} in the ER and in the cytoplasm (Fig. 8.1c). For this purpose, the ER and cytosolic Ca^{2+} probes should obviously have different spectral characteristics. In neurones, a well-established combination is based on AM-ester loading of the ratiometric, low-affinity indicator Mag-Fura-2/AM (also called furaptra; K_D in sensory neurones in vitro ~50 μM) and the patch-pipette dialysis with the water-soluble form of the non-ratiometric indicator Fluo-3 or Fluo-4 (pentapotassium salt, K_D ~325 nM – Fig. 8.1c). In addition, the technique described allows all kinds of electrophysiological recordings to be made in parallel with ER/cytoplasmic Ca^{2+} imaging. A major disadvantage of this method is the disturbance of the cytosolic environment, may be compromised of intracellular signalling pathways.

3.4. Improved ER Trapping of Synthetic Ca^{2+} Indicators by Targeted Esterases

A new strategy for targeting synthetic Ca^{2+} indicators to the endoplasmic reticulum has been developed recently (26). This method called TED (Targeted-Esterase-induced Dye loading) is based on the recombinant and targeted expression of a highly active esterase in the ER lumen, thus allowing an improved trapping of synthetic Ca^{2+} indicators in the reticulum (Fig. 8.1d). The method appears to be very useful in neurones at least in vitro as it allows a non-disruptive and specific loading of low-affinity Ca^{2+} dyes into the ER (26).

With the TED technique, a soluble ER-resident carboxylesterase is targeted genetically to the ER lumen thus providing a high level of esterase activity for rapid cleavage of the AM-dyes.

As the concentration of free Ca^{2+} in the ER is known to be in the range of 100–800 μM, the non-ratiometric Ca^{2+}-indicator dye Fluo-5N/AM is particularly useful for non-disruptive, esterase-based loading. This probe has a K_D for Ca^{2+} ~90 μM (54). Because of its low affinity for Ca^{2+}, the small amounts of Fluo-5N that may remain in the cytosol will not obscure visualization of the ER Ca^{2+} dynamics. An additional advantage is that Fluo 5N can be excited with a standard 488-nm laser line, thus allowing single-plane confocal analysis in ER-rich regions of neurones. The dye is essentially non-fluorescent in the absence of divalent cations and upon binding Ca^{2+} exhibits strong fluorescence enhancement (>100-fold) without a spectral shift (26). Thus, both dye loading and the morphology of the intraluminal Ca^{2+} store are readily visible under resting conditions when the high $[Ca^{2+}]_L$ provides a strongly fluorescent and almost background-free image (Fig. 8.4) (26).

4. Methods

4.1. AM-Ester Loading in Combination with Permeabilization Technique

The experiments described below were performed on isolated sensory neurones. Dorsal root ganglion neurones were isolated enzymatically from new-born (1–3 day-old) Sprague–Dawley rats using a conventional treatment with 0.1% protease (type XIV) in HEPES-buffered Minimum Essential Eagle Medium (MEM) for 8 min at 37°C. Individual cells were separated mechanically and plated on poly-L-ornithine (1 mg/ml) and laminin (0.01 mg/ml) covered glass cover-slips. Neurones were maintained in culture media (DMEM, supplemented with 10% horse serum, 50 U/ml penicillin/streptomycin mixture and 6 μg/ml insulin) at 37°C in an atmosphere of air supplemented with 5% CO_2 for 1–2 days prior to the experiment (12).

Neurones were incubated with 5 μM Mag-Fura-2/AM for 30 min at 37°C and washed at 37°C for 1 h prior to the experiment. Brief exposure of a Mag-Fura-2 pre-loaded neurone to saponin (10 μg/ml, 7–10 s) results in a rapid decrease in fluorescence excited at 340 nm (F_{340}) and at 380 nm (F_{380}), whereas F_{340}/F_{380} ratio (R) showed a progressive increase. These all attain steady-states in about 20–30 s (Fig. 8.2). At that stage, application of either $InsP_3$ or caffeine triggers a characteristic transient decrease in R that reflects Ca^{2+} release from the ER lumen.

The main problems of the permeabilization technique especially when employed in neurones are associated with rapid damage induced by the permeabilization agent. Therefore, great care should be taken also to keep the duration of treatment with permeabilising agent as short as possible (39).

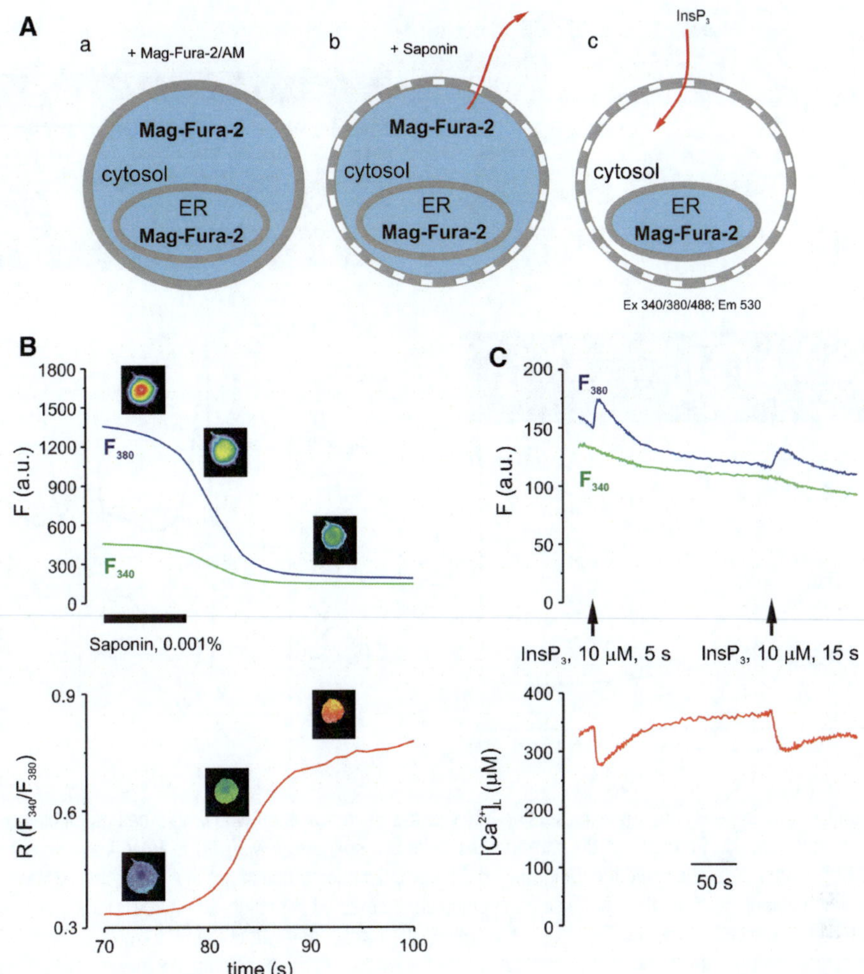

Fig. 8.2. Monitoring of intra-ER Ca^{2+} dynamics using Mag-Fura-2 in permeabilized cells (from Ref. (39) with permission). (**A**) Principle of the method. (**a**) Cells are stained by Mag-Fura-2/AM such that the probe is distributed both within the cytosol and the ER lumen. (**b**) The cellular membrane is permeabilized by brief exposure to saponin so that cytosolic Mag-Fura-2 escapes into the extracellular milieu. (**c**) The remaining Mag-Fura-2 is trapped within the ER lumen, thus allowing $[Ca^{2+}]_L$ recording. (**B**) Typical kinetics of washout of cytosolic Mag-Fura-2 after saponin permeabilization. The period of exposure to saponin is indicated by the bar. Treatment with saponin initiated rapid decrease in F_{340} and F_{380} with corresponding increase in R, which stabilized after the completion of washout. (**C**) Example of $[Ca^{2+}]_L$ changes in a DRG neurone recorded using the technique described above in response to brief (5–15 s) application of $InsP_3$ as indicated by *arrows*.

4.2. AM-Ester Loading in Combination with Patch-Pipette Dialysis and Simultaneous Imaging of Ca^{2+} Dynamics in the ER and in the Cytosol

An example of whole-cell ER/cytosole Ca^{2+} imaging experiment performed on sensory neurone is presented in Fig. 8.3 (13, 39). Here, the typical kinetics of the washout process of cytosolic Mag-Fura-2/ and wash-in of Fluo-4 are shown. The cell was initially loaded with Mag-Fura-2 by 20 min incubation with 5 μM Mag-Fura-2/AM followed by 40 min washout (both procedures carried out at 37°C).

Subsequently, the cells were transferred to the experimental chamber mounted on the inverted fluorescent microscope and whole-cell patch clamp recordings made (13). Immediately after

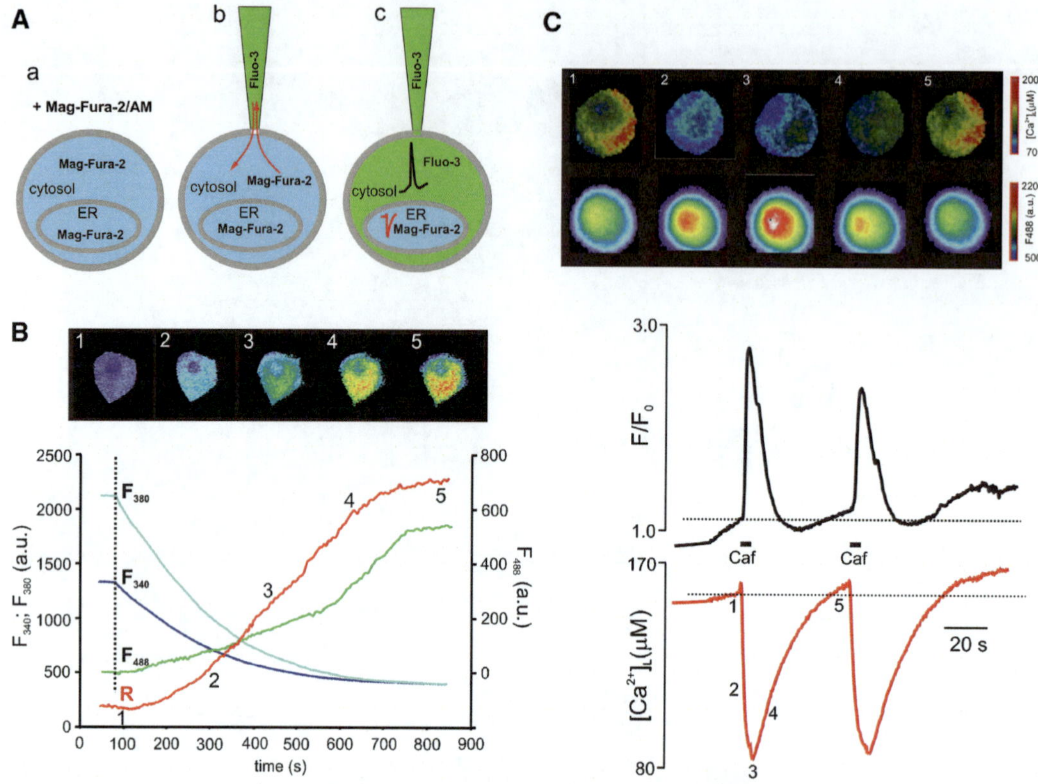

Fig. 8.3. Monitoring of intra-ER Ca^{2+} dynamics using Mag-Fura-2 in combination with whole-cell patch clamp (from Ref. (39) with permission). (**A**) Principle of the method. (**a**) Cells are stained by Mag-Fura-2/AM such that the probe is distributed both within the cytosol and the ER lumen. (**b**) The cytoplasmic portion of the dye then removed by intracellular dialysis via patch pipette. (**c**) The intrapipette solution being supplemented with high-affinity Ca^{2+} indicator Fluo-3 allows simultaneous and independent monitoring of Ca^{2+} changes in the cytosol and within the ER lumen. (**B**) Typical kinetics of washout of cytosolic Mag-Fura-2 and wash-in of Fluo-3 into the cytosol. At the beginning of experiment values of F_{340} and F_{380} are high (fluorescent values were averaged over the whole cell), and R is low, reflecting mostly cytosolic origin of the signal. After the break-through (indicated by *dotted line*) both F_{340} and F_{380} start to decrease and R rises, reflecting Mag-Fura-2 washout from the cytosol. The simultaneous increase in F_{488} reflects dialysis of the cytosol with Fluo-3K_5. All three fluorescence signals stabilize after completion of dialysis, when calculated $[Ca^{2+}]$ derived from Mag-Fura-2 approaches 250 μM, reflecting high free Ca^{2+} level in the cytosol. Selected ratio images are shown above. (**C**) Example of simultaneous $[Ca^{2+}]_L$ and $[Ca^{2+}]_i$ recordings from a DRG neurone stained and patch-clamped as described above. Images of Mag-Fura-2 ratios and Fluo-3 fluorescence are shown above; corresponding $[Ca^{2+}]_L$ and $[Ca^{2+}]_i$ traces below. The cell was challenged by 20 mM caffeine (as indicated by bars), causing a transient decrease in $[Ca^{2+}]_L$ and increase in $[Ca^{2+}]_i$.

the establishment of the whole cell configuration, the F_{340} and F_{380} values began to decrease while R increased progressively. At the same time, Fluo-4 fluorescence, excited at 488 nm (F_{488}) steadily increased reflecting dialysis of the cytosol with the intrapipette solution. F_{340}/F_{380}, R and F_{488} stabilized in about 8–10 min after break-through indicating complete exchange of cytosolic probes. The fluorescent images taken during intracellular dialysis help to visualize this washout processes: at the beginning, Mag-Fura-2 fluorescence is quite homogeneous and at the end of perfusion it clearly stains the ER leaving the nuclear part essentially fluorescence free. *R*, determined at that time corresponds to ~270 μM free Ca^{2+} a typical resting $[Ca^{2+}]_L$. At this point, simultaneous independent recordings of $[Ca^{2+}]_L$ and $[Ca^{2+}]_i$ can be made as demonstrated in Fig. 8.3C, which shows $[Ca^{2+}]_L$ and $[Ca^{2+}]_i$ dynamics in response to brief exposure to caffeine. Caffeine triggers a rapid decrease in intra-ER free Ca^{2+} mirrored by a transient increase in $[Ca^{2+}]_i$.

4.3. Targeted-Esterase Induced Dye Loading

A recombinant carboxylesterase (CES, EC 3.1.1.1) is targeted to the ER providing localized esterase activity. After AM-dye loading, this additional esterase activity results in improved trapping of Ca^{2+}-sensitive forms of the low-affinity Ca^{2+} indicators (preferentially Fluo-5N, but also Mag-Fura-2) within the ER (Table 8.1). The technique allows direct, non-disruptive measurements of Ca^{2+} dynamics in the ER lumen with high spatial and temporal resolution. The method has been successfully tested in neurones, HEK293 cells, BHK21 cells and primary astrocytes in vitro (26).

4.3.1. Cloning Strategy

Carboxylesterases (CES; EC 3.1.1.1) form a multigene family and are important for the hydrolytic transformation of a vast number of structurally diverse substances (55). These enzymes carry an ER translocation signal peptide at the N-terminus and an ER-retention and retrieval motif (KHREL* in mouseCES2 or KHVEL* in ratCES3) at the C-terminus, which identifies them as soluble ER-resident proteins. For the TED approach in neurones, the use of carboxylesterase 2 (CES2; Acc.No. BC015290) has been successful. CES2 can be cloned from liver total RNA by reverse transcription-PCR into a vector of choice (for verified CES-expressing vectors, please contact R.B.). It is recommended that a Kozak-sequence can be added to the upper primer. The following primer design is useful for full-length CES2 amplification from mouse cDNA: CES2-for: 5′-overhang-*restriction adapter*-CGCCACC **ATG** ACA CGG AAC CAA CTA CAT AAC-3′ (underlined: Kozak-sequence; bold: start codon); CES2-rev: 5′-overhang-*restriction adapter*-T AAA GCT CCC TGT GCT TGT CC-3′. This construct is targeted by its endogenous protein elements (signal peptide; ER-retention and retrieval motif).

For tagged versions, a core element of CES2 has been shown to be enzymatically active (R.B., unpublished). This allows the addition of a tag (e.g. signal peptide + GFP or signal peptide + RFP) at the N-terminus of the first amino acid of the mature protein as well as an engineered tag (e.g. myc-tag) between the CES2 core and a C-terminal optimized ER retention and retrieval motif (e.g. HKDEL*).

4.3.2. Lentiviral Transduction of Neurones

In neurones, vector transfer is critical and may be achieved through standard techniques such as transfection or electroporation. Recent years have seen the development of excellent viral vector tools that allow stable, non-toxic and highly efficient gene transfer to neurones. In cortical and hippocampal neurones in vitro, a self-inactivating lentiviral expression plasmid containing the human ubiquitin C promoter (FUGW) (56) has been used successfully for TED. Other vectors have also been developed, in which the ubiquitin promoter is replaced by neurone-specific promoters such as a modified synapsin promoter or, for use in mature neurones, a short CaM kinase II promoter (57). Lentiviral vectors can be used for the transduction of neurones in vivo, in organotypic cultures, and in vitro. For TED, transduction of cortical and hippocampal neurones in vitro is most effective on freshly dissociated cell suspensions, which are incubated with a substantial amount of lentiviral vector (example: 5×10^6 transducing units in 0.5 ml Neurobasal A/B27 medium per rat cortex, P0–P3). After 10 min incubation, the concentrated cell suspension is diluted and seeded on dry poly-L-lysine-coated glass cover-slips (12–13 mm; Menzel) at a density of 50,000 cells/cover-slip. Cells are allowed to attach for 1–2 h in a CO_2 incubator and the medium volume is then increased to 500 μl. Lentiviral transduction is stable and the ubiquitin-promoter has no known tendency to be silenced. Subsequently, the cells can be cultured for weeks to allow neuronal maturation.

4.3.3. Testing CES Expression

Western blot analysis. CES expression can be monitored readily by SDS–PAGE. In primary cortical neurones after viral transduction, 40 μg total protein lysate is sufficient to detect CES2 overexpression on PVDF membranes (Biorad) blocked with milk powder (Biorad) and labelled with rabbit anti-esterase (1/2,000–1/5,000; Abcam) using the ECL-Plus detection system (GE Healthcare).

Immunolocalization analysis. Cells are fixed with 4% paraformaldehyde in 1 × PBS for 15 min, blocked and permeabilized with 10% goat serum, 1 × PBS, 0.1% Triton X100 and then stained with antibodies against esterase (1:400; Abcam) and any appropriate secondary antibody (e.g. anti-rabbit Cy3, Jackson Laboratories). An improved protocol for detecting CES proteins is based on an additional denaturation step. In this case, fixed cells are permeabilized with 0.5% SDS, 5% β-mercaptoethanol for 30 min in 10% goat serum, 1 × PBS and then washed back to 10% goat serum, 1 × PBS, 0.1% Triton X100 (58). Afterwards, antibody labelling is performed. This protocol results in improved CES2 detection in fibroblasts. This protocol is not recommended for use in neurones in vitro since it disrupts the contact between the neurones and the cover-slip, but it might be useful for slice preparations. CES expression constructs with an immuno-reactive tag as well as a red fluorescent fusion partner are under construction (R.B.) and will be useful for improving CES detection and for distinguishing endogenous esterase expression from recombinant, engineered CES.

4.3.4. AM-Ester-Dye Loading (Fluo-5N/AM)

Since Fluo-5N is highly fluorescent when loaded to the ER, very brief excitation intensities are adequate. This allows imaging of primary neurones over long periods of up to 1 h at a resolution of 512×512 pixel at 0.2 Hz using confocal systems equipped with a high-performance objective (60×, N.A 1.4) (see Fig. 5 in (26)). To test the imaging properties of the dye, it is recommended that preliminary experiments are performed on cell lines such as BHK21 cells or on cultured astrocytes. Both are easy to grow and show strong release of Ca^{2+} from the ER and rapid refill of the ER through store-operated Ca^{2+} entry upon stimulation with ATP (perfusion stimulation, 100 μM ATP either in standard HEPES-Ringer or standard artificial cerebrospinal fluid (ACSF) with extracellular Ca^{2+} – Fig. 8.4).

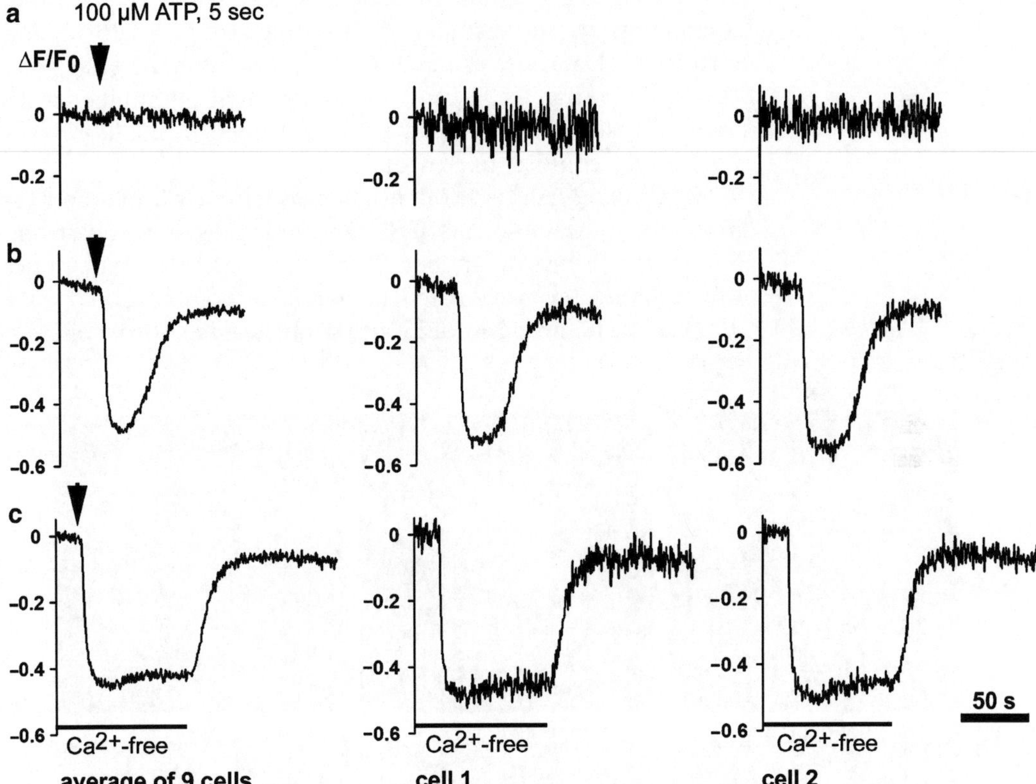

Fig. 8.4. TED control measurement in cultured astrocytes. (**a**) After Fluo-5N dye loading in non-transfected astrocytes, ER Fluo-5N loading is ineffective. Stimulation of cortical astrocytes with 100 μM ATP, a strong agonist for the release of Ca^{2+} from intraluminal Ca^{2+} stores, neither increases the Fluo-5N fluorescence (cytosolic signal) nor decreases fluorescence (luminal signal). (**b**) In astrocytes after TED and loading with Fluo-5N/AM, 100 μM ATP rapidly releases Ca^{2+} from intracellular Ca^{2+} stores. Stores are subsequently refilled, because the measurements were made in the presence of extracellular Ca^{2+} (2 mM). No increase in cytosolic fluorescence is observed in the initial phase of the stimulation. (**c**) When the same stimulation protocol is employed in the absence of extracellular Ca^{2+}, the loss of intraluminal Ca^{2+} becomes visible. Re-addition of extracellular Ca^{2+} refills the Ca^{2+} store within seconds (store-operated Ca^{2+} influx). Experimental conditions: Fluo-5N/AM loading for 10 min at 37°C in CO_2-equilibrated ACSF. Imaging conditions: Inverted microscope Olympus IX 70, confocal laser scanning with a Fluoview 300 system (Olympus), 20× objective (N.A. 0.7), excitation at 488 nm (**b**, **c**) 10 mW laser, 1.5% laser power, additionally decreased to 15%, 510 nm long pass filter, scan speed 2.3 Hz. In (**a**), laser power was fourfold increased compared to (**b**, **c**). ATP was applied by fast perfusion, ~20× chamber volume per minute (data from R. Blum).

A 5 mM stock solution of Fluo-5N/AM (Invitrogen; F14204) can be prepared in 20% Pluronic F-127 (Invitrogen) in dimethylsulphoxide (DMSO) by means of a sonifier bath for 3 min and aliquots of 0.5–1.5 µl can be stored at −20°C for months. Figure 8.5a shows primary neurones expressing CES2 labelled with Fluo-5N. Neurones were incubated with 5 µM Fluo 5N-AM in ACSF-Ringer solution. The ACSF consists of (in mM): 125 NaCl, 3 KCl, 1.25 NaH_2PO_4, 2 $CaCl_2$, 2 $MgCl_2$, 25 $NaHCO_3$, and 25 D-glucose. The solution should be saturated with 95% O_2 and 5% CO_2, resulting in a pH of 7.4. The incubation of neurones with the dye-Ringer solution works best in a cell culture incubator (37°C, 5% CO_2) for a loading time of 7–15 min.

After Fluo-5N loading, neurones should be allowed a post-loading time of about 20–30 min under continuous perfusion. Experiments are successful at room temperature (24–26°C) and up to 33°C. Neurones can be used for different imaging applications either using CCD camera-based imaging combined with monochromator-based excitation, or standard epifluorescent excitation or confocal microscopy.

Mag-Fura-2/AM is a commonly used ratiometric indicator for direct $[Ca^{2+}]_L$ measurements. CES-expressing cells show very strong accumulation of this dye in the ER lumen. When tested in neurones using a monochromator-based excitation in combination with a CCD camera for image acquisition, a strong overlap of cytosolic and

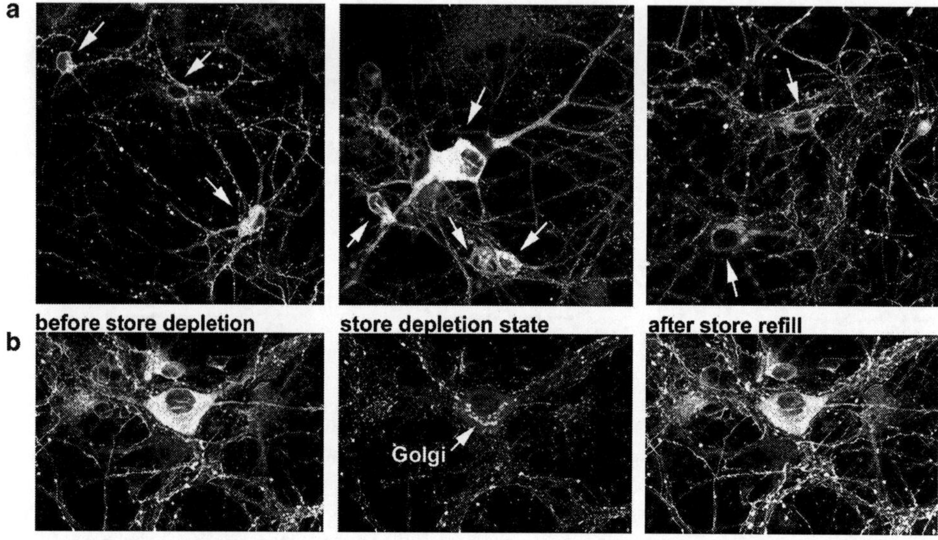

Fig. 8.5. TED labelling of cultured cortical neurones. (**a**) Visualization of Fluo-5N/Ca^{2+} complexes in cortical neurones expressing CES2. Neuronal somata are indicated by *arrows*. Images were extracted from original time-lapse experiments and represent average projections of five images, acquired at 0.2 Hz, with 512 × 512 pixel resolution and slow scan conditions. (**b**) Average projections (five images) of a store-depletion experiment. In the depleted state, Fluo-5N/Ca^{2+} complexes were lost in somatic regions, but remain in the perinuclear Golgi apparatus. In addition, fast moving vesicular–tubular structures became visible (data from Rehberg and Blum – (26)).

ER-derived signals is observed. Here, TED in combination with Mag-Fura-2-AM results in improved trapping of the dye in the ER lumen, but still requires the removal of unwanted dye from the cytosol (Fig. 8.1b, c, R. Blum unpublished observations).

4.3.5. Ca^{2+} Imaging of Neurones After TED

TED has proven successful for direct Ca^{2+} imaging of neuronal ER when used in combination with Fluo-5N/AM in confocal setups (0.2–15 Hz), as well as with CCD camera-based imaging setups (1–5 Hz) with monochromator or standard epifluorescence illumination using upright or inverted microscopes and with objectives ranging from 20× (N.A. 0.7) to 60× (N.A. 0.9 water or N.A. 1.4 oil). At rest, Fluo-5N/Ca^{2+} complexes are highly fluorescent and almost background-free. According to their spectral properties (excitation/emission maxima ~494/518 nm) Fluo-5N/Ca^{2+} complexes are best excited with a 488 nm laser line or a standard light source (xenon arc, mercury arc, or X-Cite lamps) and imaged using standard filter and/or beamsplitter sets (as used for FITC, Alexa 488 or GFP). In neurones, confocal imaging is preferable to epifluorescence microscopy. Confocal imaging allows the detection of Fluo-5N/Ca^{2+} complexes with high resolution at low sampling rates. At the cost of resolution, fast imaging is also feasible.

The TED method has been tested under two confocal imaging situations (26). First, to achieve a good resolution of the neuronal soma and neurites, an inverted confocal microscope (Fluoview 300, Olympus) equipped with a 60× objective (N.A. 1.4, oil) was used at low sampling rates (0.2 Hz). Here, the Ca^{2+}-bound form of Fluo-5N provided a strong labelling of subcellular structures in neurones, the perinuclear area, neurites and small punctuate peripheral elements (Fig. 8.5). The resolution of this slow imaging approach identified luminal Ca^{2+} dynamics in the smallest neuronal Fluo 5N/Ca^{2+}-labelled element (Fig. 8.6).

Secondly, intraluminal Ca^{2+} dynamics were monitored with a high temporal resolution (15 Hz) using an upright microscope with a 60× objective (N.A. 0.9, water). Most importantly, measurements were made in the presence of extracellular Ca^{2+} thus not affecting synaptic transmission in neuronal networks. Here, CES2-expressing primary neurones (DIV 18–25) were stimulated locally with 1 mM (*S*)-3,5-dihydroxyphenylglycine (DHPG) for 100 ms using a pressure-ejection system. DHPG is an agonist of group I metabotropic glutamate receptors (mGluRI and mGluR5) and releases Ca^{2+} from $InsP_3$-sensitive intracellular Ca^{2+} stores. The local application of DHPG to the soma of cortical neurones induced a rapid Ca^{2+} transient that lasted approximately 5 s (26).

4.3.6. Testing Ester-Derivatized Dyes for Use with TED

The TED method is not restricted to Ca^{2+} indicators, but can be employed for esterase-sensitive indicators in general. To determine whether CES proteins are able to activate any dye that needs a

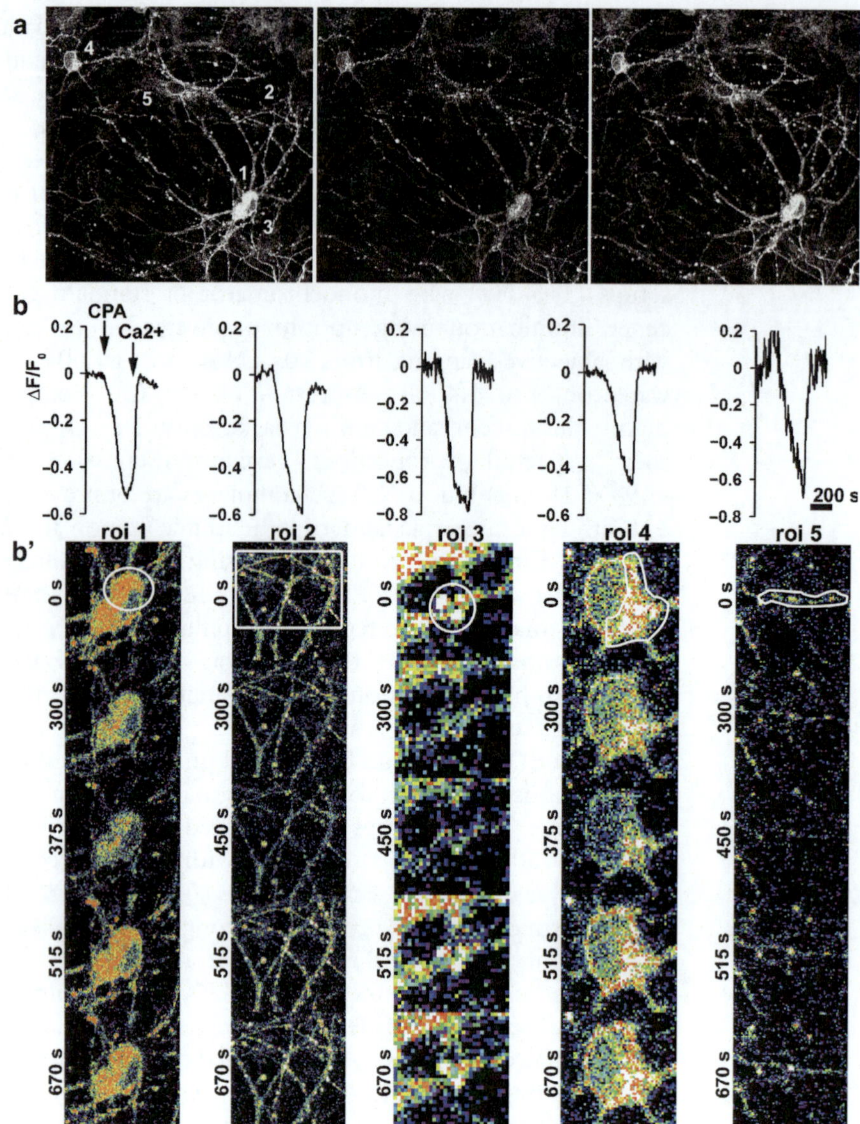

Fig. 8.6. Direct imaging of ER Ca^{2+} release signals after TED dye loading in cultured cortical neurones. Confocal image of neurones and their processes (**a**) and Ca^{2+}-release transients (**b**) in selected regions-of-interest (roi 1–roi 5, **b′**). The scanning rate was 0.2 Hz. *Arrow tips* indicate changes in the perfusion buffer. To induce Ca^{2+} release from intracellular stores, the SERCA-pump blocker CPA was used (30 μM in Ca^{2+}-free buffer). The re-uptake of Ca^{2+} to the ER was triggered by re-addition of 2 mM Ca^{2+} (data from Rehberg and Blum – (26)).

hydrolytic esterase activity, a commercially available carboxylesterase preparation (porcine liver carboxylesterase, EC 3.1.1.1., Sigma) can be used to build up an in vitro dye cleavage assay. For testing non-ratiometric Ca^{2+} indicators of the Fluo family, 10 μM of the AM-ester form are incubated in 1 × PBS (without Ca^{2+} or Mg^{2+}; Invitrogen) with 10 units of liver carboxylesterase for 3 min. Using a spectrophotometer at an excitation wavelength of 488 nm, activation of the low-affinity indicator Fluo-5N by the addition of 1 mM Ca^{2+} can easily be visualized at 500–550 nm emission wavelengths.

4.3.7. Troubleshooting TED in Neurones

Toxicity. One of the hydrolysis products of the Fluo-5N/AM is highly toxic formaldehyde. An increase in dye loading therefore increases not only the concentration of an artificial intracellular Ca^{2+} buffer, but also increases the local concentration of formaldehyde. Experience to date has shown that the increase in dye loading is innocuous. Toxicity of dye loading has been observed when cells are exposed to the dye-loading mixture (5 μM Fluo-5N/AM/0.02% Pluoronic/0.1% DMSO) for more than ~25 min. As a consequence, neurones became non-responsive. The incubation of neurones with the dye/Ringer solution works well in a cell culture incubator (37°C, 5% CO_2), but loading times of 7–15 min should not be exceeded. When toxicity is observed, a mixture of 1/3 vol. growth medium with 2/3 vol. ACSF is recommended. Alternatively, either the loading time or the dye concentration can be reduced.

Bleaching. It should be noted that Fluo-5N is rather resistant to bleaching, but not as resistant as other synthetic Ca^{2+} indicators such as Oregon Green 488 BAPTA-1 or Fluo-3. Prolonged illumination with strong epifluorescent light should thus be avoided.

Cytosolic crosstalk. In somatic regions of neurones, very strong stimulation can cause mixed ER and cytosolic signals (short increase in fluorescence, followed by rapid decrease). In this case, extended washing (20–60 min) after dye loading will solve the problem. In addition, reduction of the dye concentration and loading time may be helpful.

Post-ER labelling. Luminal Fluo-5N/Ca^{2+} complexes are also visible in post-ER- structures such as the Golgi apparatus (Fig. 8.5b). Even vesicles en route are labelled and become visible during ER-store depletion.

Dye-drop. In somatic cell regions of interest, Fluo-5N/Ca^{2+} signals after stimulation show a minimal drop and do not return to the pre-stimulating levels. This indicates that some Fluo-5N/Ca^{2+} complexes are lost from the ER lumen during stimulation.

5. Perspectives

Although methodological approaches for direct ER Ca^{2+} measurements have been improved in recent years, the accurate measurements of Ca^{2+} release from or uptake into the ER remains a major issue. An ideal probe will combine the targetability of a protein with the spectral properties of synthetic Ca^{2+} indicators (38, 44, 59).

Rapid progress in the development of protein-based indicators will certainly culminate in the development of a useful ER-targeted, genetically engineered Ca^{2+} indicator. However, at least two principal problems of these indicators will remain extremely difficult to solve. Since the photon emission rate of protein Ca^{2+} indicators is low, high expression levels are required in the ER to achieve a substantial signal in synapses, axons or dendrites of neurones. This increases the risk of protein misfolding or mistargeting by saturation of the ER retention and retrieval process, and adds high amounts of additional Ca^{2+} buffer, which can influence signalling characteristics. In addition, due to their size, transport or diffusion of protein-based indicators to remote intra-ER microdomains may be limited.

A spectacular strategy for constructing an ideal Ca^{2+} probe has been reported recently by Tsien and coworkers (60, 61). Here, a small dye of 1 kDa (Calcium Green FlAsH) is bound to be targeted channel proteins and is used to monitor local Ca^{2+} dynamics in nanodomains with rapid kinetics. In the ER, such a method may be used to analyse Ca^{2+} fluctuation directly beneath the endomembrane.

The ER targeting of synthetic indicators is still a critical issue and the TED method provides a strategy for improving ER trapping of low-affinity Ca^{2+} dyes. The enrichment of the Fluo-5N-indicator relies on the proper retention of a carboxylesterase in the ER lumen while the dye itself is not the target of the ER retrieval mechanism. Thus, the method will not be useful for the analysis of nanodomain analysis. Further development of the TED method is ongoing. It will be interesting to see whether animal models for TED in combination with in vivo labelling techniques such as the multicell bolus injection loading (62) will permit the analysis of fast ER-Ca^{2+}-release events in vivo.

The ideal probe for intraluminal Ca^{2+} fluctuations is not yet available, and it is unlikely that a "one-for-all" probe will meet the demands of individual questions. Undoubtedly, continuing methodological research involving all currently available strategies, including the protein-based indicators, the synthetic indicators as well as the combinatorial use of both, will help overcome the present limitations.

References

1. Berridge MJ (2002) The endoplasmic reticulum: a multifunctional signaling organelle. Cell Calcium 32:235–249
2. Burdakov D, Petersen OH, Verkhratsky A (2005) Intraluminal calcium as a primary regulator of endoplasmic reticulum function. Cell Calcium 38:303–310
3. Petersen OH, Michalak M, Verkhratsky A (2005) Calcium signalling: past, present and future. Cell Calcium 38:161–169
4. Verkhratsky A (2005) Physiology and pathophysiology of the calcium store in the endoplasmic reticulum of neurons. Physiol Rev 85:201–279

5. Lippincott-Schwartz J, Roberts TH, Hirschberg K (2000) Secretory protein trafficking and organelle dynamics in living cells. Annu Rev Cell Dev Biol 16:557–589
6. Park MK, Choi YM, Kang YK, Petersen OH (2008) The endoplasmic reticulum as an integrator of multiple dendritic events. Neuroscientist 14:68–77
7. Petersen OH, Burdakov D, Tepikin AV (1999) Polarity in intracellular calcium signaling. Bioessays 21:851–860
8. Terasaki M, Slater NT, Fein A, Schmidek A, Reese TS (1994) Continuous network of endoplasmic reticulum in cerebellar Purkinje neurons. Proc Natl Acad Sci USA 91:7510–7514
9. Choi YM, Kim SH, Chung S, Uhm DY, Park MK (2006) Regional interaction of endoplasmic reticulum Ca^{2+} signals between soma and dendrites through rapid luminal Ca^{2+} diffusion. J Neurosci 26:12127–12136
10. Park MK, Petersen OH, Tepikin AV (2000) The endoplasmic reticulum as one continuous Ca^{2+} pool: visualization of rapid Ca^{2+} movements and equilibration. EMBO J 19: 5729–5739
11. Petersen OH, Tepikin A, Park MK (2001) The endoplasmic reticulum: one continuous or several separate Ca^{2+} stores? Trends Neurosci 24:271–276
12. Solovyova N, Verkhratsky A (2003) Neuronal endoplasmic reticulum acts as a single functional $Ca2^{+}$ store shared by ryanodine and inositol-1, 4, 5-trisphosphate receptors as revealed by intra-ER [$Ca2^{+}$] recordings in single rat sensory neurones. Pflugers Arch 446:447–454
13. Solovyova N, Veselovsky N, Toescu EC, Verkhratsky A (2002) Ca^{2+} dynamics in the lumen of the endoplasmic reticulum in sensory neurons: direct visualization of k-induced Ca^{2+} release triggered by physiological Ca^{2+} entry. EMBO J 21:622–630
14. Subramanian K, Meyer T (1997) Calcium-induced restructuring of nuclear envelope and endoplasmic reticulum calcium stores. Cell 89:963–971
15. Bouchard R, Pattarini R, Geiger JD (2003) Presence and functional significance of presynaptic ryanodine receptors. Prog Neurobiol 69:391–418
16. Ouardouz M, Nikolaeva MA, Coderre E, Zamponi GW, McRory JE, Trapp BD, Yin X, Wang W, Woulfe J, Stys PK (2003) Depolarization-induced Ca^{2+} release in ischemic spinal cord white matter involves L-type Ca^{2+} channel activation of ryanodine receptors. Neuron 40:53–63
17. Spacek J, Harris KM (1997) Three-dimensional organization of smooth endoplasmic reticulum in hippocampal CA1 dendrites and dendritic spines of the immature and mature rat. J Neurosci 17:190–203
18. Petersen OH, Petersen CC, Kasai H (1994) Calcium and hormone action. Annu Rev Physiol 56:297–319
19. Streb H, Irvine RF, Berridge MJ, Schulz I (1983) Release of Ca2+ from a nonmitochondrial intracellular store in pancreatic acinar cells by inositol-1, 4, 5-trisphosphate. Nature 306:67–69
20. Friel DD, Tsien RW (1992) Phase-dependent contributions from Ca^{2+} entry and Ca^{2+} release to caffeine-induced $[Ca^{2+}]_i$ oscillations in bullfrog sympathetic neurons. Neuron 8:1109–1125
21. Friel DD, Tsien RW (1992) A caffeine- and ryanodine-sensitive Ca^{2+} store in bullfrog sympathetic neurones modulates effects of Ca^{2+} entry on $[Ca^{2+}]_i$. J Physiol 450:217–246
22. Mogami H, Nakano K, Tepikin AV, Petersen OH (1997) Ca^{2+} flow via tunnels in polarized cells: recharging of apical Ca^{2+} stores by focal Ca^{2+} entry through basal membrane patch. Cell 88:49–55
23. Petersen OH, Verkhratsky A (2007) Endoplasmic reticulum calcium tunnels integrate signalling in polarised cells. Cell Calcium 42:373–378
24. Hofer AM, Machen TE (1993) Technique for in situ measurement of calcium in intracellular inositol 1,4,5-trisphosphate-sensitive stores using the fluorescent indicator mag-fura-2. Proc Natl Acad Sci USA 90:2598–2602
25. Montero M, Brini M, Marsault R, Alvarez J, Sitia R, Pozzan T, Rizzuto R (1995) Monitoring dynamic changes in free Ca^{2+} concentration in the endoplasmic reticulum of intact cells. EMBO J 14:5467–5475
26. Rehberg M, Lepier A, Solchenberger B, Osten P, Blum R (2008) A new non-disruptive strategy to target calcium indicator dyes to the endoplasmic reticulum. Cell Calcium 44(4):386–399
27. Solovyova N, Fernyhough P, Glazner G, Verkhratsky A (2002) Xestospongin C empties the ER calcium store but does not inhibit $InsP_3$-induced Ca^{2+} release in cultured dorsal root ganglia neurones. Cell Calcium 32:49–52
28. Hartmann J, Konnerth A (2005) Determinants of postsynaptic Ca^{2+} signaling in Purkinje neurons. Cell Calcium 37:459–466
29. Rose CR, Konnerth A (2001) Stores not just for storage. intracellular calcium release and synaptic plasticity. Neuron 31:519–522

30. Camello C, Lomax R, Petersen OH, Tepikin AV (2002) Calcium leak from intracellular stores – the enigma of calcium signalling. Cell Calcium 32:355–361
31. Hartmann J, Blum R, Kovalchuk Y, Adelsberger H, Kuner R, Durand GM, Miyata M, Kano M, Offermanns S, Konnerth A (2004) Distinct roles of Gα(q) and Gα11 for Purkinje cell signaling and motor behavior. J Neurosci 24:5119–5130
32. Hartmann J, Dragicevic E, Adelsberger H, Henning HA, Sumser M, Abramowitz J, Blum R, Dietrich A, Freichel M, Flockerzi V, Birnbaumer L, Konnerth A (2008) TRPC3 channels are required for synaptic transmission and motor coordination. Neuron 59:392–398
33. Shimizu H, Fukaya M, Yamasaki M, Watanabe M, Manabe T, Kamiya H (2008) Use-dependent amplification of presynaptic Ca^{2+} signaling by axonal ryanodine receptors at the hippocampal mossy fiber synapse. Proc Natl Acad Sci USA 105:11998–12003
34. Verkhratsky A (2002) The endoplasmic reticulum and neuronal calcium signalling. Cell Calcium 32:393–404
35. Verkhratsky A, Shmigol A (1996) Calcium-induced calcium release in neurones. Cell Calcium 19:1–14
36. Blum R, Konnerth A (2005) Neurotrophin-mediated rapid signaling in the central nervous system: mechanisms and functions. Physiology (Bethesda) 20:70–78
37. Takechi H, Eilers J, Konnerth A (1998) A new class of synaptic response involving calcium release in dendritic spines. Nature 396:757–760
38. Rudolf R, Mongillo M, Rizzuto R, Pozzan T (2003) Looking forward to seeing calcium. Nat Rev Mol Cell Biol 4:579–586
39. Solovyova N, Verkhratsky A (2002) Monitoring of free calcium in the neuronal endoplasmic reticulum: an overview of modern approaches. J Neurosci Methods 122:1–12
40. Alvarez J, Montero M (2002) Measuring $[Ca^{2+}]$ in the endoplasmic reticulum with aequorin. Cell Calcium 32:251–260
41. Miyawaki A, Llopis J, Heim R, McCaffery JM, Adams JA, Ikura M, Tsien RY (1997) Fluorescent indicators for Ca^{2+} based on green fluorescent proteins and calmodulin. Nature 388:882–887
42. Palmer AE, Tsien RY (2006) Measuring calcium signaling using genetically targetable fluorescent indicators. Nat Protoc 1:1057–1065
43. Tsien RY (1981) A non-disruptive technique for loading calcium buffers and indicators into cells. Nature 290:527–528
44. Mank M, Griesbeck O (2008) Genetically encoded calcium indicators. Chem Rev 108:1550–1564
45. Heim N, Garaschuk O, Friedrich MW, Mank M, Milos RI, Kovalchuk Y, Konnerth A, Griesbeck O (2007) Improved calcium imaging in transgenic mice expressing a troponin C-based biosensor. Nat Methods 4:127–129
46. Wallace DJ, Borgloh SM, Astori S, Yang Y, Bausen M, Kugler S, Palmer AE, Tsien RY, Sprengel R, Kerr JN, Denk W, Hasan MT (2008) Single-spike detection in vitro and in vivo with a genetic Ca^{2+} sensor. Nat Methods 5:797–804
47. Griesbeck O, Baird GS, Campbell RE, Zacharias DA, Tsien RY (2001) Reducing the environmental sensitivity of yellow fluorescent protein. Mechanism and applications. J Biol Chem 276:29188–29194
48. Ishii K, Hirose K, Iino M (2006) Ca^{2+} shuttling between endoplasmic reticulum and mitochondria underlying Ca^{2+} oscillations. EMBO Rep 7:390–396
49. Palmer AE, Jin C, Reed JC, Tsien RY (2004) Bcl-2-mediated alterations in endoplasmic reticulum Ca^{2+} analyzed with an improved genetically encoded fluorescent sensor. Proc Natl Acad Sci USA 101:17404–17409
50. Mogami H, Tepikin AV, Petersen OH (1998) Termination of cytosolic Ca^{2+} signals: Ca^{2+} reuptake into intracellular stores is regulated by the free Ca^{2+} concentration in the store lumen. EMBO J 17:435–442
51. Park MK, Tepikin AV, Petersen OH (2002) What can we learn about cell signalling by combining optical imaging and patch clamp techniques? Pflugers Arch 444:305–316
52. Thomas D, Tovey SC, Collins TJ, Bootman MD, Berridge MJ, Lipp P (2000) A comparison of fluorescent Ca^{2+} indicator properties and their use in measuring elementary and global Ca^{2+} signals. Cell Calcium 28:213–223
53. Park MK, Tepikin AV, Petersen OH (1999) The relationship between acetylcholine-evoked Ca^{2+}-dependent current and the Ca^{2+} concentrations in the cytosol and the lumen of the endoplasmic reticulum in pancreatic acinar cells. Pflugers Arch 438:760–765
54. Launikonis BS, Barnes M, Stephenson DG (2003) Identification of the coupling between skeletal muscle store-operated Ca^{2+} entry and the inositol trisphosphate receptor. Proc Natl Acad Sci USA 100:2941–2944
55. Satoh T, Hosokawa M (2006) Structure, function and regulation of carboxylesterases. Chem Biol Interact 162:195–211
56. Lois C, Hong EJ, Pease S, Brown EJ, Baltimore D (2002) Germline transmission and tissue-specific

expression of transgenes delivered by lentiviral vectors. Science 295:868–872

57. Dittgen T, Nimmerjahn A, Komai S, Licznerski P, Waters J, Margrie TW, Helmchen F, Denk W, Brecht M, Osten P (2004) Lentivirus-based genetic manipulations of cortical neurons and their optical and electrophysiological monitoring in vivo. Proc Natl Acad Sci USA 101:18206–18211
58. Blum R, Pfeiffer F, Feick P, Nastainczyk W, Kohler B, Schafer KH, Schulz I (1999) Intracellular localization and in vivo trafficking of p24A and p23. J Cell Sci 112:537–548
59. Tsien RY (2003) Breeding molecules to spy on cells. Harvey Lect 99:77–93
60. Tour O, Adams SR, Kerr RA, Meijer RM, Sejnowski TJ, Tsien RW, Tsien RY (2007) Calcium Green FlAsH as a genetically targeted small-molecule calcium indicator. Nat Chem Biol 3:423–431
61. Tsien RY (2006) Breeding and building molecules to spy on cells and tumors. Keio J Med 55:127–140
62. Stosiek C, Garaschuk O, Holthoff K, Konnerth A (2003) In vivo two-photon calcium imaging of neuronal networks. Proc Natl Acad Sci USA 100:7319–7324

Chapter 9

Ca^{2+} Imaging of Intracellular Organelles: Mitochondria

Lucía Núñez, Carlos Villalobos, María Teresa Alonso, and Javier García-Sancho

Abstract

Calcium handling by mitochondria is important both because mitochondria can shape the cytosolic Ca^{2+} signals and because changes in mitochondrial Ca^{2+} concentration ($[Ca^{2+}]_M$) are important for controlling physiological functions such as respiration or programmed cell death. Accurate measurements of $[Ca^{2+}]_M$ require selective location of the Ca^{2+} probe inside mitochondria and this is best achieved by targeting protein probes to the mitochondrial matrix. Aequorins are very adequate as Ca^{2+}probes because: (1) they allow molecular engineering for targeting or for changing the Ca^{2+} affinity; (2) do not require irradiation for measurements; (3) Ca^{2+} buffering is small; (4) have a very steep Ca^{2+}-dependence and a very wide dynamic range, which makes them ideal for detecting and quantifying Ca^{2+} microdomains. Consumption and low light output are some of its drawbacks that make calcium imaging a hard task. Here, we describe a procedure that overcomes these disadvantages by combining herpes simplex virus type 1(HSV-1)-based expression of targeted aequorins with photon-counting imaging. This methodology allows real-time resolution of changes of $[Ca^{2+}]_M$ by photon counting imaging at the single-cell level. Since HSV virus is neurotrophic, the method is adequate for measuring $[Ca^{2+}]_M$ in living neurons.

Key words: Calcium signaling, Calcium oscillations, Mitochondria, Aequorin, Microdomains, Chemiluminescence, Bioluminescence imaging, Herpes simplex virus type 1, Neurons

1. Introduction

Many physiological functions are triggered by Ca^{2+} signals, sometimes restricted to the subcellular *domain* where the molecular machinery responsible for the function under control is located. This topological organization allows simultaneous control of different functions by the same signal: the change of the local cytosolic Ca^{2+} concentration ($[Ca^{2+}]_C$) and the specificity is conferred by the microdomain where the $[Ca^{2+}]_C$ signal takes place (1, 2). Cytoplasmic organelles are able to take up and to release calcium.

A. Verkhratsky and Ole H. Petersen (eds.), *Calcium Measurement Methods*, Neuromethods, vol. 43, DOI 10.1007/978-1-60761-476-0_9, © Humana Press, a part of Springer Science+Business Media, LLC 2010

In this way, they can both modulate the $[Ca^{2+}]_C$ signals generated by other mechanisms and provoke $[Ca^{2+}]_C$ signals by themselves. In addition, the Ca^{2+} microdomains generated inside organelles are very different of the surrounding $[Ca^{2+}]_C$ changes and have a deep impact on many important cell functions (3).

It has been known for long that isolated mitochondria can, when exposed to micromolar Ca^{2+} concentrations, accumulate large amounts of this cation. Ca^{2+} is taken up through the mitochondrial calcium uniporter using the mitochondrial membrane potential as the driving force (4). This high-capacity and low-affinity mechanism does not accumulate calcium at the resting condition when $[Ca^{2+}]_C$ is well below K_D, but mitochondrial Ca^{2+} uptake increases exponentially with $[Ca^{2+}]_C$ and can be extremely fast in the high Ca^{2+} microdomains generated during cell activation near Ca^{2+} channels either at the plasma membrane or the endoplasmic reticulum (ER) (5–10). The increase in $[Ca^{2+}]$ inside mitochondrial matrix ($[Ca^{2+}]_M$) activates several mitochondrial dehydrogenases (4, 11, 12) thus coupling cell activation to increased respiration and ATP production (13, 14). On the other hand, when large $[Ca^{2+}]_M$ are attained and maintained, programmed cell death is triggered by opening of the permeability transition pore and release of pro-apoptotic mediators from the Ca^{2+}-overloaded mitochondria (15).

Selective measurements of $[Ca^{2+}]_M$ require the use of probes that locate selectively at the mitochondrial matrix. Acidic fluorescent Ca^{2+} probes such as Rhod-2 accumulate inside mitochondria, more alkaline than the surrounding cytosol, and have been used for $[Ca^{2+}]_M$ measurements (11, 16). In some cases, trapping of the probe inside mitochondria is promoted by enzymatic modification by specific mitochondrial enzymes (17). Even though these probes accumulate inside mitochondria, a substantial part of the dye remains at the cytosol contaminating the $[Ca^{2+}]_M$ measurements and making difficult the calibration of the signal. Performance is improved by dialyzing the cytosolic dye by impalement with thick patch electrodes or by permeabilizing the plasma membrane, but measurements in intact cells are best achieved by targeting protein probes to the mitochondrial matrix (3).

Two kinds of mitochondria-targeted protein probes have been used for $[Ca^{2+}]_M$ measurements, fluorescent probes such as cameleons or pericams (18–21) and photoluminescent probes such as aequorins (22, 23). Each one has its advantages and drawbacks. Fluorescence is easier to measure and should be preferred for monitoring rapid $[Ca^{2+}]_M$ changes such as those taking place during heart beating (24). Among the drawbacks, fluorescence measurements require cell irradiation with the exciting light, the fluorescence is very much pH sensitive (20) and low Ca^{2+} affinity probes are not available, and this prevents adequate monitorization of high $[Ca^{2+}]_M$, such as those involved in triggering of apoptosis. Aequorins produce little Ca^{2+} buffering, do not require irradiation for measurements, are better designed for evidencing high Ca^{2+} microdomains and have a much larger dynamic range (3).

Aequorin, a photoprotein derived from the jellyfish *Aequorea victoria*, was first used as an intracellularly injected Ca^{2+} probe by Shimomura (25). In the 1980s, aequorin was cloned almost simultaneously into two laboratories (26, 27) giving way to the use of recombinant aequorins (22, 28). Native aequorin (AEQ) has three functional EF hands which on binding of Ca^{2+} promote oxidation of the bound cofactor coelenterazine to coelenteramide and light production:

$$\text{AEQ - coelenterazine} + Ca^{2+} + O_2 \rightarrow \text{AEQ - } Ca^{2+} + CO_2 + \text{coelenteramide} + \text{light}$$

Therefore, light output is a function of $[Ca^{2+}]$. The native aequorin is adequate for $[Ca^{2+}]$ measurements in the 0.1–5 μM range. Mutations in one of the EF hands decrease Ca^{2+} affinity (29) making possible measurements in the 10^{-8}–10^{-5} range. The use of mutated aequorin together with the synthetic coelenterazine n (23) allows to extend $[Ca^{2+}]_C$ measurements to the 10^{-3} M range (7, 30, 31). Calibrations are discussed in detail below.

The idea of directing aequorin to different organelles using targeting sequences was first proposed by Campbell and co-workers (32) and quickly applied to endoplasmic reticulum (33) and mitochondria (22). Mitochondrial targeting is obtained here by using a fragment of the human cytochrome c oxydase subunit VIII (22). Expression in primary cultures including neurons was achieved using a vector derived from the neurotrophic HSV-1 (34). The performance of the probe can be enhanced by fusion to either GFP (35) or RFP (36, 37). This allowed: (1) larger and more sustained aequorin levels by stabilization of the protein; (2) easy identification of the cells expressing GFP–AEQ just by fluorescence inspection; (3) shifting the AEQ fluorescence emission, which naturally happens in the blue towards the fluorescent protein emission maximum and, (4) eventually, simultaneous double wavelength measurement of green and red aequorins targeted to different compartments (37). Combination of the high protein expression promoted by the viral vector and high amplification by photon counting imaging allowed single cell measurements (38). In the following sections, we provide details on the methodology followed by application examples of mitochondrial Ca^{2+} measurements.

2. Methods

2.1. Expression of Aequorins

For imaging measurements, high expression of aequorin is essential and not so much a high efficiency of transfection. The use of viral vectors has the double advantage of being an efficient gene transfer method in non-dividing cells such as neurones and to yield a high expression of the transduced gene. Several viral vectors have already been used to express calcium reporters in neurones including retroviruses and adenoviruses. The use of HSV-1-derived amplicons has several advantages as a gene delivery vector: (1) vector construction is relatively easy; (2) it has a large capacity for transgenes; (3) the host range of vector transduction is very wide; and (4) in the context of this review, the main advantage of the amplicons is its natural tropism to efficiently infect postmitotic neurones (34).

2.1.1. Plasmid Construction

The use of the fusion protein green fluorescent protein (GFP)–aequorin, termed here GFP–AEQ, has several advantages over the native aequorin: (1) it allows direct visualization of the fusion protein under the fluorescence microscope; (2) the protein is more stable; and (3) it gives a higher bioluminescence yield (35). All the three properties are very appropriate for imaging. Therefore, although it is possible to image subcellular calcium with aequorin itself (38), AEQ has been in the last years replaced by the fusion protein GFP–AEQ in our laboratory.

The starting chimeric fusion GFP–aequorin DNA was a generous gift from Dr. P. Brûlet (35). The GFP–AEQ was amplified by PCR with a primer containing a HindIII site at the 5′ end (CCAAAGCTTAGCAAGGGCGAGGAGCTGTTC) and an EcoRI site at the 3′ end (CCGAATTCTTAGGGGACAGCTCCACCG). The mitochondrial GFP–AEQ, termed here mitGFP–AEQ, was obtained by fusing the first 31 amino acids of the human cytochrome c oxidase subunit VIII (22) in frame with the 5′ of the GFP–AEQ in the HindIII site. The low Ca^{2+} affinity mitmutGFP–AEQ was obtained by swapping the EcoRV–EcoRI fragment of the aequorin moiety with the one containing the mutation D119A of the aequorin targeted to the endoplasmic reticulum (ER) previously reported (29, 39). The integrity of the fusions was verified by sequencing. The correct localization of the chimeric proteins was checked by confocal microscopy of the GFP-fluorescence. Both cDNAs were isolated from the original vector by EcoRI digestion and cloned in the EcoRI site of the amplicon vector pHSVpUC (34) to generate the pHSVmitGFP–AEQ and pHSVmitmutGFP–AEQ (Fig.9.1a). The transcriptional unit of the vector contains the immediately early (IE) 4/5 promoter, the mitGFP–AEQ gene, and a polyadenylation signal. The two genetic elements from HSV-1, the ori_s and the HSV packaging sequences, allow replication and packaging of the amplicon, respectively. The prokaryotic sequences contain a bacterial origin of replication and an ampicilin selection marker that allow propagation and amplification in *E. coli*.

2.1.2. Packaging of HSV Amplicons and Viral Titration

pHSVmitGFP–AEQ and pHSVmitmutGFP–AEQ DNAs were packaged into HSV-1 particles using a deletion mutant packaging system as described previously (40). In brief, 3×10^5 2-2 cells were

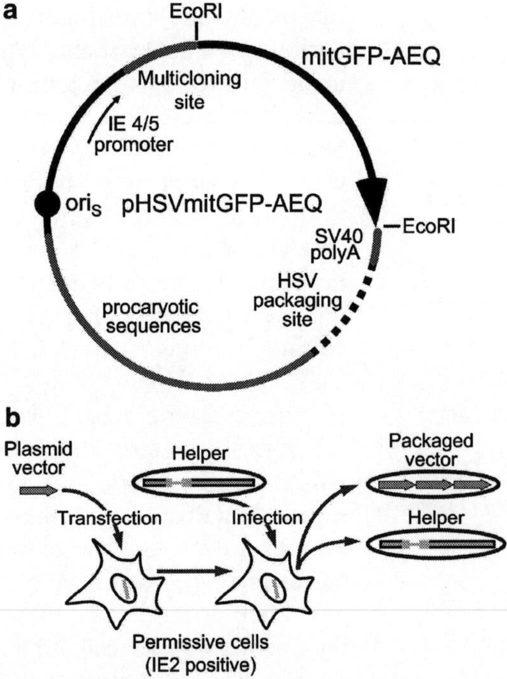

Fig. 9.1. Construction of mitochondrial aequorin amplicons. (**a**) Map of mitochondrial aequorin pHSV-mitGFP–AEQ. The transcriptional unit contains the immediately early (IE) 4/5 promoter, the corresponding mitochondrial GFP–aequorin chimeric gene cloned into the EcoRI site and a polyadenylation signal. The two genetic elements from HSV-1, the ori_s and the HSV packaging sequences, allow replication and packaging of the amplicon. The prokaryotic sequences contain a bacterial origin of replication and an ampicilin selection marker that allow propagation and amplification in *E coli.* (**b**) Overview of the packaging procedure (courtesy of Dr. F. Lim).

seeded on 60 mm-dishes and transfected with 6 μg of amplicon DNA using the lipofectamine procedure according to the manufacturer's protocol (Gibco, BRL, Madrid, Spain). One day later, the medium was replaced with 3 ml DMEM, 5% FBS and cells were infected with 2×10^6 plaque-forming units of 5dll.2 helper virus, which contains a deletion in the IE2 gene of HSV-1 strain KOS. On the following day, the virus was harvested by lysing the cells by three cycles of freeze–thawing and subsequently passaged on fresh 2-2 cells twice to increase both the ratio of vector to helper and the total amount of virus.

Titers were measured in GH_3 pituitary cells. Cells were seeded at 5×10^5 cells/well into poly-L-lysine (0.01 mg/ml) coated 24-well plates to ensure the attachment of the cells to the well surface. Two hours later, the cells were infected with various dilutions of the virus. One day later, the cells were trypsinized to determine the relative GFP fluorescence of the amplicon infected cells and the cytotoxicity by flow cytometry. Titration of the helper virus was made by inmunocytochemistry using an antibody anti-HSV capside primary antibody (1:2,000, DAKO) followed by an alkaline phosphatase-conjugated anti-rabbit secondary antibody (Sigma, Spain). Alkaline phosphatase was visualized using nitroblue tetrazolium and 5-bromo-4-chloro-3-indolyl phosphate according to the manufacturer's instructions (Sigma, Spain). Stained cells were counted by using Image J imaging software. The virus stocks have routinely titers of $0.2–1 \times 10^7$ infectious vector units (ivu) ml^{-1} and are stored at −80°C in 50–100 μl aliquots. Virus particles are thermolabile and should be thawed immediately before being added to the cells and kept on ice during manipulations. One day prior to the Ca^{2+} experiments, the cultures are infected at 0.2–20 multiplicity of infection (MOI).

2.2. Cell Preparation and Culture

2.2.1. Cerebellar Granule Neurons

Groups of two cerebella from 5- to 7-day-old rats were transferred to a Petri dish containing 1 ml of ice-cold HEPES-buffered balanced salt solution (HBSS) and cut into small pieces. The tissue was incubated in 1 ml of dispase (5 mg/ml in HBSS) for 30 min at 37°C and the cells were dispersed by gently passing the pieces through a silanized fire-polished Pasteur pipette. The cells were then washed twice by centrifugation (5 min at 100 × g) and resuspended at 10^6 cells/ml. A drop of 50 μl (5×10^4 cells) was plated on 12 mm-diameter glass coverslips coated with poly-L-lysine (0.01 mg/ml) and cultured in high-glucose Dulbecco Minimal Essential Medium (DMEM, Gibco 41966-029) supplemented with 10% fetal bovine serum (FBS), 5% horse serum, 100 units/ml penicillin and 100 μg/ml streptomycin for 2 days. Then, the culture medium was replaced by Sato´s medium (41) supplemented with 5% horse serum in order to avoid excessive proliferation of glia and culture was continued for 2–4 more days before performing the experiments. For more details see Núñez et al. (42).

2.2.2. Sympathetic Neurons from Adult Mouse Superior Cervical Ganglion

Adult (8–12 weeks) male mice were fed ad libitum and maintained under standard 12 h light/12 h dark photoregime. Groups of four to five animals were killed by cervical dislocation and their superior cervical ganglia (SCG) was rapidly removed under sterile conditions and transferred to a Petri dish containing ice-cold HBSS. Ganglia were incubated for 10 min at 37°C in HBSS containing collagenase (1.6 mg/ml) washed and incubated again in HBSS containing trypsin (1 mg/ml) for 15 min at 37°C. After gentle mechanical disruption with a silanized, fire-polished Pasteur pipette, the cells were suspended in high-glucose DMEM supplemented with 10% FBS washed twice and resuspended in the same culture medium supplemented with 50 ng/ml of NGF (Upstate). Aliquots of the cell suspension (25 μl, 10^4 cells) were plated on poly-L-lysine coated 12 mm-diameter glass coverslips. On average, we obtained the same number of cell-containing coverslips than ganglia dispersed. Cells were allowed to attach for 1 h and then additional 300 μl of culture medium were added to each well and culture was continued for 2 days before performing the experiments. For further details see Martínez et al. (43).

2.2.3. Rat Pituitary Cells

GH_3 cells (ATCC, CCL-82.1) were cultured in RPMI 1640 medium supplemented with 2.5% FBS, 15 % HS, 2 mM L-glutamine, 100 units/ml penicillin and 100 μg/ml streptomycin. The cells were grown in 75 cm^2 flasks at 37°C under a 95% air/5% CO_2 atmosphere. Cells were trypsinized once a week with 0.25% trypsin–EDTA and subcultured at 1:5 dilution. Duplication time was about 48 h. For the experiments, the cells were seeded on 12 mm diameter polylysine-coated coverslips at 3×10^4 cells/coverslip. For more details see Villalobos et al. (44).

2.3. Bioluminescence Imaging

2.3.1. The Set Up

The set up for bioluminescence imaging is made of an inverted microscope equipped with a bottom port-attached camera with photon counting ability and covered by a light-proof box (Fig. 9.2). Aequorin-expressing cells attached on 12 mm-diameter glass coverslips are mounted on a Warner platform that allows superfusion of the cells with medium pre-warmed by a Warner on-line heating system. The platform is placed on the stage of the microscope. The perfusion system is gravity-fed and includes a home-made ten valves controller and a vacuum pump for aspiration. Similar commercial systems are available from Warner (for example, VC-8 eight channel perfusion valve control system). Because bioluminescence emission is very weak, high NA objectives must be used for imaging. The light pathway must be free of unnecessary devices to minimize light loses. We use a Zeiss Axiovert S100 TV equipped with a Zeiss Fluar 40X oil/1.3 NA objective and a bottom port-attached Hamamatsu C2400-35 ICCD intensifier under the control of a Hamamatsu M431 image intensifier controller. Photonic emissions can be observed in real time using an (Argus) image processor and a RGB monitor. This device has been implemented and improved by Shigeru Uchiyama from Hamamatsu Photonics, Hamamatsu City, Japan. The new EM-CCD cameras are being tested with promising results.

The entire microscope must be covered by a light-proof box to avoid any light interference. We use a home-made box designed ad-hoc, but commercial models are also available from Hamamatsu. Alternatively, the whole set up can be located in a dark room and operated remotely from another room. The photon counting camera contains a security device that turns off the intensifier in case of excess of light. This must be taken into account when turning on any device with the camera under operation. The system can be equipped also for epifluorescence with a Xenon excitation lamp and filter wheels and cubes for GFP imaging. An antivibrational table is advisable and in any case, the table under the microscope must have a hole to hold the bottom-port attached camera (Fig. 9.2).

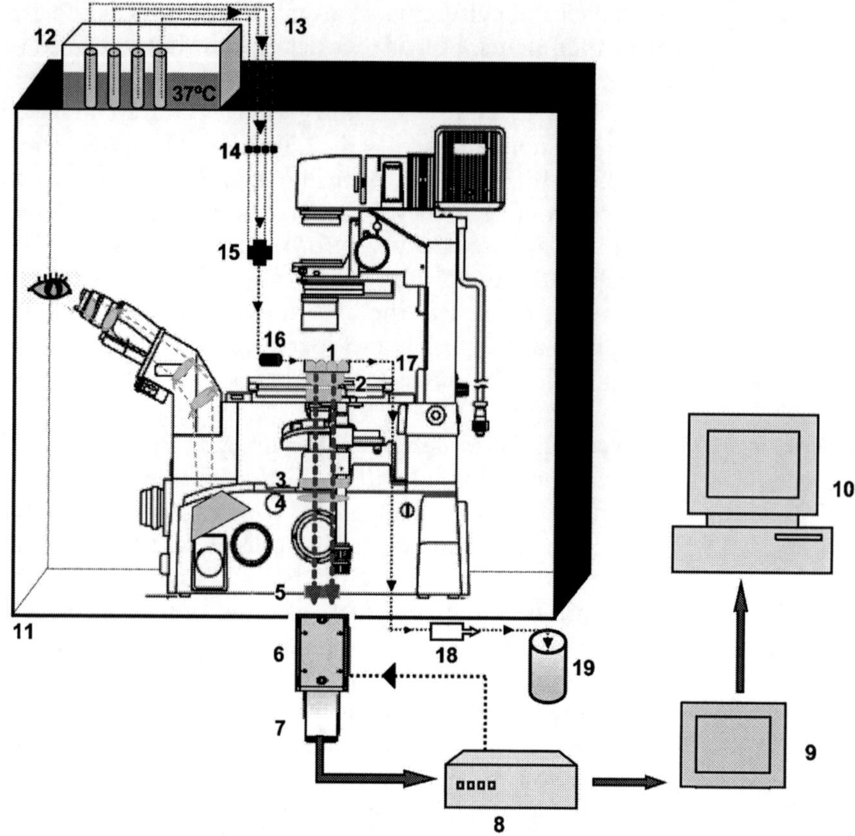

Fig. 9.2. Schematic diagram of the aequorin imaging set up. The system consists of an inverted microscope (Zeiss Axiovert S100 TV) with a thermostated platform (Warner Instruments) for cells (1), prewarmed perfusion device (Warner) and equipped with a Zeiss Fluar 40X, 1.3 NA oil objective (2). Light coming from the living neurons pass through the least possible optic devices including only an analyzer lens (3) and a tube lent (4). Light travels through the microscope's bottom port via adaptor without lent (5) and reaches the Hamamatsu C2400-35 ICCD intensifier (6) controlled by the M431 image intensifier controller, before is captured by a Hamamatsu CCD video camera (7). Signal goes then though an Argus image processor (8) and a JVC RGB monitor (9) to be finally processed by the computer system (10). The entire microscope must be covered by a light-proof box (11) to avoid light interference. Cells are perfused using a gravity-fed system. Solutions are prewarmed at 37°C (12) and driven through silicon tubes (13) by a valve-controlled perfusion system (14). All tubes converge through a manyfold system (15) into one silicon tube that passes through an on-line heating system (16) before getting to the thermostated platform (1). Medium is removed by another silicon tube (17) attached on the platform and aspirated by vacuum pump (18) to the waste bottle (19). The system should be equipped also with an excitation light source, filter wheels and filter cubes for GFP imaging.

2.3.2. Sample and System Preparation

Functional aequorin must be formed just before measurements. For this end, coverslips containing the cells or tissues expressing targeted aequorin are reconstituted with the coelenterazine. Cells are incubated in a HBBS medium containing 1 μM of the selected coelenterazine (Invitrogen) either wild type or n type (see below) for 1–2 h at room temperature in the dark. The stock solution of coelenterazine is dissolved in methanol at a concentration of 200 μM aliquoted in 30 μl portions in Eppendorf tubes in ice and gassed briefly with nitrogen before closing, wrapped with aluminum foil and store at −80 ° C (for up to 6 months). After reconstitution, the coverslips containing the cells are washed and placed in the platform on the stage of the microscope. Cells are then superfused at about 5 ml/min with pre-warmed HBBS. The microscopic field is focused and a region of interest is selected for imaging. In experiments using GFP–aequorin, cells are examined for GFP fluorescence using the FITC filter cube set. This helps selecting a region containing GFP–aequorin expressing cells. In general, because photonic emissions are scattered, it is convenient to select fields in which the aequorin-expressing cells are not too close to each other to avoid overlapping of their photonic emissions. It is convenient to take a bright field image and a GFP-fluorescence image before starting photoluminescence measurements.

Before imaging photonic emissions, any possible light source inside the light-proof box must be turned off including the microscope light and the excitation lamp. In addition, the FITC filter cube must be removed from its position in the light pathway to avoid light loss. Finally, the dark box doors are closed for complete darkness. Other putative light sources such as leads of different operators even if they are outside the dark box must be turned off. Some motorized microscopes have internal leads that produce a light leak to the camera. This is the case, for example, of the Zeiss Axiovert 200. Check for this disturbance by turning off the microscope during measurements. Because turning on and off excitation (Xenon) lamps may be damaging for these devices, a good solution is to use an external excitation lamp located outside the box with an optical fiber as a light guide. A good option that has been tested favorably is the Zeiss X-cite illumination system.

2.3.3. Image Capture

Once all the lights inside the box are turned off, capturing photonic emissions from cells perfused continuously with either control medium or the different stimuli is started. Two different imaging procedures used extensively by the authors shall be explained: (1) capturing photonic emissions via the image processor (the Argus processor), and (2) capturing via a photon counting card (the HPD-LIS system), but many other solutions are possible.

Capturing via Image Processor

In this option, the operator must pre-select a particular time interval for which the system is acquiring photonic emissions before sending the integrated image containing the accumulated photonic emissions to the PC. For this end, the intensifier is manually set to its maximum intensity and the photonic emissions are acquired for the selected time. This process can be monitored in real time in a RGB monitor. After the pre-established integration period for photon counting, the integrated image containing all the photonic emissions produced during this time interval is sent to the PC and the process of accumulation starts all over again until the end of the experiment.

This procedure is very convenient because it allows to follow in real time the effects of treatments, and this permits modifying the protocols during the experiment if required. Another advantage is that processing and analysis of the sequence of images obtained is straightforward and takes a short time. However, this procedure has some drawbacks that must be considered. For example, if the rate of photonic emissions is lower than expected, it might be more convenient to increase the integration periods. Using long integration times may be valid for monitoring luciferase activity reflecting reporter gene expression, but monitoring Ca^{2+} signals requires short times in the range of a few seconds which constrains the choices. Another drawback is that the system spends about 1 s for sending the image of accumulated photonic emissions to the memory and therefore, some photonic emissions are lost during this translocation.

The protocol we generally used is as follows: we typically choose an integration period of 10 s in a microscopic region containing between 5 and 30 individual cells. This is enough to monitor increases in $[Ca^{2+}]_M$ induced by agonists or even the faster $[Ca^{2+}]_M$ transients due to electrical activity in pituitary cells (45). Total counts per region of interest (ROI) may range between 2×10^3 and 2×10^5 for different cells. Background photonic emissions in ROIs of similar size in non-expressing cells amount typically (mean ± S.D.) 1 ± 1 counts per second (cps) per cell (about 2,000 pixels). Some examples of real experiments are shown below.

Once photonic emissions from the cells in response to the test solutions have been recorded, the experiment is finished with perfusion of the cells with a permeabilizing solution containing 0.1 mM digitonin and 10 mM $CaCl_2$. Digitonin releases the remaining aequorin counts, which must be added up to estimate the total photonic emissions, a value required for calibration (see below). Emission of all the residual counts may take up to 2–5 min. The experiment finishes once aequorin photonic emissions cease. The entire sequence of bioluminescence images is stored in the computer together with the bright field and fluorescence images captured before photon counting.

Capturing Photonic Emissions with HPD-LIS

Alternatively to use the image processor, photonic emissions can be captured with a photon counting card (HPD-LIS, developed by Karl Weinbuch, Hamamatsu Photonics, Herrsching, Germany). In this option, specially indicated for fast measurements, the intensifier is under the control of the photon counting card. The procedure allows continuous monitoring of photonic emissions from the start of the experiment to the end. A clear advantage is that, as photonic emissions are not integrated nor images are sent to the PC, there is not any loss of photonic emissions. This is important when we want to monitor rapid changes in $[Ca^{2+}]_M$. A second advantage is that the procedure generates a single file containing all the photonic emissions (instead a sequence of images) with a much smaller size (in terms of memory) and easiness to handle. The encrypted file contains the record of all the photonic emissions and when and where they were produced. The subsequent off the record "deconstruction" of the experiment allows recovering the record of photonic emissions from single pixels along the time and space. For example, after the experiment we can choose to plot integrated images representing $[Ca^{2+}]_M$ every either 10 s, 0.1 s or any other integration period, whatever is more convenient. The analysis of these recordings takes much longer than the straightforward analysis of the first procedure, but all the information is preserved. Another drawback of the HPD-LIS procedure is that it does not allow on line visualization of photonic emissions. If, for example, the first stimulus consumed most of the aequorin luminescence, further responses would hardly be observed as there are no counts left, but this would not be realized until analysis. Depending of the experiment to be carried out or the question to be addressed, the researcher must choose the best acquisition mode.

2.3.4. Calibration and Analysis

At the experiment, the results are coded either as a sequence of images (Argus capture) or an encrypted file containing the information about photonic emissions (HPD-LIS capture). In the last case, the analysis requires a previous step consisting in selecting an integration period for which the photonic emissions will added up. For example, if we have recorded photonic emissions for 15 min, we can use the encrypted file to obtain a sequence of integrated photonic emissions (images) every 10 s. This would produce a series of 90 tif images that are later converted in a sequence of images (NAF file), which is treated as the ones produced by the Argus grabber. When the HPD-LIS procedure is used for fast measurements, the resulting encrypted file can even be converted into a sequence of images at video rate (integrated every 40 ms). For a 15 min recording, this procedure would generate a sequence of 22,500 images. Neither the viewing nor the handling of such an amount of images is simple, so that it is sensible to choose just a small piece of the experiment for this kind of ultra fast analysis.

During the analysis, we want to convert the photonic emissions into $[Ca^{2+}]_M$ values. Starting from the sequence of images, we compute the rate of photonic emissions at each given time for each cell. Then, we need to compute also the total of photonic emissions released by each cell. Finally, we want to transform photonic emissions in $[Ca^{2+}]_M$ values. We have used the Aquacosmos® software from Hamamatsu to define specific regions of interest (ROIs) corresponding to every single cell. For this purpose, we draw the ROIs around each transfected cell using the GFP fluorescence image that was recorded before photon counting as a guide. It is possible that a significant part of photonic emissions exceeds the boundaries of the hot cell. In order to avoid loosing these emissions, an alternative approach is to define the ROIs over an image constructed by adding pixel by pixel all the images of the sequence. The selected ROIs are copied and pasted on every image on the sequence and finally, photonic emissions in every ROI are computed along the time to obtain the luminescence emission value (L) for each cell at each time. A few ROIs are drawn in regions devoided of cells to compute also a background, luminescence. After subtraction of the background luminescence the L(ROI, t) matrix is exported to a worksheet where all the computations are made.

The relation between aequorin luminescence and the Ca^{2+} concentration is given by the following formula derived from the model proposed by Allen et al. (46):

$$[Ca^{2+}]\,(M) = [(L/L_{TOTAL}\lambda)^{1/n} + (L/L_{TOTAL}\lambda)^{1/n} \cdot K_{TR} - 1]/[K_R - (L/L_{TOTAL}\lambda)^{1/n} \cdot K_R],$$

where L is the luminescence emitted at the time of measurement and L_{TOTAL} is the addition of the counts present in the tissue at that time, estimated by adding up all the counts from the time of measurement until the release of all the residual luminescence by lysis at the end of the experiment. For more details see Alvarez and Montero (47) and Alonso et al. (3). The values of the constants depend on the type of aequorin and coelenterazine. Native aequorin and low Ca^{2+} affinity mutated aequorin (39) can be combined with natural coelenterazine or with synthetic coelenterazine n, which decreases affinity for Ca^{2+} (23). Figure 9.3 shows calibration curves for three aequorin–coelenterazine pairs that can be used to cover a wide Ca^{2+} concentration range from 3×10^{-8} to 2×10^{-3} M (7). Values for the different parameters defining the equation are shown in the legend to Fig. 9.3. The native aequorin/coelenterazine pair is adequate for measuring concentrations up to 10^{-5} M (AEQ1 in Fig. 9.3). The pair native aequorin/coelenterazine n (AEQ2) increases the range to near 10^{-4} M and the pair mutated aequorin/coelenterazine n (AEQ3) allows measurements up to the 2×10^{-3} M range (AEQ3).

The dynamic range of aequorins is very wide (Fig. 9.3) and the changes of the luminescence with Ca^{2+} very steep (Hill number larger than 1, see values of n in Fig. 9.3 legend). This steepness sharpens the contrast of high and low $[Ca^{2+}]$ microdomains (3). Buffering of calcium by aequorin is small and this may be an

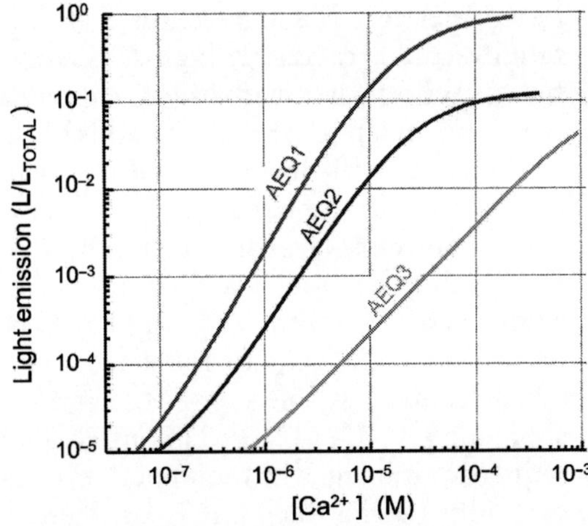

Fig. 9.3. Calibration curves of different aequorin systems. The three aequorin (AEQ) systems shown correspond to: native aequorin combined with either natural coelenterazine (AEQ1) or synthetic coelenterazine n (AEQ2) or mutated low Ca^{2+} affinity aequorin with coelentarazine n (AEQ3) (7). The function used to relate Ca^{2+} and photoluminescence (expressed as fraction of the total counts emitted at every instant, L/L_{TOTAL}), was: $[Ca^{2+}](\text{in } M) = (R + (R \times K_{TR}) - 1) / (K_R - (R \times K_R))$, where $R = (L/(L_{MAX} \cdot \lambda))^{1/n}$, using the values shown below for the constants:

Condition	K_R	K_{TR}	n	λ
AEQ1: AEQwt + COEL wt	4.8×10^7	601	2.3	1
AEQ2: AEQwt + COEL n	2.2×10^7	348	2.0	0.13
AEQ3: AEQmut + COEL n	8.5×10^7	157×10^3	1.2	0.14

This figure was redrawn with permission from Alonso et al. (3). Copyright Elsevier (2006). For more details see Montero et al. (7) and Alvarez and Montero (47).

advantage over other Ca^{2+} probes. In addition, the dependence on pH is scarce between 6.5 and 8, much smaller than for other Ca^{2+} probes (48). A serious problem that is seldom faced (both, for AEQ and for the other Ca^{2+} probes) is the dependence on Mg^{2+}. When $[Mg^{2+}]$ decreases, the apparent affinity for Ca^{2+} increases in parallel (48). The calibrations given here were derived for a Mg^{2+} concentration of 1 mM. The cytosolic Mg^{2+} concentration is assumed to be quite stable, but it may change under some conditions (49). Even more serious could be the uncertainties on the Mg^{2+} concentration in organelles and possible changes with different physiological processes.

3. Application Examples

3.1. Imaging of Mitochondrial Ca^{2+} Oscillations

Imaging of aequorin bioluminescence is difficult because of the very low light output. Using high sensitivity cameras, measurements in synchronized groups of cells were achieved (50). However, monitorization of single-cell oscillations of mitochondrial Ca^{2+} requires the exceedingly high sensitivity provided by the combined action of virus-driven expression (51) and photon counting imaging (38). Oscillations of $[Ca^{2+}]_M$ were first imaged at the single cell level in anterior pituitary (AP) cells using mitochondria-targeted aequorin as the Ca^{2+} probe (38). These oscillations were ultimately driven by spontaneous electric activity, which generated $[Ca^{2+}]_C$ oscillations. These $[Ca^{2+}]_C$ oscillations were followed and amplified by mitochondria. Ca^{2+} oscillations in AP cells are regulated by hypothalamic hormones, which control in this way AP secretion. Figure 9.4 illustrates spontaneous $[Ca^{2+}]_M$ oscillations in GH_3 pituitary cells. The sequence of images in the upper two rows shows the photonic emissions during one $[Ca^{2+}]_M$ oscillation (in pseudocolor scale, superimposed to the gray bright field image of the cell; each image corresponds to an integration period of 10 s). The trace in the lower part of Fig. 9.4 shows the oscillatory pattern during a 10 min period. The oscillations of $[Ca^{2+}]_M$ were generally in the submicromolar level (Fig. 9.4), but these changes are enough to activate mitochondrial dehydrogenases (4) and the respiratory rate (11, 12, 38). Hypothalamic hormones greatly stimulated $[Ca^{2+}]_M$ oscillations (38). Similar $[Ca^{2+}]_M$ oscillations have also been reported in pancreatic β cells in response to high glucose (52) or in adrenal chromaffin cells in response to veratridine (unpublished results; see Ref. (53)).

As remarked before, physiological mitochondrial oscillations are usually in the submicromolar or low micromolar range. The native aequorin reconstituted with the natural coelenterazine is therefore adequate for such measurements. Intense and/or prolonged stimulation of Ca^{2+} entry through voltage-gated Ca^{2+} channels can produce much larger mitochondrial load. For example, depolarization of adrenal chromaffin cells with high K^+ for 15–30 s can promote an increase of $[Ca^{2+}]_M$ as high as 300–400 μM (7). The uptake is very fast (V_{MAX} over 6 mmol/l cells/s) and it is quickly cleared through the mitochondrial Ca^{2+}/Na^+ and Ca^{2+}/H^+ exchanger with no apparent mitochondrial damage (10). If the mitochondrial Ca^{2+} overload is maintained for several minutes, then programmed cell death is triggered (54). Native aequorin is inadequate for measurements of increases of $[Ca^{2+}]_M$ above 5–10 μM, as it is massively consumed in less than 1 min (3) and measurements become meaningless. The combination of the mutated low Ca^{2+} affinity aequorin and coelenterazine n should be used instead (see Fig. 9.3). Even with this combination substantial consumption of aequorin would happen in 5–10 min at Ca^{2+} concentrations near 1 mM. Owning to these reasons, aequorin is not adequate for measuring large Ca^{2+} overloads in long-term experiments.

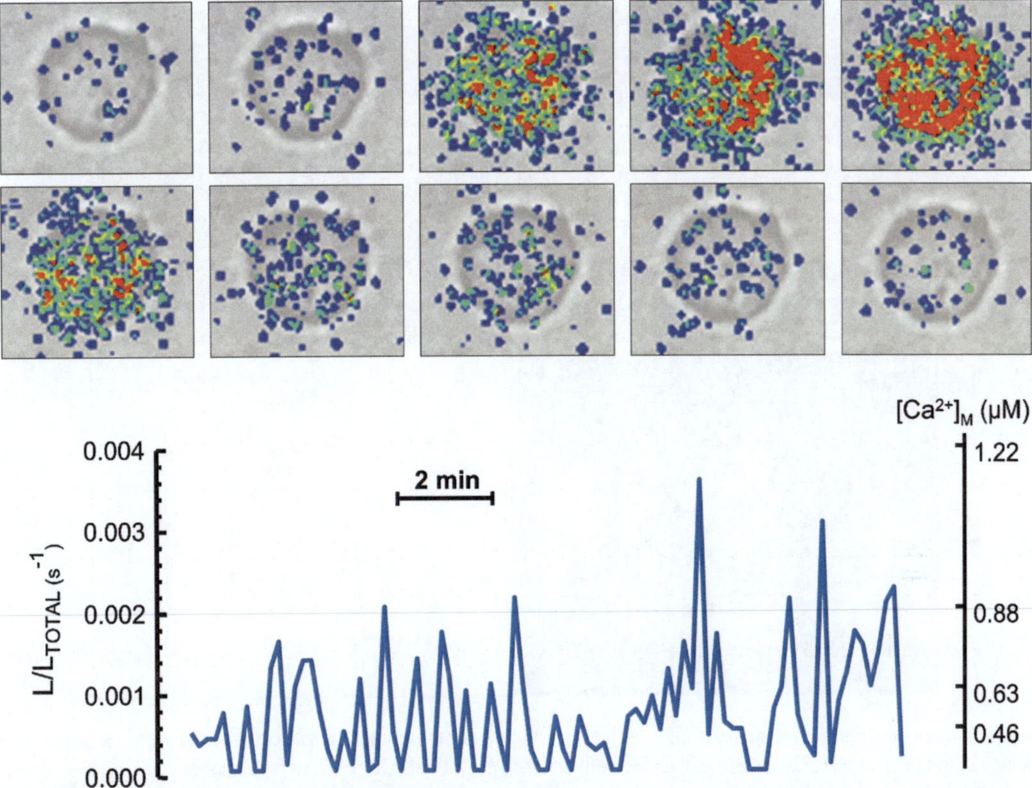

Fig. 9.4. Mitochondrial Ca^{2+} oscillations in GH_3 pituitary cells. The trace represents the spontaneous oscillations of $[Ca^{2+}]_M$ in a single cell expressing mitochondria-targeted aequorin. The images corresponding to one single oscillation are shown on top. Photonic emissions, coded in pseudocolor, have been superimposed to the bright field image taken at the beginning of the experiment. Time sequence goes from left to right and from top to bottom. The interval between images is 10 s. The size of each image box is 10 × 10 μm. Unpublished results by Villalobos C, Nuñez L, Chamero P, Alonso MT, and García-Sancho J. For more details see Villalobos et al. (38).

3.2. Aequorin Reveals Functional Heterogeneity of Subcellular Compartments

The specificity for subcellular distribution of aequorin relies on its targeting sequence rather than on the optical resolution of the imaging device. This specificity is superior to the Ca^{2+} dyes, whose distribution rests on physicochemical differences among different cellular compartments that are not absolute. The optical spatial resolution provided by aequorin is on the contrary very poor because of the low light output. Therefore, aequorin is not generally adequate to evidence spatial differences within the single cell surrounding at least in small and spherical cells. Figure 9.5 shows an exception, which is possible because of the large size and long ramifications of primary sympathetic neurons. Mitochondria-targeted aequorin was expressed using the viral vector and then the responses of $[Ca^{2+}]_M$ in the soma and the neurites were compared. Two different stimuli were tested, depolarization by high K^+, producing Ca^{2+} entry through voltage-gated Ca^{2+} channels and caffeine releasing Ca^{2+} from the intracellular Ca^{2+} stores by

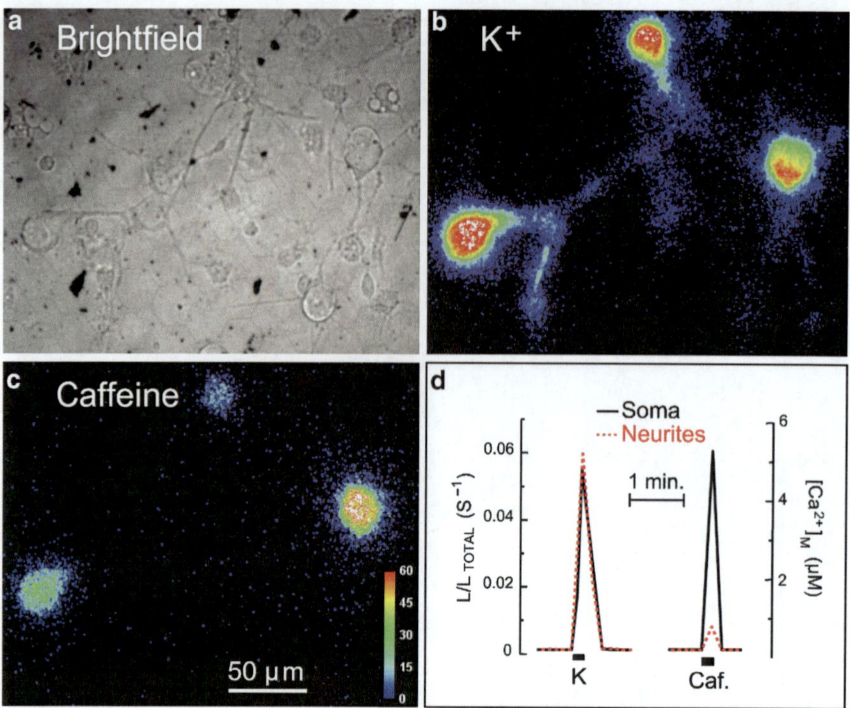

Fig. 9.5. Mitochondrial Ca^{2+} responses differ in the cell body and the neurites of sympathetic neurons. (**a**) Bright field image of the field studied. Cells were stimulated with either high (150 mM) K^+ (**b**) or with caffeine 50 mM (**c**). $[Ca^{2+}]_M$ was measured in mouse sympathetic neurons infected with mitochondrial aequorin. (**d**) Compares the mean responses of body (*continuous line*) and dendrites (*dotted line*) of ten neurons.

activation of ryanodine receptors. Surprisingly, whereas high-K^+ produced a comparable increase of $[Ca^{2+}]_M$ in soma and neurites (Fig. 9.5b), caffeine increased $[Ca^{2+}]_M$ in the soma, but had little or no effect in neurites (Fig 9.5c). These results were not due to the absence or low density of either ER or mitochondria in neurites as the distribution of these organelles evidenced by mitotracker and ER-tracker dyes was similar to the one found in the soma (55). Panel D in Fig 9.5 compares the average responses in ten sympathetic neurons. The $[Ca^{2+}]_M$ increase produced by high-K^+ and caffeine was similar in the soma, but very different in the neurites.

Topological information provided by aequorin on subcellular structure is not very accurate, but it contains subtle clues about history of $[Ca^{2+}]$ changes in the cell compartment under study, which can be learned from aequorin consumption. This information can reveal functional inhomogeneities of a given subcellular pool. For example, we find that depolarization with high K^+ of adrenal chromaffin cells produces quick consumption of about 50% of the mitochondrial aequorin pool (These experiments are best performed with native aequorin). Further stimulation with high K^+ does promote significant further consumption of mitochondrial AEQ even though fura-2 reveals the same Ca^{2+} entry as before. In contrast, membrane permeabilization with digitonin makes accessible the remaining 50% of the aequorin (7). We interpret these results in terms of two different mitochondrial pools, one easily accessible to the Ca^{2+} entering through the plasma membrane Ca^{2+} channels and the other one sensing much smaller $[Ca^{2+}]_C$ changes. These two pools, also supported by other evidences, do probably correspond to mitochondria located either just below the plasma membrane or at the cell core (7, 10). Aequorin consumption in chromaffin cells also suggests partial overlap of the mitochondrial pools sensing both Ca^{2+} entry through the plasma membrane and caffeine-induced Ca^{2+} release from the intracellular calcium stores. This topology of plasma membrane Ca^{2+} channels, ryanodine receptors and mitochondria would optimize amplification of subplasmalemmal Ca^{2+} signals by Ca^{2+}-induced Ca^{2+} release while dampening the propagation of the Ca^{2+} wave towards the cell core (7, 10). The organization of mitochondria in two different pools seems to apply also to other excitable cells (52, 55).

3.3. Improving Aequorins by Fusion to Green Fluorescent Protein

Another inconvenient for aequorin imaging is that the cells expressing luminescence cannot be identified until the end of the experiment, which, when combined with low rate of expression decreases very much efficiency of the experimental work. It would be very convenient then to have a tracer allowing identification under the microscope of the cells expressing aequorin before starting luminescence measurements. A fusion protein of GFP with aequorin (GFP–AEQ) was prepared and characterized by P. Brulêt and coworkers in 2000 (35). GFP–AEQ was not only easily traced by fluorescence microscopy, but also gave a higher luminescence yield, which facilitates imaging (56). In addition, the emission of photoluminescence is shifted from blue towards green by bioluminescence resonance energy transfer (BRET) between aequorin and GFP. Figure 9.6 shows $[Ca^{2+}]_M$ measurements in cerebellar granule neurons using GFP–AEQ. Fluorescence inspection before luminescence measurements allows selection of neurons expressing aequorin beforehand (Fig. 9.6 left). Mitochondrial calcium load has been proposed to play a role in the excytotoxic injury induced by glutamate in neurons. Then, we followed the changes in $[Ca^{2+}]_M$ induced by stimulation of ionotropic NMDA glutamate receptors. As expected, Ca^{2+} entry was quickly followed by an increase of $[Ca^{2+}]_M$ (Fig. 9.6 right). The increase of $[Ca^{2+}]_M$ was easily reverted if the treatment with NMDA was suspended before 10 min (results not shown), which is consistent with the reversibility of short treatments with excytotoxic amino acids (57).

Fusion of aequorin with red fluorescent protein (RFP) has also been achieved recently (36, 37). Fusion to RFP (RFP–AEQ) also stabilized aequorin expression and produced a red shift in

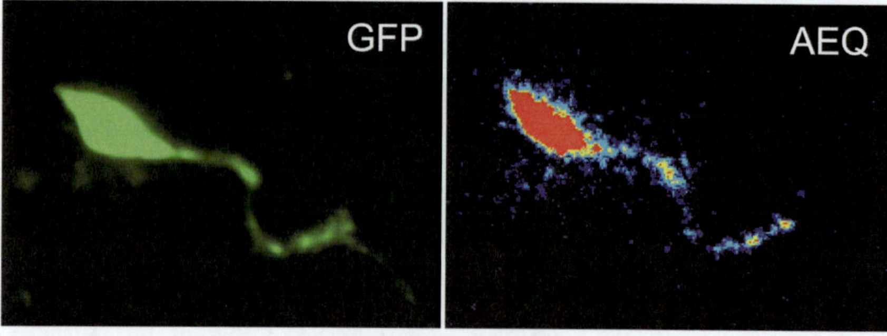

Fig. 9.6. Glutamate receptor stimulation induces mitochondrial Ca^{2+} uptake in cerebellar granule neurons. The GFP–aequorin fusion protein was used for these experiments. (**Left**) GFP fluorescence; (**Right**) Stimulation with 100 μM NMDA + 10 μM glycine on $[Ca^{2+}]_M$ for 1 min. The luminescence image shown was taken during the last 10 s of the stimulation period. It is pseudocolor-coded similarly to Fig. 9.5b.

luminescence. This increased tissue transmission of the emitted light and therefore improved in vivo measurements of luminescence (36, 58). On the other hand, co-expression of GFP–AEQ and RFP–AEQ targeted to different organelles permits simultaneous and independent monitoring of $[Ca^{2+}]$ in different subcellular domains of the same cell (37). Thus, photon counting imaging of the new aequorins may provide new insights as to the handling of Ca^{2+} by mitochondria in health and disease.

Acknowledgments

Financial support from the Spanish Ministerio de Educación y Ciencia (grants BFU2007-60157, BFU2005-02078 and BFU2006-05202), Instituto de Salud Carlos III (PI07/0766) and the Junta de Castilla y León (VA-088/A06) is gratefully acknowledged.

References

1. Alvarez J, Montero M, Garcia-Sancho J (1999) Subcellular Ca^{2+} dynamics. News Physiol Sci 14:161–168
2. Petersen OH, Tepikin AV (2008) Polarized calcium signaling in exocrine gland cells. Annu Rev Physiol 70:273–299
3. Alonso MT, Villalobos C, Chamero P, Alvarez J, Garcia-Sancho J (2006) Calcium microdomains in mitochondria and nucleus. Cell Calcium 40:513–525
4. Gunter TE, Pfeiffer DR (1990) Mechanisms by which mitochondria transport calcium. Am J Physiol 258:C755–C786
5. Csordas G, Thomas AP, Hajnoczky G (1999) Quasi-synaptic calcium signal transmission between endoplasmic reticulum and mitochondria. EMBO J 18:96–108
6. Hajnoczky G, Csordas G, Madesh M, Pacher P (2000) The machinery of local Ca^{2+} signalling between sarco-endoplasmic reticulum and mitochondria. J Physiol (London) 529:69–81
7. Montero M, Alonso MT, Carnicero E, Cuchillo-Ibáñez I, Albillos A, García AG, García-Sancho J, Alvarez J (2000) Chromaffin-cell stimulation triggers fast millimolar mitochondrial Ca^{2+} transients that modulate secretion. Nat Cell Biol 2:57–61

8. Rizzuto R, Brini M, Murgia M, Pozzan T (1993) Microdomains with high Ca^{2+} close to IP3-sensitive channels that are sensed by neighboring mitochondria. Science 262: 744–747
9. Rizzuto R, Pinton P, Carrington W, Faym FS, Fogarty KE, Lifshitz LM, Tuft RA, Pozzan T (1998) Close contacts with the endoplasmic reticulum as determinants of mitochondrial Ca^{2+} responses. Science 280:1763–1766
10. Villalobos C, Nuñez L, Montero M, García AG, Alonso MT, Chamero P, Alvarez J, García-Sancho J (2002) Redistribution of Ca^{2+} among cytosol and organella during stimulation of bovine chromaffin cells. FASEB J 16:343–353
11. Hajnoczky G, Robb-Gaspers LD, Seitz MB, Thomas AP (1995) Decoding of cytosolic calcium oscillations in the mitochondria. Cell 82:415–424
12. Pralong WF, Spat A, Wollheim CB (1994) Dynamic pacing of cell metabolism by intracellular Ca^{2+} transients. J Biol Chem 269:27310–27314
13. Jouaville LS, Pinton P, Bastianutto C, Rutter GA, Rizzuto R (1999) Regulation of mitochondrial ATP synthesis by calcium: evidence for a long-term metabolic priming. Proc Natl Acad Sci USA 96:13807–13812
14. Maechler P, Wollheim CB (2000) Mitochondrial signals in glucose-stimulated insulin secretion in the beta cell. J Physiol (London) 529:49–56
15. Szabadkai G, Duchen MR (2008) Mitochondria: the hub of cellular Ca^{2+} signaling. Physiology (Bethesda) 23:84–94
16. Voronina S, Sukhomlin T, Johnson PR, Erdemli G, Petersen OH, Tepikin A (2002) Correlation of NADH and Ca^{2+} signals in mouse pancreatic acinar cells. J Physiol (London) 539:41–52
17. Gunter TE, Restrepo D, Gunter KK (1988) Conversion of esterified fura-2 and indo-1 to Ca^{2+}-sensitive forms by mitochondria. Am J Physiol 255:C304–C310
18. Giepmans BN, Adams SR, Ellisman MH, Tsien RY (2006) The fluorescent toolbox for assessing protein location and function. Science 312:217–224
19. Miyawaki A, Llopis J, Heim R, McCaffery JM, Adams JA, Ikura M, Tsien RY (1997) Fluorescent indicators for Ca^{2+} based on green fluorescent proteins and calmodulin. Nature 388:882–887
20. Nagai T, Sawano A, Park ES, Miyawaki A (2001) Circularly permuted green fluorescent proteins engineered to sense Ca^{2+}. Proc Natl Acad Sci USA 98:3197–3202
21. Palmer AE, Tsien RY (2006) Measuring calcium signaling using genetically targetable fluorescent indicators. Nat Protoc 1:1057–1065
22. Rizzuto R, Simpson AW, Brini M, Pozzan T (1992) Rapid changes of mitochondrial Ca^{2+} revealed by specifically targeted recombinant aequorin. Nature 358:325–327
23. Shimomura O, Musicki B, Kishi Y, Inouye S (1993) Light-emitting properties of recombinant semi-synthetic aequorins and recombinant fluorescein-conjugated aequorin for measuring cellular calcium. Cell Calcium 14:373–378
24. Robert V, Gurlini P, Tosello V, Nagai T, Miyawaki A, Di Lisa F, Pozzan T (2001) (2001) Beat-to-beat oscillations of mitochondrial [Ca^{2+}] in cardiac cells. EMBO J 20:4998–5007
25. Shimomura O, Johnson FH, Saiga Y (1963) Microdetermination of calcium by aequorin luminescence. Science 140:1339–1340
26. Inouye S, Noguchi M, Sakaki Y, Takagi Y, Miyata T, Iwanaga S, Miyata T, Tsuji FI (1985) Cloning and sequence analysis of cDNA for the luminescent protein aequorin. Proc Natl Acad Sci USA 82:3154–3158
27. Prasher D, McCann RO, Cormier MJ (1985) Cloning and expression of the cDNA coding for aequorin, a bioluminescent calcium-binding protein. Biochem Biophys Res Commun 126:1259–1268
28. Knight MR, Campbell AK, Smith SM, Trewavas AJ (1991) Recombinant aequorin as a probe for cytosolic free Ca^{2+} in Escherichia coli. FEBS Lett 282:405–4088
29. Kendall JM, Dormer RL, Campbell AK (1992) Targeting aequorin to the endoplasmic reticulum of living cells. Biochem Biophys Res Commun 189:1008–1016
30. Alonso MT, Barrero MJ, Michelena P, Carnicero E, Cuchillo I, García AG, García-Sancho J, Montero M, Alvarez J (1999) Ca^{2+}-induced Ca^{2+} release in chromaffin cells seen from inside the ER with targeted aequorin. J Cell Biol 144:241–254
31. Barrero MJ, Montero M, Alvarez J (1997) Dynamics of [Ca^{2+}] in the endoplasmic reticulum and cytoplasm of intact HeLa cells. A comparative study. J Biol Chem 272: 27694–27699
32. Kendall JM, Sala-Newby G, Ghalaut V, Dormer RL, Campbell AK (1992) Engineering the Ca^{2+}-activated photoprotein aequorin with reduced affinity for calcium. Biochem Biophys Res Commun 187:1091–1097
33. Kendall JM, Sala-Newby G, Ghalaut V, Dormer RL, Campbell AK (1992) Engineering aequorin to measure Ca^{2+} in defined compartments of living cells. Biochem Soc Trans 20:144S

34. Geller AI, Breakfield XO (1988) Defective HSV-1 vector expresses *Escherichia coli* β-galactosidase in cultured peripheral neurons. Science 241:1667–1669
35. Baubet V, Le Mouellic H, Campbell AK, Lucas-Meunier E, Fossier P, Brulet P (2000) Chimeric green fluorescent protein-aequorin as bioluminescent Ca^{2+} reporters at the single-cell level. Proc Natl Acad Sci USA 97:7260–7265
36. Curie T, Rogers KL, Colasante C, Brulet P (2007) Red-shifted aequorin-based bioluminescent reporters for in vivo imaging of Ca^{2+} signaling. Mol Imaging 6:30–42
37. Manjarrés IM, Chamero P, Domingo B, Molina F, Llopis J, Alonso MT, García-Sancho J (2008) Red and green aequorins for simultaneous monitoring of Ca^{2+} signals from two different organelles. Pflugers Arch 455:961–970
38. Villalobos C, Núñez L, Chamero P, Alonso MT, García-Sancho J (2001) Mitochondrial $[Ca^{2+}]$ oscillations driven by local high $[Ca^{2+}]$ domains generated by spontaneous electric activity. J Biol Chem 276:40293–40297
39. Montero M, Brini M, Marsault R, Alvarez J, Sitia R, Pozzan T, Rizzuto R (1995) Monitoring dynamic changes in free Ca^{2+} concentration in the endoplasmic reticulum of intact cells. EMBO J 14:5467–5475
40. Neve R, Lim F (2001) Overview of gene delivery into cells using HSV-1-based vectors. Curr Protoc Neurosci 4, Unit 4.12
41. Bottenstein JE, Sato GH (1979) Growth of a rat neuroblastoma cell line in serum-free supplemented medium. Proc Natl Acad Sci USA 76:514–517
42. Núñez L, Sánchez A, Fonteriz RI, García-Sancho J (1996) Mechanisms for synchronous calcium oscillations in cultured rat cerebellar neurons. Eur J Neurosci 8:192–201
43. Martínez JA, Lamas JA, Gallego R (2002) Calcium current components in intact and dissociated adult mouse sympathetic neurons. Brain Res 951:227–236
44. Villalobos C, Nuñez L, Frawley LS, García-Sancho J, Sánchez A (1997) Multiresponsiveness of single anterior pituitary cells to hypothalamic-releasing hormones: a cellular basis for paradoxical secretion. Proc Natl Acad Sci USA 94:14132–14137
45. Schlegel W, Winiger BP, Mollard P, Vacher P, Wuarin F, Zahnd GR, Wollheim CB, Dufy B (1987) Oscillations of cytosolic Ca^{2+} in pituitary cells due to action potentials. Nature 329:719–721
46. Allen DG, Blinks JR, Prendergast FG (1977) Aequorin luminescence: relation of light emission to calcium concentration a calcium-independent component. Science 195:996–998
47. Alvarez J, Montero M (2002) Measuring $[Ca^{2+}]$ in the endoplasmic reticulum with aequorin. Cell Calcium 32:251–260
48. Moisescu DG, Ashley CC (1977) The effect of physiologically occurring cations upon aequorin light emission. Determination of the binding constants. Biochim Biophys Acta 460:189–205
49. Szanda G, Rajki A, Gallego-Sandín S, Garcia-Sancho J, Spät A (2009) Effect of cytosolic Mg^{2+} on mitochondrial Ca^{2+} signaling. Pflugers Arch 457(4):941–954
50. Rutter GA, Burnett P, Rizzuto R, Brini M, Murgia M, Pozzan T, Tavaré JM, Denton RM (1996) Subcellular imaging of intramitochondrial Ca^{2+} with recombinant targeted aequorin: significance for the regulation of pyruvate dehydrogenase activity. Proc Natl Acad Sci USA 93:5489–5494
51. Alonso MT, Barrero MJ, Carnicero E, Montero M, Garcia-Sancho J, Alvarez J (1998) Functional measurements of $[Ca^{2+}]$ in the endoplasmic reticulum using a herpes virus to deliver targeted aequorin. Cell Calcium 24:87–96
52. Quesada I, Villalobos C, Nunez L, Chamero P, Alonso MT, Nadal A, Garcia-Sancho J (2008) Glucose induces synchronous mitochondrial calcium oscillations in intact pancreatic islets. Cell Calcium 43:39–47
53. Lopez MG, Garcia AG, Artalejo AR, Neher E, Garcia-Sancho J (1995) Veratridine induces oscillations of cytosolic calcium and membrane potential in bovine chromaffin cells. J Physiol (London) 482:15–27
54. Jambrina E, Alonso R, Alcalde M, del Carmen Rodríguez M, Serrano A, Martínez AC, García-Sancho J, Izquierdo M (2003) Calcium influx through receptor-operated channel induces mitochondria-triggered paraptotic cell death. J Biol Chem 278:14134–14145
55. Núñez L, Senovilla L, Sanz-Blasco S, Chamero P, Alonso MT, Villalobos C, García-Sancho J (2007) Bioluminescence imaging of mitochondrial Ca^{2+} dynamics in soma and neurites of individual adult mouse sympathetic neurons. J Physiol (London) 580:385–395
56. Rogers KL, Stinnakre J, Agulhon C, Jublot D, Shorte SL, Kremer EJ, Brûlet P (2005) Visualization of local Ca^{2+} dynamics with genetically encoded bioluminescent reporters. Eur J Neurosci 21:597–610
57. Greenwood SM, Connolly CN (2007) Dendritic and mitochondrial changes during glutamate excitotoxicity. Neuropharmacology 53:891–898
58. Rogers KL, Picaud S, Roncali E, Boisgard R, Colasante C, Stinnakre J, Tavitian B, Brûlet P (2007) Non-invasive in vivo imaging of calcium signaling in mice. PLoS One 2:e974

Chapter 10

Ca^{2+} Imaging of Dendrites and Spines

Knut Holthoff

Abstract

The intracellular calcium concentration is one key parameter triggering numerous intracellular signalling pathways in neuronal cells. The development of optical techniques like fast confocal or 2-photon microscopy has made it possible to measure calcium dynamics even in sub-cellular compartments like dendrites and dendritic spines at high temporal and spatial resolution. This chapter provides experimental and technical details for different imaging techniques appropriate for calcium measurements in sub-cellular compartments, discusses specific advantages and limitations and calls attention to possible pitfalls.

Key words: 2-photon microscopy, Calcium imaging, Confocal, Dendrite, Dendritic spine

1. Introduction

The change in intracellular Ca^{2+}-concentration is one key parameter initiating numerous intracellular signalling cascades in neuronal cells. Therefore, it has been of pivotal interest to identify Ca^{2+}-signalling mechanisms in different sub-cellular compartments like somata, dendrites or dendritic spines. The invention of fluorescent indicator dyes (1) was the basis of the challenging approach to use optical means to investigate Ca^{2+} signalling mechanisms at high resolution in space and time. During the last decades, several technical improvements extended the range of possible applications of optical recording techniques. Firstly, confocal microscopy with 1-photon or 2-photon excitation (2) made it possible to image fluorescence at the fractional limit even in strongly scattering preparations in vitro and in vivo. Secondly, the still ongoing development of a wide range of fluorescent probes with different properties in affinity, sensitivity or spatial distribution

A. Verkhratsky and Ole H. Petersen (eds.), *Calcium Measurement Methods*, Neuromethods, vol. 43,
DOI 10.1007/978-1-60761-476-0_10, © Humana Press, a part of Springer Science+Business Media, LLC 2010

has allowed a custom-made experimental design. Thirdly, new scanner technologies improved the time resolution and together with better detection sensors increased the sensitivity of fluorescence imaging techniques leading to real-time applications for 1-photon and 2-photon excitation.

This chapter will provide experimental and technical specifications of different optical techniques for the recording of Ca^{2+}-concentration changes in sub-cellular neuronal compartments like dendrites and dendritic spines. It will present specific advantages and limitations of the different methods and will call attention to possible pitfalls.

2. Methods

2.1. Fluorescent Indicator Dyes

The selection of the appropriate fluorescent indicator dye for the determination of the intracellular calcium concentration changes is one of the first decisions to make, depends on several parameters and often is a trade-off. First of all, each dye is characterized by its specific excitation and emission spectrum. For example, one of the most popular indicator dye families excited in the near ultra-violet (UV) is the group of Fura dyes. They belong to the so-called ratiometric dyes and allow the determination of absolute Ca^{2+}-concentrations by analyzing the ratio of the fluorescence signals stimulated at two different excitation wavelengths in the near UV. These dyes are often not suited for commercially available laser-based confocal microscopes because these imaging systems provide only a limited number of excitation wavelengths and in most cases do not allow excitation in the UV. In this case e.g. arc-lamp-based system and with limitations 2-photon microscopes are advantageous, because they provide flexible excitation over a wide range of wavelengths. Therefore, already the specifications of the imaging system in use can limit the available indicator dyes.

One of the most important properties of an indicator dye is its affinity to the ion of interest. The affinity (specified by the dissociation constant K_d) is defined as the ion concentration at which half of the indicator dye is bound to its specific ion. In case of calcium ions, indicator dyes with affinities between hundreds of nM, called high affinity dyes, and affinities from several up to hundreds of μM, called low affinity dyes, are available. Why is the affinity so important for the choice of the appropriate indicator dye? The relation between the fluorescence intensity of the indicator dye and the concentration of the ion of interest is typically a sigmoid curve. It is obvious that there is only a small concentration range around the K_d of the indicator dye, in which the change in fluorescence is linearly correlated to the concentration change of the ion of interest. For example, if the concentration of the ion of interest increases far beyond the K_d

of the indicator dye, the concentration change will be underestimated because the indicator dye runs into saturation. On the other hand, low affinity indicator dyes cannot detect concentration changes which happen far below their K_d. It is important to note that not only the amplitude of the expected calcium transients but also the level of intracellular calcium concentration at rest needs to be considered in this context. If the resting calcium concentration is already far above the K_d of the indicator dye no additional fluorescence change can be expected upon calcium concentration transients because the dye is already saturated. Taken together, the indicator dye should be chosen depending on the resting calcium concentration and depending on the expected amplitude of the monitored concentration transient in a way that its K_d is as high as possible to prevent dye saturation and as low as necessary to detect the smallest transient of interest.

Another important property of an indicator dye is its dynamic range defined as the maximal relative change in fluorescence upon subsequent exposure to zero and saturating concentrations of the ion of interest. The larger the dynamic range of the indicator the better the potential signal-to-noise ratio of the resulting optical signals. But as always, there is a trade-off for indicator dyes with a very large dynamic range. These dyes (e.g. Fluo dyes) tend to be very dim at low ion concentrations making it difficult e.g. for calcium indicators to identify the stained structure at resting Ca^{2+} concentration. A second complication of these dyes is that the accurate measurement of the relative change in fluorescence upon a change in ion concentration (see also 2.2) can easily be falsified by the wrong determination of the baseline fluorescence. To circumvent this problem, Svoboda and colleagues invented a new protocol by introducing a second dye of different colour which is not sensitive to changes in the ion concentration of interest (3). The fluorescence level of this ion concentration-insensitive dye is used to normalize the fluorescence change of the indicator dye by analyzing the ratio between both fluorescence intensities (4).

Once an indicator dye is chosen, the appropriate intracellular concentration of the indicator during the experiment needs to be defined. It is important to note that on top of a potentially present endogenous buffer system of the ion of interest any indicator dye acts as an additional so-called exogenous buffer system. The addition of this exogenous buffer strength leads to a decrease in the measured amplitude of a concentration transient and slows down its kinetics (5). The strength of the exogenous buffer and thereby its impact on the measured quantity is dependent on to the affinity of the indicator dye and its concentration. The differential buffer-capacity κ_B reflects the buffer strength at a given free calcium concentration and can be expressed as the rate in change of buffer-bound calcium concentration with respect to the free calcium (6):

$$\kappa_B = \frac{\partial[\mathrm{BCa}]}{\partial[\mathrm{Ca}^{2+}]} = \frac{K_d \cdot [\mathrm{B}]}{\left(K_d + [\mathrm{Ca}^{2+}]\right)^2}$$

Because the impact of the indicator dye on the measured quantity is supposed to be as little as possible, the concentration of the indicator dye and its affinity should be chosen as low as possible. For the measurement of synaptically induced Ca^{2+} transients in dendrites and dendritic spines, we recently used the low affinity dyes Oregon Green BAPTA-6F and Oregon Green BAPTA-5N (Invitrogen, Carlsbad, USA) at concentrations of 100–200 μM. In cortical pyramidal neurons and at resting calcium concentrations of 50 nm (7), the indicator dye Oregon Green BAPTA-5N adds under these conditions an exogenous buffer capacity κ_B between 2.5 and 5 to the endogenous buffer capacity κ_S determined to be between 100 and 200 in main apical dendrites of these cells (5). Under these conditions, the impact of the exogenous buffer is negligible and Ca^{2+} transients show large amplitudes and fast kinetics, most likely reflecting almost physiological conditions. In comparison, high affinity dyes like e.g. Oregon Green BAPTA-1 at the same concentration would add a more than 60 times higher exogenous buffer capacity and would therefore strongly reduce the amplitude and would slow down the kinetics of the Ca^{2+}-concentration gradients.

2.2. Cell Loading

In many cases, fluorometric Ca^{2+} measurements in dendrites and dendritic spines are combined with somatic or dendritic patch-clamp recordings of the same cell. Therefore, a very common way of loading neuronal cells with indicator dyes is to fill them via the patch pipette in whole cell configuration experiments. The intracellular solution used for the staining is very similar to standard patch-clamp recordings. Our standard pipette solution comprises (in mM) 140 K-gluconate, 10 NaCl, 4 Mg-ATP, 2 Na_2-ATP, 0.4 Na-GTP, 10 K-HEPES, 0.1–0.2 Ca^{2+}-indicator dye. Please note that different from standard patch-clamp experiments no Ca^{2+} chelators like ethyleneglycotetraacetic acid (EGTA) or 1,2-bis(o-aminophenoxy)ethane-N,N,N ,N -tetraacetic acid (BAPTA) are included because they would substantially buffer Ca^{2+}-concentration transients leading to smaller and slowed-down fluorometric signals. All intracellular solutions are prepared using double-processed water (Sigma, No. W-3500, ST. Louis, MO) and can be kept frozen at –20°C for up to 2 weeks. In our hands, water from ion starvation systems, even if serviced at prescribed time intervals, often causes problems when used for the preparation of intracellular solutions.

In whole cell recording experiments, we wait at least 20 min after the break-in for the dye to equilibrate in the whole dendritic tree before we start with fluorometric recordings. The access resistance should be monitored during the loading period because a low access resistance (<15 MΩ) is crucial for sufficient dye loading. One often expressed disadvantage of whole cell patch-clamp recordings during fluorometric Ca^{2+} measurements is the potential wash-out of mobile endogenous buffer systems. To test for potential wash-out, we used to perform control experiments using a bulk loading protocol (6). Cells were patch-clamped in whole cell configuration for 2 min with regular intracellular pipette solution containing 2 mM of the indicator dye. Afterwards, the pipette was carefully pulled off the cell and we waited 20 min for the dye to equilibrate. The biggest disadvantage of this protocol is that depending on variations of the access resistance in different experiments the intracellular dye concentration cannot be controlled quantitatively and might influence the calcium transients in a non-predictable way.

2.3. Calibration

In addition to the concentration of the ion of interest, the absolute fluorescence intensity of a non-ratiometric indicator dye depends on several experimental conditions like the dye concentration, the brightness of the excitation light source, the sensitivity of the detector and the thickness of the excited volume. The final goal must be to compensate for all these experimental conditions and to obtain a fluorescence signal which is only dependent on the ion concentration. A commonly used method to correct for these experimental conditions is to analyze the relative change in fluorescence (ΔF/F) rather than the change of absolute values (ΔF). Although using this procedure changes in different parts of the preparation can be compared semi-quantitatively, non-ratiometric dyes do not provide absolute numbers about the concentration changes of the ion of interest. To get a first estimate of absolute changes in ion concentration, we established a two-step calibration procedure for the low affinity dye Oregon Green BAPTA-6F (OG-6F) (8), which is also applicable for any other non-ratiometric indicator dye. In a first step in vitro and by means of calibration kit solutions (Molecular Probes, Cat # C-3723), we obtained a calibration curve using a micro-cuvette and the pipette solution used in our whole-cell patch-clamp experiments. Because the dynamic range of an indicator dye and its dissociation constant K_d depends on the ionic background, it is very important to perform the calibration procedure under conditions as similar as possible to the experimental conditions. The dissociation constant K_d and the dynamic range $R_F = F_{max}/F_{min}$ were determined by fitting the data points with a sigmoid curve (IgorPro, Wavemetrics). We obtained a R_F of 6 and a K_d value of 2.2 µM under these conditions. Please note that these values differ significantly from the data provided by the manufacturer of the indicator dye (Invitrogen, Carlsbad, USA). In a second step and to get even closer to the experimental conditions, we determined in whole cell recordings the maximum change of fluorescence between resting and saturating Ca^{2+}-concentrations in neurons containing 200 μM Oregon Green BAPTA-6F. After measuring the fluorescence level at rest (F_0) we saturated the dye with Ca^{2+} that is necessary for obtaining the maximal fluorescence value (F_{max}), by either disrupting the seal by slightly tapping on the recording stage or by local application of 200 μM of the Ca^{2+} ionophore Ionomycine. With both procedures, we obtained similar ratio values of about $F_{max}/F_0 = 4.5$ (8). It is important to note that this procedure underestimates the full dynamic range ($R_F = F_{max}/F_{min}$) of the indicator dye because only the ratio of F_{max}/F_0 is determined. Assuming that the resting Ca^{2+}-concentration is 50 nM (7), we constructed the calibration curve from these results and used it for quantifying absolute changes in Ca^{2+}-concentration.

2.4. Scanning Systems

The first attempts to measure calcium with optical means in neuronal dendrites and spines were done using widefield fluorescence microscopy (9–11). Although possible, these types of measurements suffer from poor spatial resolution of small subcellular structures like spines in strongly scattering tissue. Different from imaging systems using whole-field illumination, a scanner driven imaging system scans the preparation subsequently with one or several spots of excitation light using a fixed scanning pattern. The main advantage of the usage of scanning systems is the possibility to implement optical sectioning methods namely confocal and 2-photon microscopy (12, 13). The combination of confocal microscopy and elaborate fluorescence probes has revolutionized the field of imaging tools for both morphological and functional studies. For the first time, high resolution measurements even in highly scattering tissue were possible. Because of the principle of subsequent scanning individual image pixels, one serious drawback of many scanning systems is their poor time resolution. Many attempts have been made to overcome this limitation in time resolution and some of them will be described below.

2.4.1. Galvano-Scanner

Most commercially available scanning systems are based on two moving mirrors each mounted on a galvanometer and are therefore called galvano-scanners. The typical scanning pattern of these systems consists out of parallel straight lines scanned subsequently from the top to the bottom of the covered area forming one image or frame. One mirror is responsible for the horizontal movement of the illumination spot along the lines of the images and the other moves the spot perpendicular to the line orientation and takes care of the vertical position of the subsequently scanned lines. The maximal repetition rate of a full frame of these scanning systems crucially depends on the maximal scan speed of the mirror which is responsible for scanning the lines and ends up typically around 1 Hz. To increase the time resolution, these

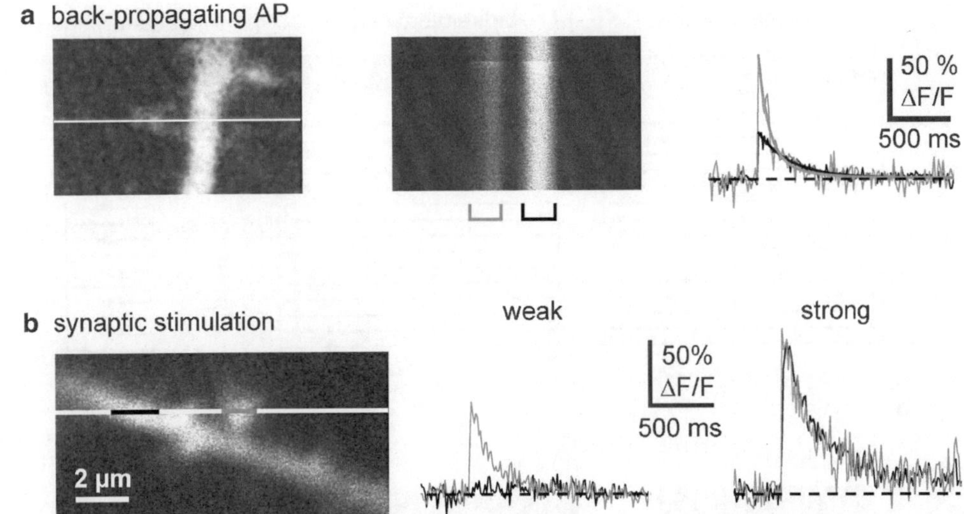

Fig. 10.1. 2-Photon dendritic calcium imaging using line-scan mode. A: B: *Left panel* shows part of a dendrite which was imaged during experiment. Calcium transients were monitored using the line-scan mode of a 2-photon microscope. The *white line* labels the line scanned during experiment. *Colored segments* represent the analyzed regions of interest namely the spine head (*red*) and parent dendrite (*black*). *Traces* represent calcium dynamics in individual spine (*red trace*) and parent dendrite (*black trace*) to single shock synaptic stimulation with weak (*middle panel*) and strong (*right panel*) intensity. Note that after weak stimulation calcium transient is restricted to spine head. After strong stimulation, a dendritic spike was initiated and both the spine and parent dendrite respond with a larger calcium transient. (Modified from (27)).

systems are often equipped with the so-called line-scan mode, where only one line of the image is scanned repetitively (Fig. 10.1). At the expense of the loss of spatial information, the gain in time resolution is remarkable because line repetition rates of up to 1 kHz can be obtained with a typical system. To increase flexibility in orientation at the expense of speed, different companies offer line-scan modes with straight lines at different angles, free user-drawn lines or the so-called multiple line-scan modes (14).

If detailed spatial information is required, full frames need to be scanned as mentioned above. In most cases, the x-mirror is driven by a saw-tooth like command signal to keep the scan speed constant along the line (Fig. 10.2a). This scanning pattern also makes sure that the illumination time (called dwell time) for each pixel of the image is the same, which can be important for quantitative measurements. This type of scanner allows typically full frame rates of about 1 Hz, which sometimes is too slow for functional measurements, but often fast enough for morphological studies. One way to increase the scanning speed is to scan only a part of the image. Because the movement of the horizontal scanning mirror is the limiting step, scanning fewer lines is the most effective way to improve the time resolution of these systems. Another way to increase the time resolution of a galvano-scanner is to implement a so-called bidirectional scan mode (Fig. 10.2a). Here, subsequent

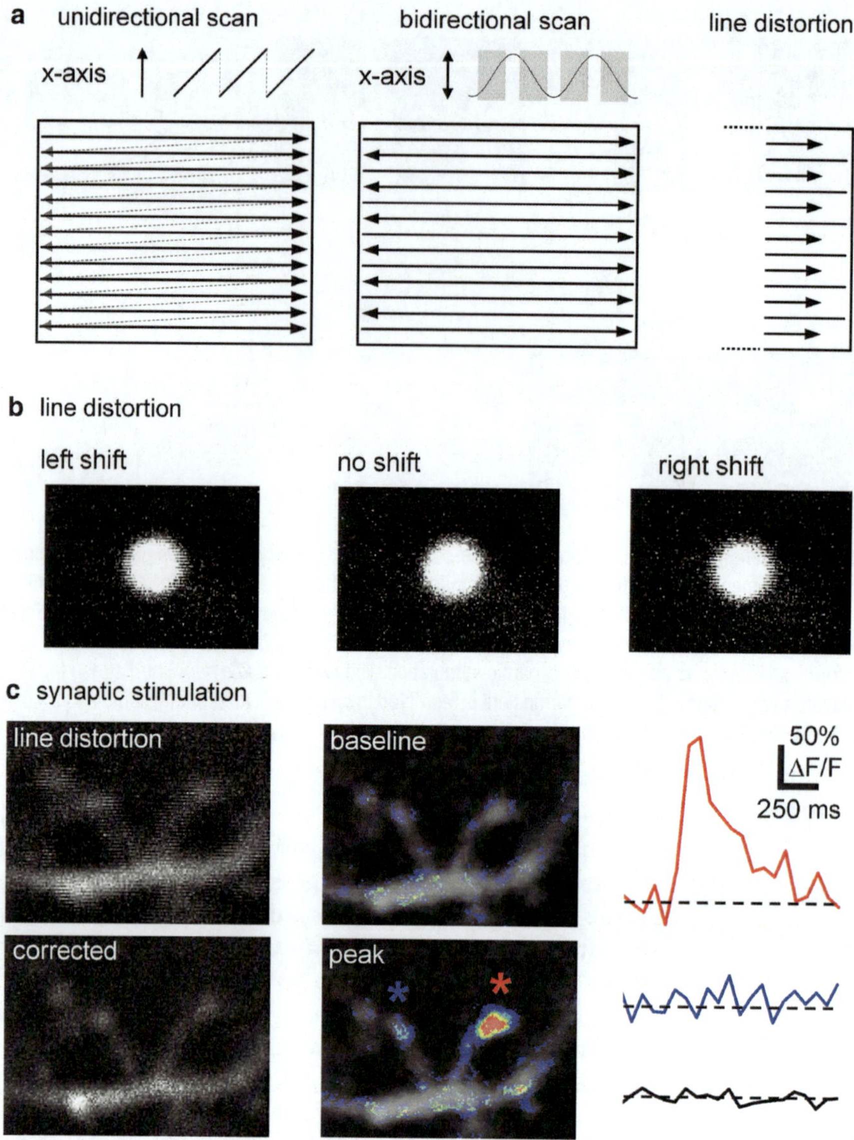

Fig. 10.2. 2-Photon calcium imaging using galvanometer-scanner. (**a**) Scanning patterns of galvanometer-scanners and possible scanning artefact. *Left panel* represents scanning pattern during unidirectional scan mode. *Horizontal lines* are scanned subsequently from left to right with a constant speed. *Trace* represents the command signal over time which drives the mirror responsible for horizontal deflection. Vertical deflection is caused by a second mirror mounted perpendicular to the former one. *Middle panel* represents scanning pattern of the same scanner type in bidirectional scan mode. Note that subsequent lines are scanned in opposite direction. *Trace* represents the sinusoidal-shaped command signal over time which drives the mirror responsible for horizontal deflection in bidirectional mode. Data are collected only during the gray shaded time periods during which scanning speed of the mirror is almost constant. *Right panel* illustrates a frequent scanning artefact during bidirectional scanning mode. Because scanning direction of subsequent lines alternates, it is technically difficult to ensure that data acquisition of one line stops at the exact same horizontal position where it starts for the next line. Any deviation causes scanning artefacts of the recorded image, which are similar to motion artefacts known from interlaced television signals and which we call line distortion artefact. (**b**) Line distortion artefact during bidirectional scan mode. *Panels* show images of a fluorescent bead with and without line distortion artefacts at the left and right side of the bead. *Left and right panels* show clear line distortion artefacts because odd lines are laterally shifted to the left or to the right regarding to

lines are scanned in opposite direction to use also the backward movement of the horizontally swinging mirror for data acquisition (Figs. 10.2a, b). Different from unidirectional scanning mode and to improve the scanning speed even more, the position command signal for the x-mirror changes from a saw-tooth like wave to a sinusoidal wave. To keep the scanning speed along the line and therefore the pixel dwell times almost constant only the time interval around the zero-crossing of the sinusoidal wave is used for data acquisition (Fig. 10.2a). It is important to note that the horizontal position of even and odd lines to each other are often not exactly aligned causing a deterioration of image quality (Fig. 10.2b). In combination with a reduction in the scanned area, maximal frame rates of about 15 Hz can be accomplished.

A further development of the regular galvano-scanner driven by a sinusoidal waveform is the so-called resonance scanner. Here, the regular galvanometer for the horizontal line scan is replaced by a much faster galvanometer swinging in its resonance frequency. Resonance scanners are always used in the bidirectional scan mode and manage to accomplish full frame rates of about 30 Hz, which is also called video-rate because 30 Hz represents the standard video frame rate. Because the resonance frequency has a fixed value for each galvanometer-mirror pair, resonance scanners have an unchangeable scan speed and for a given pixel a fixed dwell time. Nevertheless, in between one line the dwell time of the pixels is slightly different. Because of the sinusoidal moving pattern of the resonance mirror, the pixel dwell time is shortest in the centre of the image, where the speed of the mirror peaks and is getting longer for pixels located at the left and right side, where the mirror is slowed down to finally reverse its direction. The differences in scan speed also cause distortions at the left and right side of the image, which need to be compensated either mathematically or by appropriate timing of the pixel clock. Because the scan rate of an individual line is determined by the resonance frequency, the frame rate of a resonance scanner can only be alternated by changing the number of scanned lines per image. Therefore by scanning, only half or a quarter of the lines frame repetition rates of 60 or 120 Hz can be achieved respectively.

A very recent development extended the application of galvano-scanners to the fast registration of neuronal activity in

even lines respectively. *Middle panel* shows image without any line distortion artefact when odd lines show no lateral shift to even lines. (**c**) 2-Photon calcium imaging after synaptic stimulation using bidirectional scanning mode. *Upper left panel* illustrates the quality of original image deteriorated by line distortion artefact. *Lower left panel* shows the same image corrected off-line using image processing software (ImageJ, http://rsb.info.nih.gov/ij/index). *Middle panels* show colour-coded changes in fluorescence representing calcium concentration increase after single-shock synaptic stimulation. *Traces* display the analyzed relative change in fluorescence in right spine head (*red trace*), left spine head (*blue trace*) and parent dendrite (*black trace*) after synaptic stimulation. Note compartmentalization of calcium transient to individual spine head.

3-dimensional volumes over time. The group of Fritjof Helmchen developed a 3-dimensional line-scan mode utilizing a combination of a regular galvano-scanner with piezo-driven focussing device, which allows the recording of either neuronal network activity or changes in calcium concentration in dendrites in 3-dimensional brain volumes over time (15, 16).

In conclusion, scanners based on galvanometers are versatile and technically sophisticated systems and are able to fit different needs in terms of resolution in time and space. Because the excitation beam is positioned by mirrors which are reflective for a wide range of wavelengths, galvano-scanners are easily applicable for 1-photon or 2-photon excitation fluorescence microscopy.

2.4.2. AOD Scanner

To overcome the speed limitations of a galvano-scanner described above, one possibility is to substitute one or both scanning mirrors with a much faster scanning device, a so-called acousto-optical-deflector (AOD). The AOD utilizes the acousto-optic effect, which describes the change in the refractive index of a medium caused by an acoustic wave. If an acoustic wave is enforced on a crystal by a piezoelectric transducer, it causes a moving periodic change in the refractive index inside of the crystal representing an optical grating. Incoming light is being scattered off this grating comparable to Bragg's diffraction. The angle of diffraction is only dependent on the frequency of the acoustic wave. It is obvious that the scan speed of such a device is only dependent on the maximal speed at which the frequency of the acoustic wave can be changed in the crystal, which is practically feasible in tens of nanoseconds.

We use the Odyssey XL confocal microscope (Noran Instruments, Middleton, WI) for fast imaging applications at video frame rate (8) (Fig. 10.3). It is equipped with an AOD responsible for the horizontal deflection of the excitation beam and enables real-time recordings up to 240 Hz.

Originally built for 1-photon excitation confocal fluorescence microscopy, AOD scanners can principally be used also for 2-photon excitation. A serious complication is that the AOD crystal adds a large amount of spatial and temporal dispersion if used with 2-photon lasers providing femtosecond pulses. Therefore, either adequate compensation mechanisms need to be incorporated or the use of 2-photon lasers using pulse-widths in the picoseconds range, which show much less dispersion is necessary. Different labs demonstrated successfully the combination of the fast scanning principle of AOD-scanners with the benefits of non-linear 2-photon excitation (17–19).

2.4.3. Nipkow-Disc Scanner

The scanning principle of our Nipkow-disc imaging system (CSU 10, Yokogawa) is fundamentally different from all laser scanning systems described above. Whereas in most other systems, a single laser beam is scanned over the preparation utilizing a set of mirrors

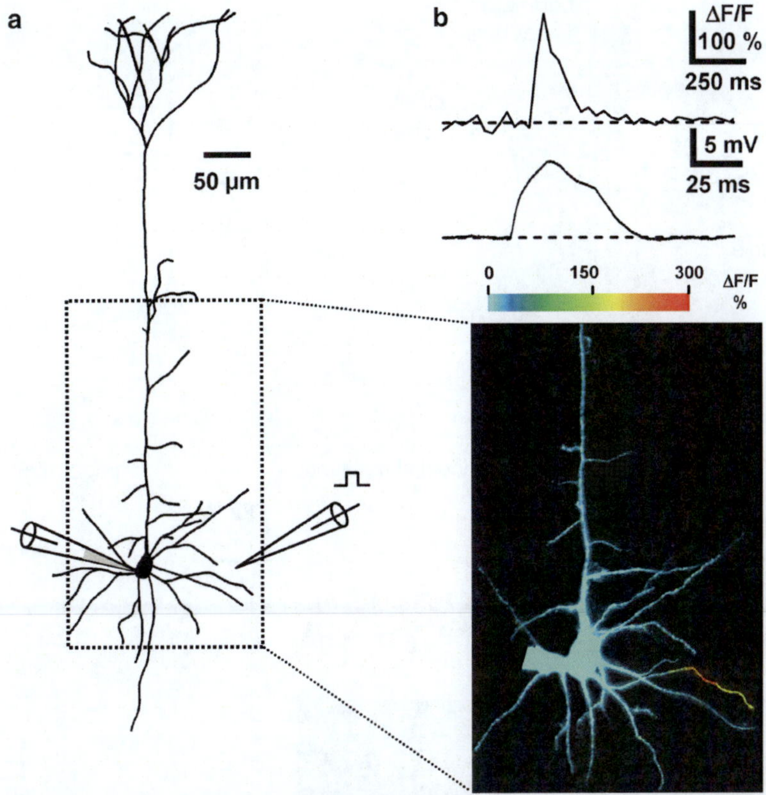

Fig. 10.3. Fast confocal calcium imaging using AOD scanner after dendritic spike initiation. (**a**) Drawing of a layer V pyramidal neuron. Cell was patch-clamped in whole cell configuration and filled via the patch pipette with the calcium indicator dye Oregon Green BAPTA-6F. Synaptic stimulation of basal dendrite was performed using an extracellular stimulation pipette. (**b**) *Upper trace* displays relative change in fluorescence after strong synaptic stimulation in part of the basal dendrite close to the stimulation pipette reflecting dendritic calcium concentration transient. *Lower trace* represents somatic voltage recording at the same time. Note complex shape of excitatory postsynaptic potential reflecting the initiation of a dendritic action potential. *Colour-coded image* represents peak change in fluorescence after synaptic stimulation of the dendritic action potential. Note that calcium transient is restricted to only one branch of the basal dendritic tree (Modified from (28)).

or crystals, the Nipkow-disc imaging system scans the preparation with multiple light beams in parallel (Fig. 10.4a). To produce an array of light beams, a widened and collimated laser beam is projected onto a rotating disc which contains thousands of pinholes oriented in spiral lines. The design of the pinhole pattern on the disc allows capturing twelve full frames per rotation. With a rotational speed of 1800 turns/min this system produces a maximal frame rate of 360 Hz. By increasing the rotational speed to 5000 turns/min, a more recent version of the Nipkow-disc imaging system (CSU-22, Yokogawa) accomplishes a maximal frame rate of 1 kHz. Because the pinholes cover only 1–2% of the disc surface, the excitation light would be diminished by 98–99%. Therefore and as a specific feature, the Yokogawa scanner contains an additional disc in front of the pinholes containing an array of micro-lenses

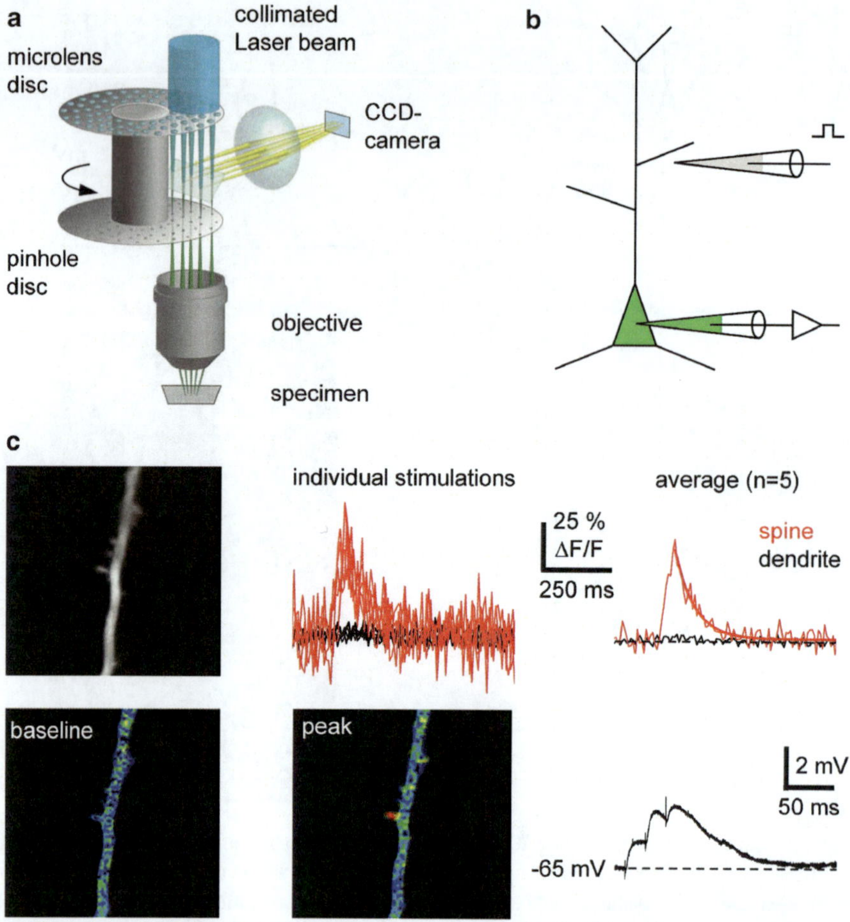

Fig. 10.4. Fast confocal calcium imaging using Nipkow-disc scanner. (**a**) Scheme of Nipkow-disc scanner. A collimated laser beam is projected on a rotating disc containing thousands of pinholes arranged in spiral lines. To increase effectiveness of light transmission through the pinhole disc, opposed to each pinhole a microlense is positioned. Because of this arrangement, the preparation is scanned in parallel by an array of light spots allowing full-frame rates of scanning up to 360 Hz. To preserve the spatial information of the emission light a fast CCD-camera is used as detector. (**b**) Experimental arrangement for recording synaptically induced dendritic calcium concentration transients in cortical pyramidal neurons. In acute brain slices, individual cell is patch clamped in whole cell configuration and filled via the patch pipette with a fluorescent indicator dye (*green color*). To stimulate the cell synaptically under visual control, an extracellular pipette is positioned close (~10 μm) to the dendrite of interest and short voltage pulses are applied. (**c**) Calcium transients in individual spine upon extracellular synaptic stimulation. *Black and white image* shows at high spatial resolution the stretch of the stimulated dendrite with individual spines. *Color-coded images* display fluorescence before (*baseline*) and maximal change after (*peak*) synaptic stimulation (triple stimulation at 50 Hz). *Traces* represent quantified relative changes in fluorescence upon synaptic stimulation in spine (*red trace*) and parent dendrite (*black traces*) of the same experiment. Electrophysiological trace displays somatic recording during synaptic triple stimulation.

each positioned opposite to a pinhole. The focal point of each lens is positioned in the centre of the opposing pinhole. Due to this configuration, the transmission of the pinhole disc is increased to 70% of the initial laser intensity. The emission light takes the same way back through the same pinholes producing the confocality and gets reflected on the detector by a dicroic mirror (Fig. 10.4a).

Because of the multi-beam excitation mode, a CCD-camera needs to be used for detection to preserve the spatial information. To make use of the high frame repetition rate of the scanning system, we chose as detector a high speed CCD camera (NeuroCCD, RedShirt Imaging) with low spatial resolution (80×80 pixel) but minimal readout and dark noise. To resolve the morphology of the cells of interest in detail, we attached a high spatial resolution CCD-camera (Pixelfly QE, PCO imaging) at the second port of the Nipkow-scanner.

Drawbacks of the Nipkow-scanner are that a cross-talk between adjacent pinholes produces some optical noise and that the pinholes size cannot be adapted to different objectives. The most important advantages of this imaging system for calcium imaging (20) are its high full frame repetition rate of 360 Hz or even 1 kHz, the extraordinary signal-to-noise ratio because of the use of a high-speed and low-noise CCD-camera and the very low photo-toxicity and bleaching due to low light levels necessary for excitation (21).

3. Conclusions

High resolution fluorescence imaging is nowadays the gold standard for detecting intracellular calcium concentration changes in cellular and sub-cellular compartments like neuronal dendrites and spines. Nevertheless, a large and often confusing variety of different fluorescent probes or scanner and laser technologies are available making it difficult for the normal user to find the optimized combination for a given scientific question. On the other hand, in these days, several companies provide turnkey imaging systems including sophisticated software control allowing to benefit from the extraordinary advantages of optical means also for the non-specialized scientist.

The combination of new low affinity and high dynamic range calcium indicator dyes with high speed imaging systems has taught us that physiological intracellular calcium transients can be very fast and are often spatially compartmentalized (22, 23).

Nevertheless, the physical limit inherent in all optical imaging techniques, the diffraction spatial resolution barrier seemed to be unconquerable for many decades. But very recently, even this theoretical limit in spatial resolution has been overcome and extended to the nanometre scale by the development of the Stimulated-Emission-Depletion (STED) microscopy (24). STED microscopy has already been proven to be very useful for the high resolution and high speed investigation of sub-cellular neuronal structures (25, 26). Because the physical principle of this new optical technique is relatively simple, it is conceivable that also functional fluorescence microscopy like calcium imaging will benefit from the improvement in spatial resolution.

Acknowledgements

This work was supported by grants from the German Research Foundation (DFG). I would like to thank Friedrich Johenning for comments to the manuscript.

References

1. Tsien RY (1989) Fluorescent probes of cell signaling. Annu Rev Neurosci 12:227–253
2. Denk W, Strickler JH, Webb WW (1990) Two-photon laser scanning fluorescence microscopy. Science 248:73–76
3. Oertner TG, Sabatini BL, Nimchinsky EA, Svoboda K (2002) Facilitation at single synapses probed with optical quantal analysis. Nat Neurosci 5:657–664
4. Yasuda R, Sabatini BL, Svoboda K (2003) Plasticity of calcium channels in dendritic spines. Nat Neurosci 6:948–955
5. Helmchen F, Imoto K, Sakmann B (1996) Ca2+ buffering and action potential-evoked Ca2+ signaling in dendrites of pyramidal neurons. Biophys J 70:1069–1081
6. Holthoff K, Tsay D, Yuste R (2002) Calcium dynamics of spines depend on their dendritic location. Neuron 33:425–437
7. Garaschuk O, Yaari Y, Konnerth A (1997) Release and sequestration of calcium by ryanodine-sensitive stores in rat hippocampal neurones. J Physiol 502:13–30
8. Holthoff K, Kovalchuk Y, Yuste R, Konnerth A (2004) Single-shock LTD by local dendritic spikes in pyramidal neurons of mouse visual cortex. J Physiol 560:27–36
9. Connor JA (1986) Digital imaging of free calcium changes and of spatial gradients in growing processes in single, mammalian central nervous system cells. Proc Natl Acad Sci USA 83:6179–6183
10. Tank DW, Sugimori M, Connor JA, Llinas RR (1988) Spatially resolved calcium dynamics of mammalian Purkinje cells in cerebellar slice. Science 242:773–777
11. Muller W, Connor JA (1991) Dendritic spines as individual neuronal compartments for synaptic Ca^{2+} responses. Nature 354:73–76
12. Conchello JA, Lichtman JW (2005) Optical sectioning microscopy. Nat Methods 2: 920–931
13. Helmchen F, Denk W (2005) Deep tissue two-photon microscopy. Nat Methods 2:932–940
14. Lorincz A, Rozsa B, Katona G, Vizi ES, Tamas G (2007) Differential distribution of NCX1 contributes to spine-dendrite compartmentalization in CA1 pyramidal cells. Proc Natl Acad Sci U S A 104:1033–1038
15. Gobel W, Helmchen F (2007) New angles on neuronal dendrites in vivo. J Neurophysiol 98:3770–3779
16. Gobel W, Kampa BM, Helmchen F (2007) Imaging cellular network dynamics in three dimensions using fast 3D laser scanning. Nat Methods 4:73–79
17. Roorda RD, Hohl TM, Toledo-Crow R, Miesenbock G (2004) Video-rate nonlinear microscopy of neuronal membrane dynamics with genetically encoded probes. J Neurophysiol 92:609–621
18. Salome R, Kremer Y, Dieudonne S, Leger JF, Krichevsky O, Wyart C, Chatenay D, Bourdieu L (2006) Ultrafast random-access scanning in two-photon microscopy using acousto-optic deflectors. J Neurosci Methods 154:161–174
19. Duemani RG, Kelleher K, Fink R, Saggau P (2008) Three-dimensional random access multiphoton microscopy for functional imaging of neuronal activity. Nat Neurosci 11:713–720

20. Gundlfinger A, Bischofberger J, Johenning FW, Torvinen M, Schmitz D, Breustedt J (2007) Adenosine modulates transmission at the hippocampal mossy fibre synapse via direct inhibition of presynaptic calcium channels. J Physiol (Lond) 582:263–277
21. Wang E, Babbey CM, Dunn KW (2005) Performance comparison between the high-speed Yokogawa spinning disc confocal system and single-point scanning confocal systems. J Microsc 218:148–159
22. Yuste R, Majewska A, Holthoff K (2000) From form to function: calcium compartmentalization in dendritic spines. Nat Neurosci 3:653–659
23. Sabatini BL, Oertner TG, Svoboda K (2002) The life cycle of Ca^{2+} ions in dendritic spines. Neuron 33:439–452
24. Klar TA, Jakobs S, Dyba M, Egner A, Hell SW (2000) Fluorescence microscopy with diffraction resolution barrier broken by stimulated emission. Proc Natl Acad Sci U S A 97:8206–8210
25. Nagerl UV, Willig KI, Hein B, Hell SW, Bonhoeffer T (2008) Live-cell imaging of dendritic spines by STED microscopy. Proc Natl Acad Sci U S A 105:18982–18987
26. Westphal V, Rizzoli SO, Lauterbach MA, Kamin D, Jahn R, Hell SW (2008) Video-rate far-field optical nanoscopy dissects synaptic vesicle movement. Science 320:246–249
27. Holthoff K (2004) Regenerative dendritic spikes and synaptic plasticity. Curr Neurovasc Res 1:381–387
28. Holthoff K, Kovalchuk Y, Konnerth A (2006) Dendritic spikes and activity-dependent synaptic plasticity. Cell Tissue Res 326:369–377

Chapter 11

In Vivo Ca^{2+} Imaging of the Living Brain Using Multi-cell Bolus Loading Technique

Gerhard Eichhoff, Yury Kovalchuk, Zsuzsanna Varga, Alexei Verkhratsky, and Olga Garaschuk

Abstract

This chapter overviews the use of acetoxymethyl (AM) ester-based multi cell bolus loading (MCBL) technique for in vivo Ca^{2+} imaging of neural circuits. This technique provides anatomically targeted, rapid and non-invasive staining of both neurones and glia with small molecule Ca^{2+} indicators thus allowing real-time imaging of brain activity. We describe the protocols for staining cells in newborn, juvenile, adult and aged tissue; discuss critical steps and possible pitfalls and introduce the multicolor imaging approach for functional characterization of specific cell types. We show, furthermore, that the use of MCBL can be extended to label other elements within the brain tissue as, for example, amyloid plaques in a mouse model of Alzheimer's disease (AD). Finally, we illustrate the use of MCBL for two-photon calcium imaging of neurones and glia in the aged mouse cortex as well as the mouse cortex in an animal model of AD.

Key words: In vivo two-photon microscopy, Calcium indicator dye, Staining technique, Calcium imaging, Ageing, Alzheimer's disease

1. Introduction

Monitoring physiological activity of neural circuits is fundamental for understanding of the brain function. Mammalian brain is the extremely complex and dense cellular network. Its grey matter is composed of two closely associated cellular circuits, the neuronal net and the glial web. Closely apposed neuronal and glial membranes compartmentalize the extracellular space into narrow subregions and clefts, where neurotransmitters, neuromodulators and hormones diffuse and undergo complex processes of enzymatic degradation and active uptake. The three-dimensional organization of the brain is therefore critically important for the integration within neural circuits, and the function of such

A. Verkhratsky and Ole H. Petersen (eds.), *Calcium Measurement Methods,* Neuromethods, vol. 43
DOI 10.1007/978-1-60761-476-0_11, © Humana Press, a part of Springer Science+Business Media, LLC 2010

three-dimensionally organized computing space can only be studied in vivo.

Because activity of many individual cells has to be studied simultaneously, high resolution imaging techniques are required. Over last decades, two-photon microscopy has become a method of choice for analyzing neuronal circuits in the intact mammalian brain (1–4). This technique utilizes fluorescent ion-selective probes, which allow real-time imaging of cellular movements of physiologically relevant ions, the Ca^{2+} ions being at the very focus of these indicators (5, 6). Fluorescent Ca^{2+} probes come in two flavors: (1) membrane-impermeant (salt) and (2) membrane-permeant (acetoxymethyl (AM) ester) forms (7) and therefore require different techniques of cell staining. The staining of individual cells can be either achieved by direct injection of the membrane-impermeant form into the cytosol (either through microelectrodes/patch-pipettes or by single-cell electroporation), or can be accomplished by introducing the membrane-permeant AM form of the dye into the extracellular space. In the later case, the dye is freely diffusing through the cell membrane and is cleaved in the cytosol by intracellular esterases. This traps the probe within the cell of interest. Both loading techniques were successfully adapted for the in vivo brain imaging (4, 8–10).

In this chapter, we shall overview the use of AM ester-based multi cell bolus loading technique (4, 11) for in vivo Ca^{2+} imaging of the neural circuits. We will discuss its applicability in animals of different ages as well as the approaches used to identify different cell types. We will articulate the critical steps and possible pitfalls and will describe several possible applications of the technique either alone or in combination with electrophysiological recordings.

2. Multi-cell Bolus Loading Technique

The technique for multi-cell bolus loading (MCBL) of neural circuits in the living brain was initially developed for in vivo staining of cells in the mouse cortex (4). Subsequently, it was successfully employed in different species and different brain regions (2, 3, 12–15). This technique provides rapid non-invasive staining of both neurones and glia with small molecule Ca^{2+} indicators thus allowing real-time imaging of brain activity.

2.1. Surgery

The animals are anesthetized by inhalation of isoflurane (1–1.5% in pure O_2), or by intraperitoneal injection of ketamine/xylazine or urethane (0.1/0.01 mg/g and 1.9 mg/g body weight, respectively). The degree of anesthesia as well as the physiological status of the animal (e.g., breathing/hard beat rate, body temperature, etc.) is closely monitored using the anesthesia monitoring system (for example, from AD Instruments (Sydney, Australia)). Subsequently, the animal is fixed in the stereotactic device and the skin is removed (after subcutaneous injection of a local anesthetic agent (e.g., 2% lidocaine)) giving an access to the skull. Optical imaging can be performed either through the thinned skull (which improves the stability of recording by smoothing heart beat and breathing artifacts) or through the window created by removing a piece of the skull (10). According to our experience, imaging through the open-skull window allows better image quality and provides increased imaging depth (4). It has to be noticed, however, that the results obtained during chronic imaging (days to weeks) through the open-skull window may be compromised by activation of glial cells (16).

For imaging through the skull, the skull is thinned with drill bits of increasingly smaller size (Ø of 0.7–0.5 mm) to the thickness of ~10–20 μm and then the remaining bone is treated with a felt polisher (e.g., from Dr. Ihde Dental, Munich, Germany) – (4)). This is a critical step as the procedure of thinning/polishing may damage the brain tissue and should therefore be performed with maximal gentleness (11). Alternatively, the skull is opened and the window is formed, though which the "open access" imaging can be performed. The open access window should be <1 mm in diameter. Increase in the size of craniotomy affects the stability of the recording and increases mechanical artifacts (especially because of heartbeat). The stability of the recordings can be improved by covering the opening with 2% agarose and a coverglass (10, 11).

When the skull is prepared for imaging, the recording chamber made from standard plastic tissue culture dish 35 mm in diameter (11) is glued (using the cyanoacryl glue) to the bone. Subsequently, the animal is transferred to the recording set-up (consisting of the two-photon microscope, micromanipulators and, if required for combined recordings, some electrophysiological equipment) where it is placed onto the warming blanket that is kept at 38°C. The recording chamber affixed to the skull is perfused with standard external saline (125 mM NaCl, 4.5 mM KCl, 26 mM $NaHCO_3$, 1.25 mM NaH_2PO_4, 2 mM $CaCl_2$, 1 mM $MgCl_2$, 20 mM glucose, pH 7.4, when bubbled with 95% O_2 and 5% CO_2) at physiological temperature (37°C).

Throughout the experiment, the blood pressure, body temperature, respiratory and pulse rate of the animal (as well as any other parameters relevant to the particular experiment) have to be monitored and kept at physiological levels (e.g., using anesthesia monitoring system (see above)). The status of the animal is critical for obtaining high quality long-lasting recordings. Furthermore, all the manipulations with the skull should be done with utmost precision and tenderness as the damage inflicted to the blood vessels or the brain parenchyma directly determines the outcome of both staining procedure and ensuing recordings.

2.2. Staining Protocol

The AM form of the Ca^{2+} dye is dissolved in DMSO containing 20% Pluronic F-127 (w/v) to the final concentration of 10 mM. The injection pipette is prepared using the standard protocol for patch-clamp electrodes. As a rule, the resistance of the staining pipette varies between 6 and 9 MΩ when filled with physiological saline. The pipette is filled with the pipette solution of the following composition (in mM): 150 NaCl, 2.5 KCl and 10 HEPES supplemented with 0.1–1 mM of the Ca^{2+} dye (11). Note that the pipette solution does not contain any divalent cations. This is done on purpose in order to minimize precipitation of the dye and clogging of the injection pipette. To avoid clogging, it is important to ensure that all solutions are freshly made immediately before the experiment. In addition, the staining solution can be filtered using for example a Millipore filter (pore diameter of 0.45 μm) if necessary. Even when imaging through the thinned skull, a tiny skull opening has to be made lateral to the imaging window to provide an access for the dye injection pipette. The staining pipette is inserted into the brain tissue using a micromanipulator. The resistance of the pipette is constantly monitored with a patch-clamp amplifier. This resistance can transiently increase up to 15 MΩ during penetration of dura mater. It has to be stressed that removal of dura mater is not advised; the intact dura does not affect the staining quality while improving recording stability. Stable increase of the pipette resistance above 20 MΩ indicates blockage of the pipette, which may render the dye ejection procedure impossible. In this case, the staining pipette should be replaced. When the pipette reaches the desired depth, the dye is pressure ejected into the brain by applying the positive pressure (70 kPa for 1 min). With the pipette tip located 150–200 μm below the cortical surface, the dye ejection stains both neurones and glia up to the depth of ~400 μm from the cortical surface (Fig. 11.1, (4)). When deeper cortical layers are targeted, the tip of the pipette is positioned at 650–700 μm below the pia (2). Following dye ejection, the pipette is removed and the experiment commences approximately 1 h later to ensure complete deesterification of the dye.

Interestingly, while all cortical cells are stained in the middle of the injection area (Fig. 11.1c), on the periphery, some 400 μm apart from the injection spot, only astrocyte cell bodies are stained (Fig. 11.1d). This finding suggests that mechanisms other than diffusion are involved in the astrocyte staining. For example, astrocytes might take up the dye at the injection side and pass it to the neighboring cells most probably through gap junctions. The same mechanisms are likely to be employed by other astrocyte-specific labeling techniques as for example, topic application of the indicator dyes (17, 18).

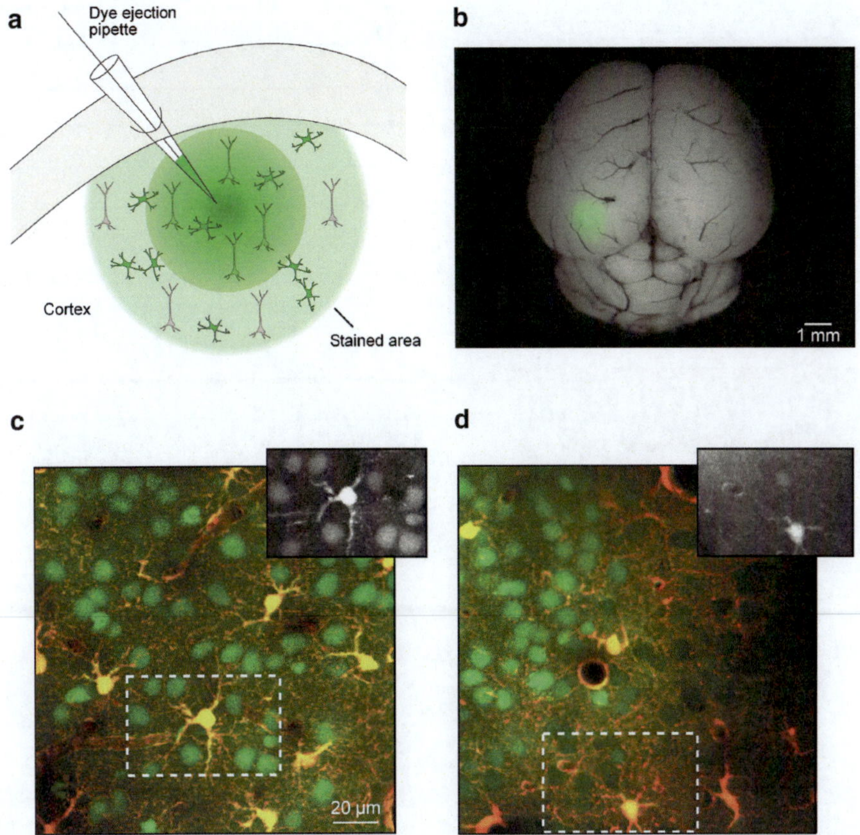

Fig. 11.1. Staining of defined brain areas with Ca^{2+} indicators using multi cell bolus loading technique. (**a**) A schematic drawing of the stained area. The stained area consists of two concentric spheres. In the *inner sphere* both neurones and glial cells are stained. In the *outer sphere* only glia cells are labeled. (**b**) A microphotograph of the in vivo stained brain. The staining pipette was targeted to the primary visual cortex. (**c**, **d**) High resolution images of the labeled area at the center of the injection spot (**c**) and appr. 400 μm to the periphery (**d**). Fluorescence of Oregon Green BAPTA-1 (OGB-1) was directed to the *green channel* and the fluorescence of the glial marker sulforhodamine 101 to the *red channel*. Glia cells contain both dyes and therefore appear *yellow*. All experiments illustrated here and below were performed in accordance with institutional guidelines and were approved by the local government.

To label the neural cells in the aged brain (see Figs. 11.2 and 11.3), the protocol described above was slightly modified in that the dye concentration within the staining pipette was reduced to 0.5 mM and the delivery was made as three consecutive 1-min-long dye injections with 1–2 min intervals between them. This protocol yielded a stained area with the dimensions of 500 × 700 μm and a maximal imaging depth of 300–350 μm (20).

Originally, the MCBL procedure employed high concentrations of the indicator (1 mM), which in turn required high concentrations of DMSO (10%) in the staining solution. Although the 400 fl of the staining solution injected into the brain (4) are very likely to be rapidly diluted by the extracellular saline, there still remained some worries that high concentrations of detergent

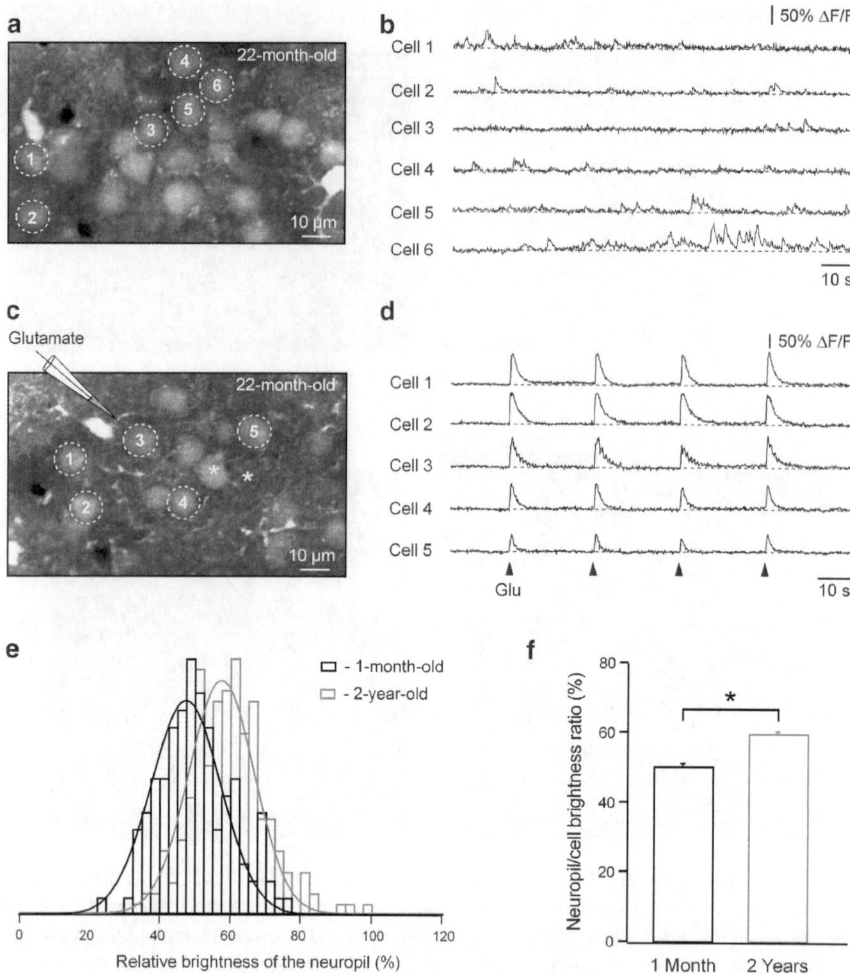

Fig. 11.2. In vivo Ca^{2+} imaging of the aged mouse cortex. (**a**) Cells in the layer 2/3 in the visual cortex of a 22-month-old mouse stained with OGB-1 AM using MCBL. (**b**) Spontaneous Ca^{2+} transients recorded from cells, marked with respective numbers in (**a**). (**c**, **d**) Neuronal Ca^{2+} transients (**d**) caused by local iontophoretic glutamate application (100 mM within the application pipette) to the OGB-1-stained cells, marked with respective numbers in (**c**). The location of the application pipette is schematically shown in (**c**). In this experiment spontaneous activity was blocked by a "bath" (19) application of 2.5 μM TTX. (**e**, **f**) Histograms (**e**) and a corresponding bar graph (**f**) showing the distributions (**e**) and the mean values (**f**) of the neuropil/cell brightness ratio in the juvenile (*black*) and aged (*red*) mouse cortex. For analyses, intensity values recorded from the regions of interest drawn within large blood vessels were considered as a background and were subtracted in all experiments. To measure brightness of the neuropil the region of interest identical to the one used to analyze the corresponding cell was placed in the immediate neighborhood of the cell of interest (see *asterisks* in (**c**)). Reproduced, with permission, from Ref. (20).

may cause damage to the tissue. As was demonstrated in the subsequent study (11), the dye concentration used can be reduced to 0.1 mM thus also lowering the DMSO concentration to 1%. The quality of cell labeling was very little affected by lowering the concentration of the dye as judged by both high-resolution images and parameters of Ca^{2+} signals. This finding suggests

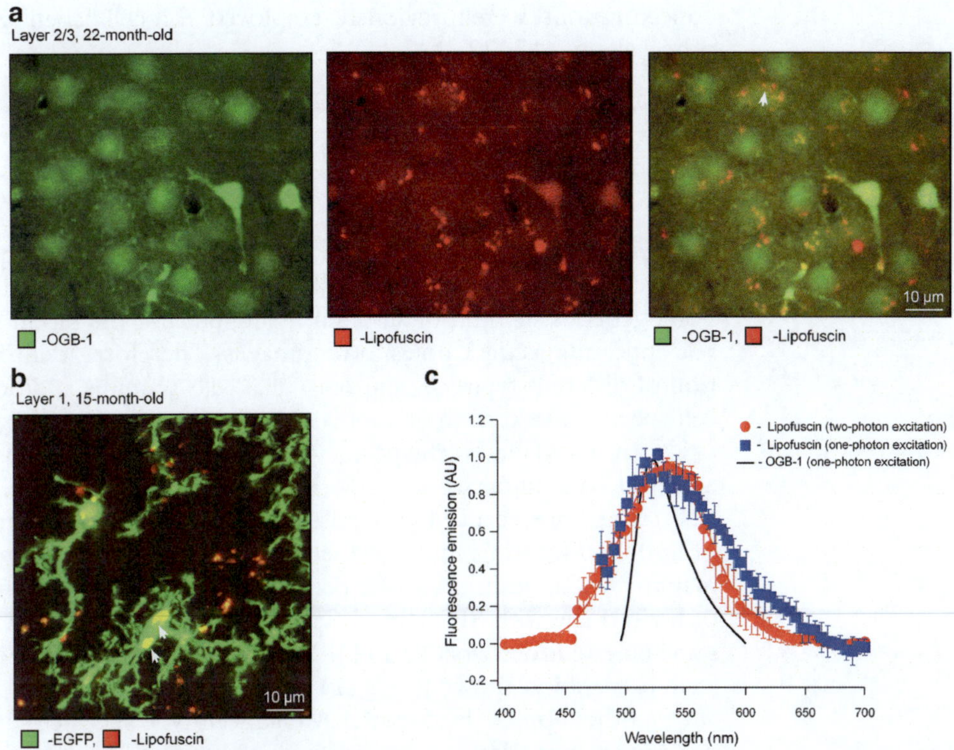

Fig. 11.3. In vivo two-photon imaging of lipofuscin autofluorescence. (**a**) An image of cells in cortical layer 2/3 stained with OGB-1 (*left*). Simultaneously recorded autofluorescence of lipofuscin is shown in the *middle* and a merged image to the *right*. The emission light was split into two channels at 570 nm. (**b**) An experiment, similar to the one in (**a**) conducted in a CX3CR1-EGFP transgenic mouse in which microglia is labeled with EGFP (21). Note large lipofuscin deposits within the microglia cell body (*arrows*). Similar, albeit smaller, lipofuscin accumulations are also seen in neurones (an *arrow* in (**a**), *right*). (**c**) Emission spectra of lipofuscin (step 5 nm, spectral width 10 nm, obtained with Olympus Fluoview 1000MPE, Tokyo, Japan). Single photon excitation was performed at 458 nm and two-photon excitation at 800 nm. For comparison the emission spectrum of OGB-1 (taken from Molecular Probes catalogue) is included. Reproduced, with permission, from Ref. (20).

that the dye injected into the extracellular space gets washed out very quickly most likely by the microcirculation (19), and therefore similar intracellular dye concentrations are achieved when adding either 1.0 or 0.1 mM of the dye to the staining solution. The use of the lower dye concentration however, decreased the robustness of staining procedure and made it more susceptible to the variations in the amount of applied dye due for example to the variability in the pipette resistance and/or partial pipette clogging. Therefore it seems advisable to increase other parameters of dye application (e.g., duration or the pipette tip size) when using the lower indicator concentrations within the staining pipette. The experience accumulated over last years however strongly suggests that the use of a DMSO-containing staining solution (as described in the original paper) does not cause much neuronal damage. Furthermore, similar detergent concentrations were previously employed for cell labeling by other groups (17, 22, 23).

3. Cell-Specific Staining Using MCBL

Inherently, the MCBL technique does not provide for cell type-specific staining. As illustrated in Fig.11.1c it results in relatively homogeneous staining of all brain tissue, but not the blood vessels appearing as dark holes in the images. Therefore, identification of different types of neuronal/glial cells requires additional cell-specific markers in combination with multicolor imaging (19). In particular, cell-specific labeling of astroglia can be achieved with fluorescent marker sulforhodamine 101 (SR101, (24)). The fluorescence light emitted by SR101 has a wavelength of 550–750 nm thus much longer than the light emitted by the majority of Ca^{2+} indicators. The protocol for simultaneous staining of brain tissue with SR101 and Ca^{2+} indicators injected from the same pipette is described in (11). The same protocol was used to stain preparation shown in Fig .11.1c and d.

Another approach to confer MCBL cell type specificity is to use transgenic animals carrying fluorescent proteins in defined populations of neuronal and/or glial cells. The light emitted by red fluorescent proteins (see, for example, (25)) can be easily separated from that of many calcium indicators enabling two-color imaging of Ca^{2+} signals in identified cells. When using eGFP-labeled cells, it is practical to use the technique described by Sohya et al. (26). Briefly, the cells are stained with a calcium indicator dye Fura-2 using MCBL. The eGFP expressed in the particular cell type is excited at 950 nm (where the fluorescence of Fura-2 is negligible) whereas the Ca^{2+} imaging is performed at 800 nm excitation wavelength (which produces very little excitation of eGFP). This approach was successfully used to characterize responses of GAD-positive cortical interneurones to orientation of the visual stimuli (26). Similar approach albeit using Ca^{2+}-sensitive dye X-Rhod-1 was used by Petzold et al. (14) to study in vivo Ca^{2+}-signaling in olfactory bulb astrocytes.

4. MCBL is Applicable at Any Age from Newborn to Old

The MCBL technique has been successfully used for staining neural cells in the brains of animals of very different age – from newborns to adult and old. This represents a specific advantage of the bolus-loading technique, as hitherto staining neurones or glia in the in vivo/in vitro preparations obtained from old brains has been exceedingly difficult. In contrast, the MCBL technique applied with minor modifications allowed Ca^{2+} imaging from rodents of any age (newborn animals – (27); juvenile/adult – (3, 4, 11, 19, 28); old – (20)).

Neurones stained using MCBL in adult/aged brains easily showed spontaneous as well as glutamate-evoked Ca^{2+} transients (Fig. 11.2a–d; (20)). The properties of these Ca^{2+} transients are similar to those recorded in juvenile preparations (compare Fig. 11.2 to Figs. 6, 7 in (19)). Nonetheless, the background fluorescence created mostly by the neuropil was higher in adult/aged tissues. We quantified this finding (Fig. 11.2e, f) by measuring pairwise the absolute brightness of the cell bodies and that of the surrounding neuropil (marked by asterisks in Fig. 11.2c). On average, the relative brightness of the neuropil (background corrected neuropil/cell brightness ratio) was 60% in 2-year-old mice and thus slightly but significantly higher than the value (50%) obtained for 1 month-old animals (20). This increased background fluorescence decreases contrast of acquired images. It is most likely caused by the slower/less effective removal of the indicator dye from the extracellular space during the staining procedure.

Another critical issue when imaging the adult/aged brain is the endogenous "background" fluorescence of the tissue. The brain tissue contains various "biological" fluorophores such as flavins, porphyrins and a reduced form of the nicotine-amide adenine dinucleotide (29), which may contribute to the Ca^{2+}-independent background fluorescence. Many of these fluorophores are distributed homogeneously and their fluorescence can be measured before staining and subtracted from subsequent images. This is unfortunately not the case for the "aging pigment" lipofuscin (30). The latter is a product of iron-catalyzed oxidation/polymerization of proteins and lipid residues and may bind various metals including mercury, aluminum, iron, copper and zinc (31). The role for lipofuscin remains unclear although the rate of its accumulation is increased in neurodegenerative diseases (31). Furthermore, lipofuscin may interfere with autophagic process thus hampering intrinsic cellular regeneration (31, 32). Within the cells, lipofuscin accumulates in grain-like structures associated with lysosomes and therefore produces punctuate fluorescent image (Fig. 11.3a, b). Lipofuscin is easily excited by red light (790–980 nm) and emits fluorescence over the entire visible light spectrum (450–700 nm, Fig. 11.3c). However, we noted during experiments that the cross-section (33) of lipofuscin is smaller than that of commonly used Ca^{2+} indicators and hence, in certain conditions (for example when imaging layer 1 of the cerebral cortex), the intensity of the two-photon light required for excitation of the Ca^{2+} probe (e.g., Oregon Green BAPTA-1, OGB-1) is not sufficient to excite lipofuscin. In our experiments,

10 mW of the excitation light (at 800 nm wavelength) was sufficient to excite layer 1 cells stained with OGB-1, whereas 21.5 mW were required to visualize lipofuscin. Still, when lipofuscin is excited, it is not possible to filter its fluorescence out. However, it is possible to make use of the long-wavelength part of lipofuscin emission spectrum to visualize the lipofuscin granules (Fig. 11.3a) and to exclude them from regions of interest during image analyses.

5. Special Applications

5.1. Targeted Patching Using MCBL

MCBL-stained preparation can be used to screen the functional responses of many cells and to identify cells with particular/

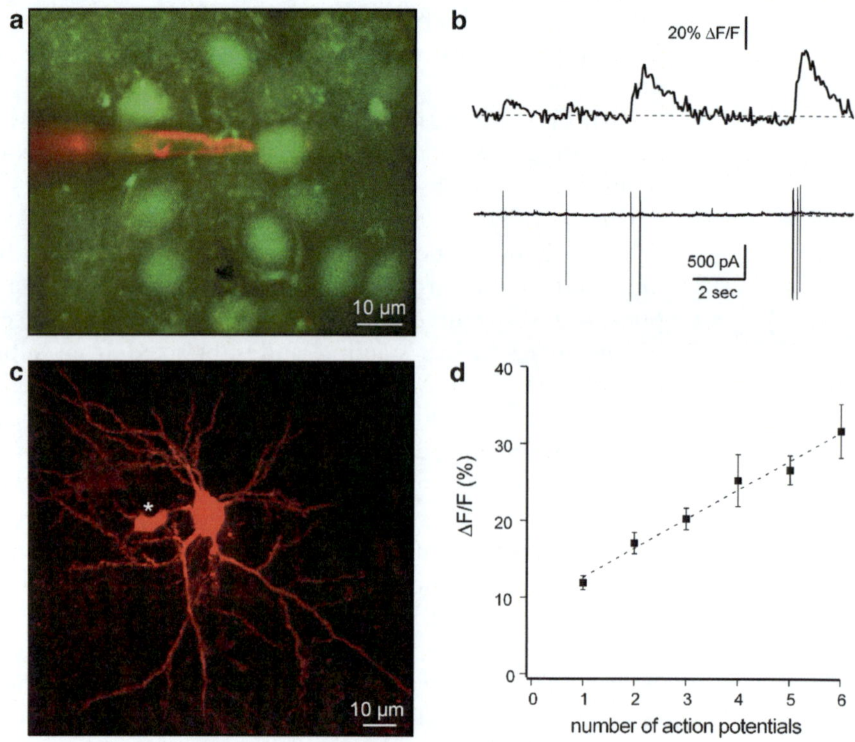

Fig. 11.4. Targeted loose-patch recording of a neurone stained using MCBL. (**a**) A microphotograph of layer 2/3 neurones in the frontal cortex of an 8-month-old mouse. Cells are stained with OGB-1, recording pipette is filled with a pipette solution containing in mM: 150 NaCl, 2.5 KCl, 10 HEPES, 2 $CaCl_2$, 1 $MgCl_2$, 0.025 Alexa Fluor 594. (**b**) Spontaneous intracellular Ca^{2+} transients (*upper*) and underlying action potentials (*lower*) recorded simultaneously. (**c**) A morphological reconstruction of the recorded cell revealed by a subsequent electroporation with Alexa Fluor 633 (1 mM in the electroporation pipette). The image represents a maximal projection of 73 μm of the recorded volume. Note that a small satellite cell (marked with an *asterisk*) was also filled with the dye during electroporation. (**d**) The relationship between the amplitude of a Ca^{2+} transient and the number of underlying action potentials ($n = 15$ layer 2/3 neurones).

desired properties. These cells can then be subjected to detailed electrophysiological analyses using MCBL-guided targeted patching. Figure 11.4 illustrates such an experiment. The cortical cell in the brain of a mouse model of Alzheimer's disease was selected based on the frequency of its spontaneous Ca^{2+} transients (normal cell, see Ref. (34)) and loose-seal recording mode was established. Simultaneous electrical and Ca^{2+} recordings revealed linear relation between the amplitude of Ca^{2+} transients and the number of underlying action potentials (Fig. 11.4b, d; similar data were obtained by (2) in the rat motor and somatosensory cortex). The morphology of the neurone was assessed by an electroporation with Alexa Fluor 633 at the end of the experiment (Fig. 11.4c). When conducting these recordings, we realized that whereas multiple spikes caused robust Ca^{2+} transients in MCBL-stained cells, Ca^{2+} signals caused by a single action potential were small and of variable amplitude (Fig. 11.4b). We extracted Ca^{2+} transients from the recorded Ca^{2+} trace using a template-matching algorithm recognizing sharply rising transients with amplitude three times larger than the standard deviation of the corresponding baseline noise. The analyzed trace was then compared with the underlying electrical recordings to determine the signal detection probability. In 8-month-old mice (e.g., Fig. 11.4), Ca^{2+} recordings reliably reported firing of four and more action potentials (detection probability of 100%). For triplets the detection probability was 88.5%, for doublets 79.1% and for single action potentials it was 60% ($n=21$ cells). Also in young tissue (postnatal days 13–16), the probability of detection of a single action potential was 62% ($n=6$ cells). Thus, under our experimental conditions majority but not all single action potentials can be detected in vivo when imaging neurones stained using MCBL. To improve detection sensitivity one can (1) increase signal amplitudes by using better dyes, (2) decrease movement artifacts contributing to background noise, (3) increase brightness of the dye to minimize shot noise in the system, (4) increase the sampling rate of the system (in order to decrease filtering of fast Ca^{2+} signals) and (5) improve signal extraction algorithms.

5.2. In Vivo Imaging of Alzheimer's Disease

Understanding the early cellular events in the development of Alzheimer's disease (AD – (35)) is fundamental for the development of therapeutic strategies against this devastating neurodegenerative pathology. The early events in the pathogenesis of AD are largely unknown (36, 37). The advanced disease is marked by aberrant protein deposition in the form of β-amyloid containing senile plaques and tau-containing neurofibrillary tangles associated with extensive neuronal and synaptic loss in vulnerable parts of the brain. The cognitive disability specific to AD is associated with the progressive loss of neurones and synapses in the networks involved in memory and higher cognitive functions such as the nucleus of Meynert and the hippocampus. This loss of specific neuronal populations is paralleled with the accumulation of amyloid plaques and neurofibrillary tangles. Several hypotheses have been proposed over the years to explain the pathogenesis of the AD one of these concentrating on dysregulation of neuronal Ca^{2+} homeostasis (38–40). Accumulation of senile plaques and tau-pathology can be mimicked in numerous mouse models, which are overexpressing AD-related proteins with mutations isolated from the family form of the disease (41, p. 344, 42, 43).

As shown in Fig. 11.5, MCBL can be combined with specific staining of amyloid plaques with thioflavin-S (20, 44, 45). Thioflavin-S can either be applied topically or injected into the brain in a MCBL-like fashion. As described in (20), topical application results in reduced depth penetration of the dye and therefore pressure injection is the method of choice if aiming at deeper cortical regions. A tri-color staining in Fig. 11.5b illustrates neurones, astrocytes and amyloid-plaques in a mouse model of AD (34). Cells and plaques are stained with OGB-1 as well as specific markers (SR101 and thioflavin-S) and are visualized using two different dichroic beamsplitters (515 nm to separate fluorescence of thioflavin-S and 570 nm to distinguish between OGB-1 and SR101). In our resent study (34), we examined the pattern of spontaneous cortical activity in neurones located in the vicinity of amyloid plaques (e.g., Fig. 11.5c–e). It turned out that many of these neurones become hyperactive in the course of the disease. Hyperactive neurones not only show increased frequencies of spontaneous Ca^{2+} transients, they also tend to fire in synchrony and therefore under certain conditions may act as a seed for epileptiform activity.

6. Advantages and Disadvantages of MCBL

The core principle of the MCBL lies in a targeted delivery of the membrane-permeant indicators directly to the region of the interest for subsequent in vivo imaging. This technique has several advantages: (1) it requires a rather minor surgery thus being relatively harmless to the experimental animal; (2) the staining is confined to the target region (may be because connective tissues present the effective barrier for AM dyes; in our experiments injecting the indicator in the hippocampus never resulted in staining the cortex); (3) targeted administration of the indicator minimizes the amount of the dye used as compared to the bath application and reduces the loss of the dye due to diffusion or uptake by non-neural cells; (4) in combination with cell type-specific markers, it allows to study defined neuronal populations.

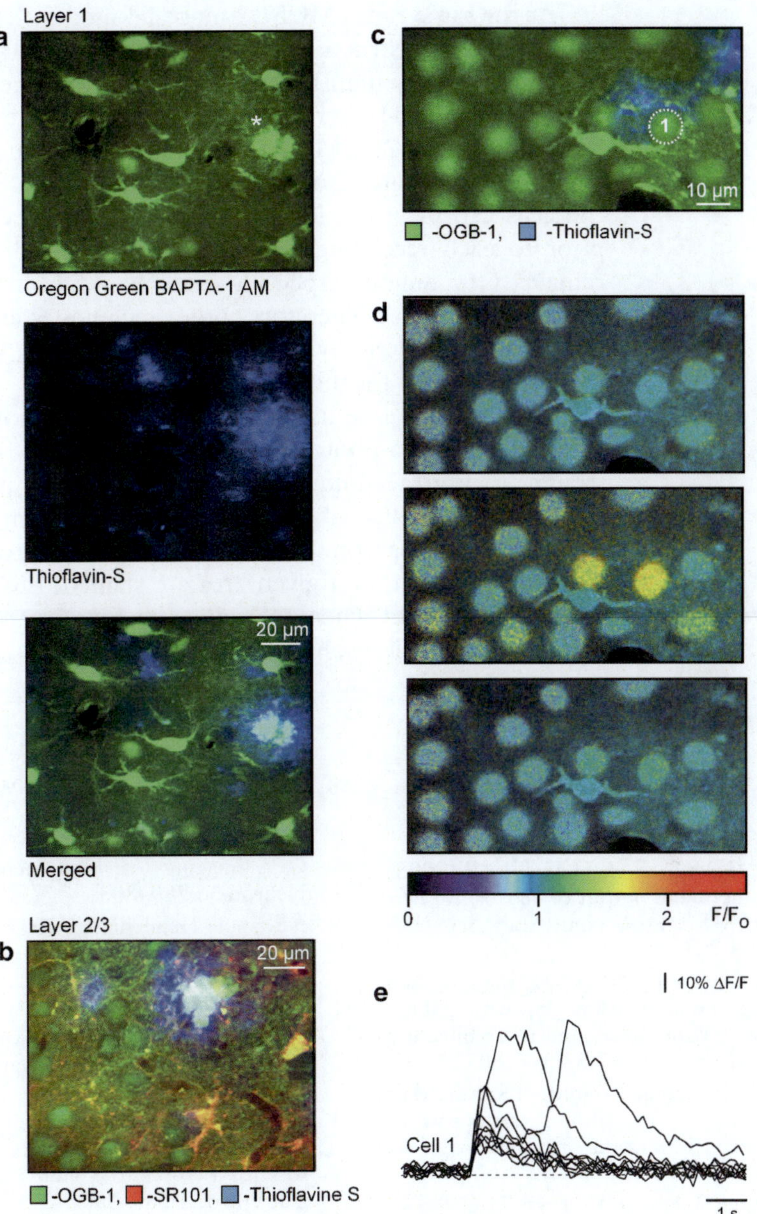

Fig. 11.5. In vivo two-photon imaging of neuronal function in a mouse model of Alzheimer's disease. (**a**) Images of the cortical layer 1 stained with OGB-1 (*upper*) and with the fluorescent plaque binding substance thioflavin-S (*middle*). The merged image is shown in the *lower panel*. Note that OGB-1 also binds to large amyloid plaques (marked with an *asterisk* in the *upper panel*). (**b**) A microphotograph of the layer 2/3 stained with OGB-1 (*green*) Sulforhodamine 101 (*red*) and thioflavin-S (*blue*). (**c**) An image of the vicinity of two amyloid plaques. (**d**) Pseudo color images taken before (*upper*) during (*middle*) and after (*lower*) spontaneous Ca^{2+} transients occurring simultaneously in two cells. The same cells as in (**c**). (**e**) Ten consecutive Ca^{2+} transients recorded in vivo from a neurone, marked with a corresponding number in (**c**).

Moreover, MCBL is the only known technique so far which allows effective loading of various Ca^{2+} indicators into neural cells in adult and old brain tissue.

At the same time, MCBL cannot be used for the recordings from subcellular structures. The reasons for that are associated with (1) relatively high background staining of the surrounding neuropil and (2) a rather low concentration of the dye attained in the cells (around 20 μM (4)). Continuous chronic recordings of cellular function are also difficult when using MCBL. Although the cells can in principle be re-stained with the Ca^{2+} indicator prior to each recording session, in practice such a procedure remains very tedious especially under conditions of implanted glass window (46). Therefore, other techniques as for example expression of genetically encoded Ca^{2+} indicators (47, 48) seem to be more suited for this kind of recordings.

In conclusion, the use of MCBL in combination with multicolor in vivo two-photon imaging is a versatile and relatively straight-forward technique which allows real time monitoring of the activity of neuronal and glial cells in different areas of the brain and in brains from animals of different age groups. Because all cell somata within a given area are stained, this technique is especially powerful when analyzing the function of neural networks at single cell resolution.

References

1. Denk W, Svoboda K (1997) Photon upmanship: why multiphoton imaging is more than a gimmick. Neuron 18:351–357
2. Kerr JN, Greenberg D, Helmchen F (2005) Imaging input and output of neocortical networks in vivo. Proc Natl Acad Sci USA 102:14063–14068
3. Ohki K, Chung S, Ch'ng YH, Kara P, Reid RC (2005) Functional imaging with cellular resolution reveals precise micro-architecture in visual cortex. Nature 433:597–603
4. Stosiek C, Garaschuk O, Holthoff K, Konnerth A (2003) In vivo two-photon calcium imaging of neuronal networks. Proc Natl Acad Sci USA 100:7319–7324
5. Grynkiewicz G, Poenie M, Tsien RY (1985) A new generation of Ca^{2+} indicators with greatly improved fluorescence properties. J Biol Chem 260:3440–3450
6. Tsien RY (1980) New calcium indicators and buffers with high selectivity against magnesium and protons: design, synthesis, and properties of prototype structures. Biochemistry 19: 2396–2404
7. Tsien RY (1981) A non-disruptive technique for loading calcium buffers and indicators into cells. Nature 290:527–528
8. Nagayama S, Zeng S, Xiong W, Fletcher ML, Masurkar AV, Davis DJ, Pieribone VA, Chen WR (2007) In vivo simultaneous tracing and Ca^{2+} imaging of local neuronal circuits. Neuron 53:789–803
9. Nevian T, Helmchen F (2007) Calcium indicator loading of neurons using single-cell electroporation. Pflugers Arch 454:675–688
10. Svoboda K, Denk W, Kleinfeld D, Tank DW (1997) In vivo dendritic calcium dynamics in neocortical pyramidal neurons. Nature 385:161–165
11. Garaschuk O, Milos RI, Konnerth A (2006) Targeted bulk-loading of fluorescent indicators for two-photon brain imaging in vivo. Nat Protoc 1:380–386
12. Brustein E, Marandi N, Kovalchuk Y, Drapeau P, Konnerth A (2003) "In vivo" monitoring of neuronal network activity in zebrafish by two-photon Ca^{2+} imaging. Pflugers Arch 446:766–773
13. Niell CM, Smith SJ (2005) Functional imaging reveals rapid development of visual response properties in the zebrafish tectum. Neuron 45:941–951
14. Petzold GC, Albeanu DF, Sato TF, Murthy VN (2008) Coupling of neural activity to

blood flow in olfactory glomeruli is mediated by astrocytic pathways. Neuron 58:897–910

15. Sullivan MR, Nimmerjahn A, Sarkisov DV, Helmchen F, Wang SS (2005) In vivo calcium imaging of circuit activity in cerebellar cortex. J Neurophysiol 94:1636–1644
16. Xu HT, Pan F, Yang G, Gan WB (2007) Choice of cranial window type for in vivo imaging affects dendritic spine turnover in the cortex. Nat Neurosci 10:549–551
17. Hirase H, Qian L, Bartho P, Buzsaki G (2004) Calcium dynamics of cortical astrocytic networks in vivo. PLoS Biol 2:E96
18. Wang X, Lou N, Xu Q, Tian GF, Peng WG, Han X, Kang J, Takano T, Nedergaard M (2006) Astrocytic Ca^{2+} signaling evoked by sensory stimulation in vivo. Nat Neurosci 9:816–823
19. Garaschuk O, Milos RI, Grienberger C, Marandi N, Adelsberger H, Konnerth A (2006) Optical monitoring of brain function in vivo: from neurons to networks. Pflugers Arch 453:385–396
20. Eichhoff G, Busche MA, Garaschuk O (2008) In vivo calcium imaging of the aging and diseased brain. Eur J Nucl Med Mol Imaging 35(Suppl 1):S99–S106
21. Jung S, Aliberti J, Graemmel P, Sunshine MJ, Kreutzberg GW, Sher A, Littman DR (2000) Analysis of fractalkine receptor CX_3CR1 function by targeted deletion and green fluorescent protein reporter gene insertion. Mol Cell Biol 20:4106–4114
22. Ikegaya Y, Aaron G, Cossart R, Aronov D, Lampl I, Ferster D, Yuste R (2004) Synfire chains and cortical songs: temporal modules of cortical activity. Science 304:559–564
23. Wachowiak M, Cohen LB (2001) Representation of odorants by receptor neuron input to the mouse olfactory bulb. Neuron 32: 723–735
24. Nimmerjahn A, Kirchhoff F, Kerr JN, Helmchen F (2004) Sulforhodamine 101 as a specific marker of astroglia in the neocortex in vivo. Nat Methods 1:31–37
25. Hirrlinger PG, Scheller A, Braun C, Quintela-Schneider M, Fuss B, Hirrlinger J, Kirchhoff F (2005) Expression of reef coral fluorescent proteins in the central nervous system of transgenic mice. Mol Cell Neurosci 30: 291–303
26. Sohya K, Kameyama K, Yanagawa Y, Obata K, Tsumoto T (2007) GABAergic neurons are less selective to stimulus orientation than excitatory neurons in layer II/III of visual cortex, as revealed by in vivo functional Ca2+ imaging in transgenic mice. J Neurosci 27:2145–2149
27. Adelsberger H, Garaschuk O, Konnerth A (2005) Cortical calcium waves in resting newborn mice. Nat Neurosci 8:988–990
28. Mrsic-Flogel TD, Hofer SB, Ohki K, Reid RC, Bonhoeffer T, Hubener M (2007) Homeostatic regulation of eye-specific responses in visual cortex during ocular dominance plasticity. Neuron 54:961–972
29. Aubin JE (1979) Autofluorescence of viable cultured mammalian cells. J Histochem Cytochem 27:36–43
30. Porta EA (2002) Pigments in aging: an overview. Ann N Y Acad Sci 959:57–65
31. Terman A, Brunk UT (2004) Lipofuscin. Int J Biochem Cell Biol 36:1400–1404
32. Brunk UT, Terman A (2002) Lipofuscin: mechanisms of age-related accumulation and influence on cell function. Free Radic Biol Med 33:611–619
33. Xu C (2000) Two-photon cross sections of indicators. In: Yuste R, Konnerth A (eds) Imaging: a laboratory manual. Cold Spring Harbor Press, Cold Spring Harbor, NY, pp 11–19
34. Busche MA, Eichhoff G, Adelsberger H, Abramowski D, Wiederhold KH, Haass C, Staufenbiel M, Konnerth A, Garaschuk O (2008) Clusters of hyperactive neurons near amyloid plaques in a mouse model of Alzheimer's disease. Science 321:1686–1689
35. Alzheimer A (1907) Über eine eigenartige Erkrankung der Hirnrinde. Allg Z Psychiat Psych-Gericht Med 64:146–148
36. Maudsley S, Mattson MP (2006) Protein twists and turns in Alzheimer disease. Nat Med 12:392–393
37. Van Broeck B, Van Broeckhoven C, Kumar-Singh S (2007) Current insights into molecular mechanisms of Alzheimer disease and their implications for therapeutic approaches. Neurodegener Dis 4:349–365
38. Bezprozvanny I (2009) Calcium signaling and neurodegenerative diseases. Trends Mol Med 15:89–100
39. Stutzmann GE, Smith I, Caccamo A, Oddo S, Laferla FM, Parker I (2006) Enhanced ryanodine receptor recruitment contributes to Ca^{2+} disruptions in young, adult, and aged Alzheimer's disease mice. J Neurosci 26:5180–5189
40. Verkhratsky A (2005) Physiology and pathophysiology of the calcium store in the endoplasmic reticulum of neurons. Physiol Rev 85:201–279

41. Sturchler-Pierrat C, Abramowski D, Duke M, Wiederhold KH, Mistl C, Rothacher S, Ledermann B, Burki K, Frey P, Paganetti PA, Waridel C, Calhoun ME, Jucker M, Probst A, Staufenbiel M, Sommer B (1997) Two amyloid precursor protein transgenic mouse models with Alzheimer disease-like pathology. Proc Natl Acad Sci USA 94:13287–13292
42. Gotz J, Streffer JR, David D, Schild A, Hoerndli F, Pennanen L, Kurosinski P, Chen F (2004) Transgenic animal models of Alzheimer's disease and related disorders: histopathology, behavior and therapy. Mol Psychiatry 9:664–683
43. Oddo S, Caccamo A, Shepherd JD, Murphy MP, Golde TE, Kayed R, Metherate R, Mattson MP, Akbari Y, LaFerla FM (2003) Triple-transgenic model of Alzheimer's disease with plaques and tangles: intracellular Abeta and synaptic dysfunction. Neuron 39:409–421
44. Bacskai BJ, Kajdasz ST, Christie RH, Carter C, Games D, Seubert P, Schenk D, Hyman BT (2001) Imaging of amyloid-beta deposits in brains of living mice permits direct observation of clearance of plaques with immunotherapy. Nat Med 7:369–372
45. Christie RH, Bacskai BJ, Zipfel WR, Williams RM, Kajdasz ST, Webb WW, Hyman BT (2001) Growth arrest of individual senile plaques in a model of Alzheimer's disease observed by in vivo multiphoton microscopy. J Neurosci 21:858–864
46. Trachtenberg JT, Chen BE, Knott GW, Feng G, Sanes JR, Welker E, Svoboda K (2002) Long-term in vivo imaging of experience-dependent synaptic plasticity in adult cortex. Nature 420:788–794
47. Mank M, Santos AF, Direnberger S, Mrsic-Flogel TD, Hofer SB, Stein V, Hendel T, Reiff DF, Levelt C, Borst A, Bonhoeffer T, Hubener M, Griesbeck O (2008) A genetically encoded calcium indicator for chronic in vivo two-photon imaging. Nat Methods 5:805–811
48. Wallace DJ, Borgloh SM, Astori S, Yang Y, Bausen M, Kugler S, Palmer AE, Tsien RY, Sprengel R, Kerr JN, Denk W, Hasan MT (2008) Single-spike detection in vitro and in vivo with a genetic Ca^{2+} sensor. Nat Methods 5:797–804

Chapter 12

Ca^{2+} Imaging of Glia

Christian Lohr and Joachim W. Deitmer

Abstract

Glial cells are besides neurons the second major cell type of nervous systems and are either of neuroectodermal (macroglia) or mesodermal (microglia) origin. As electrically non-excitable cells, they employ calcium signals in response to most external stimuli, which initiate cellular activity. A variety of techniques are described here, which have been developed to monitor cellular calcium in different glial cell preparations from cell culture, acute tissue slices to in vivo measurements. New optical innovations in the last two decades, as e.g. generations of new calcium-sensitive fluorescent dyes, genetically-encoded calcium sensors and multiple applications of laser scanning microscopy have allowed novel experimental approaches and have provided important results for our understanding of nervous systems. The study of calcium signalling in glial cells from both vertebrate and invertebrate preparations has also significantly contributed to our understanding of the significance of cytosolic calcium elevations in different cell types and subcellular compartments.

Key words: Calcium imaging, Glia, Astrocyte, Cell culture, Brain slices, In vivo imaging, Invertebrates

1. Introduction

Cytosolic calcium signalling in electrically non-excitable cells as are most types of glial cells in central and peripheral nervous systems often plays a dominant role in mediating primary signals to cellular activity. Glial cells in vertebrate animals comprise astrocytes the major type of macroglial cells, oligodendrocytes and Schwann cells, the myelinating glial cells in the central and peripheral nervous system respectively, and microglial cells, immune-competent cells with macrophage-like origin. In addition, there is a number of glial cell types in invertebrate animals often with similar functions as either astrocytes or microglia in vertebrate nervous systems. Astrocytes themselves are by no means a homogeneous group of glial cells, but comprise a number of different

A. Verkhratsky and Ole H. Petersen (eds.), *Calcium Measurement Methods,* Neuromethods, vol. 43
DOI 10.1007/978-1-60761-476-0_12, © Humana Press, a part of Springer Science+Business Media, LLC 2010

cell types and we assume today that there are many types of astrocytes in different brain regions. Mammalian astrocytes were originally divided into either protoplasmic type I or fibrous type II astrocytes and more recently astrocytes with ionotropic glutamate receptors, but no glutamate transporters (GluR-type astrocytes) were distinguished from astrocytes with glutamate transporters but no ionotropic glutamate receptors (GluT-type astrocytes) respectively (1). These latter types of astrocytes also differ in gap junctional communication with the GluT-type astrocytes being strongly coupled by gap junctions, in particular connexion 43 and connexion 30 and with a linear current–voltage relationship and the GluR-type astrocytes being not coupled to neighbouring astrocytes with a high membrane resistance, and a non-linear current–voltage relationship (1, 2).

As with all of these cell type classifications established so far, there are known exemptions, such as the Bergmann glial cells in the cerebellum. Bergmann glia is a radial type of macroglial cell, which is a specialized astrocyte, which has both EAATs and ionotropic alpha-amino-3-hydroxy-5-methyl-4-isoxazolepropionic acid (AMPA)-type glutamate receptors. There are radial-type astrocytes also in the developing cortex, which extend from the ventricular surface to the pial surface. Furthermore, Müller glial cells the principal glial cell type in the vertebrate retina extend across the entire retina from the photoreceptors to the inner retinal surface, and have a variety of properties and functions which go beyond those of classical astrocytes.

Single and repetitive rises of cytosolic Ca^{2+} play a complex role for initiating intracellular signalling cascades modulating cellular functions and intercellular interaction in the majority of glial cells in particular astrocytes. Most of these functions and interactions involve other neighbouring astrocytes and neurons, but there are also signalling pathways to oligodendrocytes and microglial cells. Astrocytes are endowed with a large number of metabotropic receptors in their cell membrane, most of which are coupled to the release of Ca^{2+} from the endoplasmic reticulum (ER) via phospholipase C (PLC)-mediated formation of inositol-trisphosphate (IP_3). Astrocytic Ca^{2+} signalling can consist of a single Ca^{2+} transient with or without a shoulder or plateau phase, repetitive Ca^{2+} transients, the so-called Ca^{2+} oscillations, or irregular Ca^{2+} rises, depending on the species of primary messenger (neurotransmitter, hormone, cytokine, growth factor) and its concentration. These Ca^{2+} signals may spread along the cell processes and across cell boundaries to neighbouring astrocytes in form of Ca^{2+} waves and can be evoked or modulated by neuronal activity. The spatial and temporal properties of these Ca^{2+} signalling modes reflect the versatility of this intracellular messenger system. The signalling pathway leading to a cytosolic Ca^{2+} rise common to many cell types may be regarded as a type of "excitation" in electrically non-excitable cells like astrocytes. Cytosolic Ca^{2+} transients may initiate Ca^{2+}-dependent release of transmitters (gliotransmitters) and modulators affecting neuronal excitability or vasoconstriction/vasodilation of blood vessels in the brain and may hence modulate the supply of energy to neurons .

In the present chapter, we discuss some approaches of studying cytosolic Ca^{2+} signalling in astrocytes in culture, in acute tissue slices (in situ), and in vivo, as well as in some other types of glial cells. We shall focus our attention to the application of measuring cytosolic Ca^{2+} in glial cells including dye loading, identification of glial cells and special features of Ca^{2+} signalling in astrocytes such as Ca^{2+} oscillations and propagation of Ca^{2+} waves using different microscopic techniques.

2. Cultured Glial Cells

2.1. Culture Conditions

Fundamental properties of glial Ca^{2+} signalling can often best be studied in glial cell cultures. The advantage of glial cell cultures as compared to brain slices is the monolayer and thus "2-dimensionality" of cells and the lack of neuronal influence on glial performance, which has to be considered in a tissue preparation. Using suitable culture techniques, virtually pure glial cell cultures can be obtained. We have used newborn rats to produce cultures of cerebellar astrocytes of >95% purity (3). Puppies are decapitated and the cerebellum is dissected in ice-cold saline (120 mM NaCl, 5.4 mM KCl, 0.8 mM $MgCl_2$, 25 mM Tris–HCl, 15 mM glucose; pH 7.2). Particular care should be taken removing the meninges, because meningal fibroblasts are the main source of contamination in astrocyte cultures. After removal of the meninges, pieces of cerebellar tissue are digested with 1% trypsin for 5 min. The enzymatic treatment is stopped by addition of 10% foetal calf serum. The tissue is triturated and then centrifuged. The cell pellet is resuspended in medium (per litre: 13.4 g DMEM, 1 g BSA, 10 μg EGF, 10 mg insulin, 10 ml penicillin/streptomycin, 3.3 ml transferrin; pH 7.2) and transferred into poly-d-lysine-coated culture flasks. Cells are maintained at 37°C and 7% CO_2. After 10 days, the cells are subcultivated by plating 4×10^6 cells/ml on poly-d-lysine-coated glass coverslips in Petri dishes. Subcultivated astrocytes are maintained at 37°C and 7% CO_2 and can be used for up to 3 weeks. A confluent monolayer of cells is usually obtained after 4–8 days in culture when an astrocyte culture is often used for measuring cytosolic Ca^{2+}.

The purity of the astrocyte culture can be tested by co-labelling of formaldehyde-fixed cells with a nuclear stain such as DAPI or propidium iodide to obtain the total cell number, and with anti-GFAP (glial fibrillar acidic protein) antibodies the number of

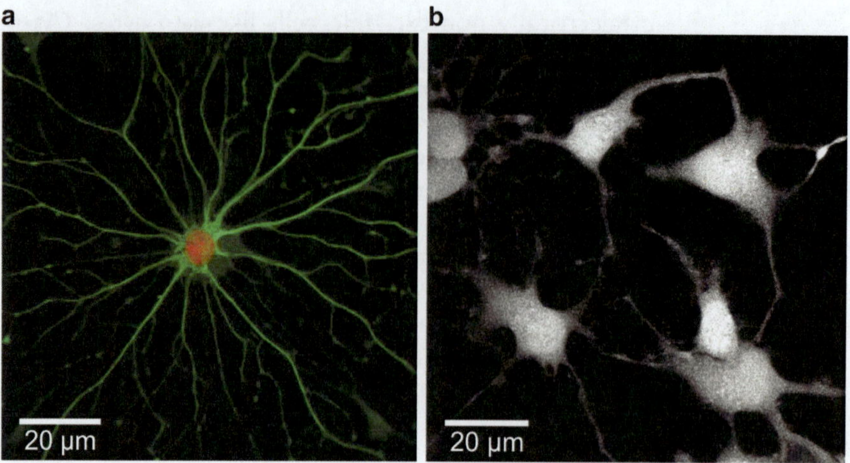

Fig. 12.1. Cultured astrocytes. (**a**) Astrocyte labelled with anti-GFAP antibody (*green*) and the nuclear stain propidium iodide (*red*). (**b**) Cultured astrocytes loaded with Fura-2 AM.

mature GFAP-expressing cells (Fig. 12.1) and/or with anti-S100B antibodies to obtain the number of S100B-expressing cells. For other types of glial cells, antibodies against the respective marker proteins have to be used such as galactocerebroside for oligodendrocytes (4) and OX-42 for microglia (5). For immunolabelling of cultured astrocytes, cell cultures on coverslips are fixed with 4% formaldehyde in PBS (pH 7.4) for 1 h. The coverslips are rinsed with PBS, and unspecific binding sites are blocked with 10% normal goat serum in PBS with 0.1% Triton-100. The primary antibody (rabbit anti-glial fibrillary acidic protein, 1:1,000; Z 0334; DakoCytomation, Glostrup, Denmark) in PBS with 0.05% Triton-100 is added to the fixed cells and incubated overnight at 4°C. After rinsing with PBS the secondary antibody (Alexa Fluor 488 goat-anti-rabbit, 1:1,000; Invitrogen) and propidium iodide (20 μM) in PBS are incubated for 2 h. The coverslips with the immuno-stained cells are then mounted on slides with mounting medium. The number of propidium iodide-stained nuclei and the number of GFAP-positive and/or S100B-positive cells are counted to obtain the relative number of astrocytes. We usually achieve cell cultures with >95% astrocytes.

Neuron–glia co-cultures are used to study neuron–glia interactions under controlled conditions. For co-cultivation, mainly two methods are used: First, glial cells and neurons are grown simultaneously in the same culture (6), and second, glial cells are first grown to a confluent monolayer and neurons are seeded on top of the glial cells thereafter (7).

2.2. Dye Loading

A wide range of Ca^{2+} indicator dyes is available for Ca^{2+} imaging in cultured glial cells (see Chap. 3). Cell cultures are usually studied using conventional fluorescence microscopy and therefore no limitation with respect to the excitation wavelengths of the Ca^{2+} indicators has to be considered in contrast to laser scanning microscopy, where the excitation is limited by the available wavelengths of the lasers. Excitation of Ca^{2+} indicators ranges from UV light such as for Fura-2 and Indo-1 to yellow light such as for Calcium Crimson and X-Rhod-1. Thus, Ca^{2+} indicators can be used in combination with other dyes, e.g. to label organelles, or fluorescent proteins that are used to identify cells or to tag a given protein. In addition to the excitation and emission wavelengths of the Ca^{2+} indicator, the Ca^{2+} indicator's affinity for Ca^{2+} given as the binding constant K_D has to be taken into consideration. The K_D represents the Ca^{2+} concentration at which half of the Ca^{2+} indicator molecules are bound to Ca^{2+} and half of the Ca^{2+} indicator molecules are Ca^{2+}-free. The K_D also represents the Ca^{2+} concentration at which the Ca^{2+} indicator responds most sensitively to changes of free Ca^{2+} ions; hence, a Ca^{2+} indicator with a K_D within the range of the expected Ca^{2+} concentrations should be chosen. In case of glial cells, the cytoplasmic Ca^{2+} concentration ranges from 50 to 100 nM at resting conditions to several hundred nM up to 1 μM upon stimulation, e.g. with a receptor agonist. The most frequently used Ca^{2+} indicators such as Fura-2, Fluo-3, Fluo-4, Calcium Green-1, and Oregon Green BAPTA-1 have K_D values between 100 and 400 nM and are therefore well suited to measure Ca^{2+} increases in the physiological range. However, these high-affinity Ca^{2+} indicators bind a considerable amount of Ca^{2+} and thus buffer intracellular Ca^{2+} changes resulting in Ca^{2+} transients with reduced amplitude and slowed kinetics (8). Therefore, low-affinity Ca^{2+} indicators such as Fluo-5, Furaptra, Calcium Green-2 and Oregon Green BAPTA-5N are used in some cases to minimize the impact of Ca^{2+} indicators as Ca^{2+} buffers on intracellular Ca^{2+} signalling (9, 10).

Bulk loading of a given Ca^{2+} indicator is an efficient technique to reliably introduce the Ca^{2+} indicator into the entire cell population (Fig. 12.1). Ca^{2+} indicators are polyanionic – therefore they bind several Ca^{2+} – and are hence membrane-impermeable. In order to enable uptake of the Ca^{2+} indicator into cells, the indicator molecules are derivatized as acetoxymethyl (AM) esters resulting in a nonpolar Ca^{2+}-insensitive molecule that can permeate the cell membrane (Fig. 12.2). Once in the cell, intracellular, intrinsic esterases hydrolyze the ester bounds releasing the polar membrane-impermeable Ca^{2+}-sensitive indicator molecules thereby trapping the dye in the cell. We have adopted the bulk loading technique to load astrocyte cell cultures with Fura-2 AM or Fluo-4 AM (3, 11). Stock solutions of the AM ester of the Ca^{2+} indicator are prepared in 20% pluronic acid/80% DMSO at a concentration of 1 mM and can be stored at –20°C for up to 6 months. The stock solution of the Ca^{2+} indicator is

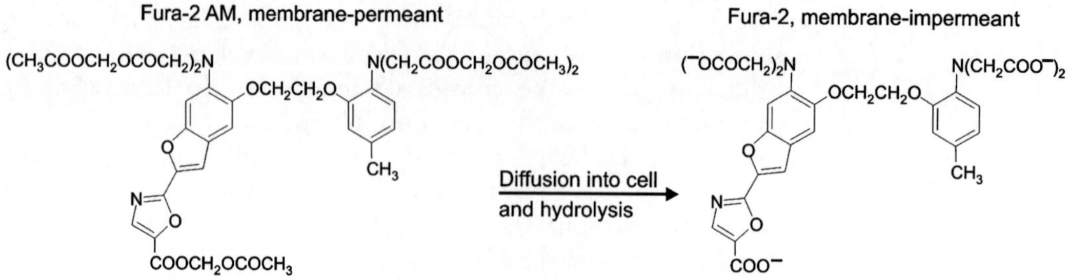

Fura-2 AM, membrane-permeant

$(CH_3COOCH_2OCOCH_2)_2N$ $N(CH_2COOCH_2OCOCH_3)_2$ OCH_2CH_2O CH_3 $COOCH_2OCOCH_3$

Diffusion into cell and hydrolysis

Fura-2, membrane-impermeant

$(^{-}OCOCH_2)_2N$ $N(CH_2COO^{-})_2$ OCH_2CH_2O CH_3 COO^{-}

Fig. 12.2. Fura-2 chemical formula and principle of bulk loading with Fura-2 AM (acetomethylether of Fura-2), which diffuses freely across cell membranes and is cleaved by esterases and thus trapped in the cells .

diluted to 1–2 µM in HEPES-buffered ACSF (145 mM NaCl, 5 mM KCl, 2 mM $CaCl_2$, 1 mM $MgCl_2$, 10 mM d-glucose, 10 mM HEPES, pH 7.4, adjusted by addition of NaOH) and cell cultures are incubated in this solution for 30 min at room temperature. Cells are then briefly rinsed two to three times with ACSF and can immediately be used for Ca^{2+} imaging experiments. By varying the final dye concentration and the duration of incubation, the amount of dye loaded in the cells can be adjusted according to need (e.g. to get a larger signal and/or better signal-to-noise ratio).

2.3. Ratiometric Ca^{2+} Imaging Using Conventional Microscopy

We have used Fura-2 to measure Ca^{2+} in cultured glial cells. Fura-2 is brightly fluorescent has a low photobleaching rate and can readily be bulk-loaded into cultured cells. Fura-2 has a K_D of 135 and 224 nM as measured in KCl solution and physiological buffer respectively (12), but this value may differ in the intracellular environment and both smaller (13, 14) as well as larger values (15) have been determined intracellularly. The most important advantage of Fura-2 is the ability to perform ratiometric measurements thereby eliminating measurement errors due to photobleaching and differences in Ca^{2+} indicator uptake efficacy. Fura-2 exhibits a shift in the excitation wavelength from near 360 nm under Ca^{2+}-free conditions to near 340 nm under Ca^{2+}-saturated conditions. The ratio is build of the fluorescence measured at 340 nm excitation divided by the fluorescence measured at 360 nm excitation or to improve the dynamic range 380 nm excitation.

Since a given ratio value corresponds to a given Ca^{2+} concentration, the Ca^{2+} concentration in cultured glial cells can be calculated from the ratio measurements after in vivo calibration of the system. For calibration, Fura-2-loaded cells are permeabilized with 20 µM 4-bromo-A23187 and/or 10 µM ionomycin to allow for equilibration of the cytoplasmic Ca^{2+} concentration with the extracellular Ca^{2+} concentration, and hence the adjustment of

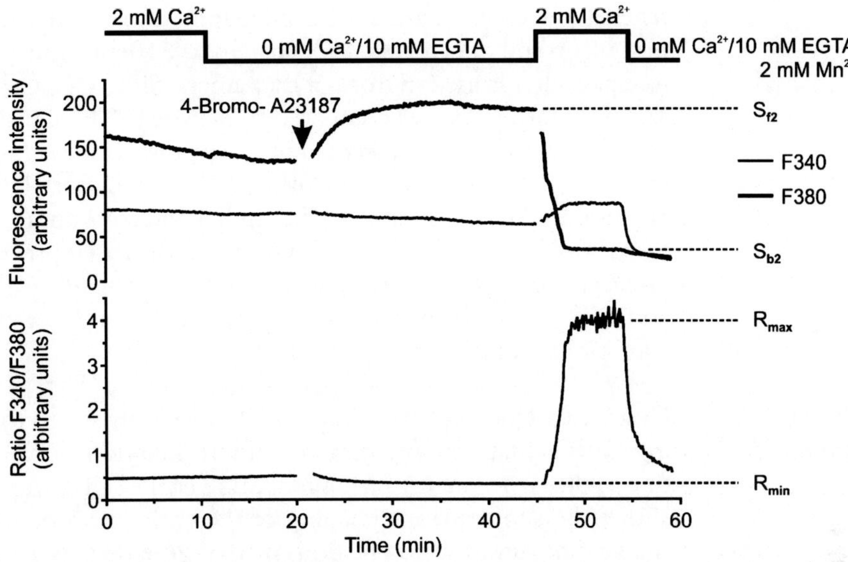

Fig. 12.3. Calibration of the Fura-2 imaging system using astrocyte cultures. Cells are bulk-loaded with Fura-2 AM and the fluorescence at excitation with 340 nm (F340) and 380 nm (F380) is measured to calculate the ratio (F340/F380). External Ca^{2+} is washed out for 10 min (0 mM Ca^{2+}/10 mM EGTA) before the ionophore 4-bromo-A23187 (20 μM) is added to permeabilize the cell membrane. After intracellular and external Ca^{2+} have equilibrated (after approximately 20 min), 2 mM Ca^{2+} are added. After another 10 min, a Ca^{2+}-free, 2 mM Mn^{2+}-containing solution is washed in. Mn^{2+} enters the cells and quenches the Fura-2 fluorescence revealing the Fura-2-independent background fluorescence, which has to be subtracted from all measured fluorescence values for background correction. R_{min}, R_{max}, S_{f2} and S_{b2} are read from the calibration curves, background-corrected and implemented into the formula given by Grynkiewicz et al. (12) to calculate Ca^{2+} concentrations from the ratio values (see text for more details).

defined cytoplasmic Ca^{2+} concentrations (Fig. 12.3). Fluorescence and ratio values are measured under Ca^{2+}-free conditions (Ca^{2+}-free ACSF with 0.5 mM EGTA added) and Ca^{2+}-saturated conditions (ACSF containing 2–10 mM Ca^{2+}). These values are implemented into the equation given by Grynkiewicz et al. (12) to calculate the free cytoplasmic Ca^{2+} concentration $(Ca^{2+})_i$:

$$[\mathrm{Ca}^{2+}]_\mathrm{i} = K_\mathrm{D}(R - R_\mathrm{min}) / (R_\mathrm{max} - R) \times (S_\mathrm{f2} / S_\mathrm{b2})$$

R is the ratio of fluorescence values measured with excitation wavelengths of 340 nm versus 380 nm throughout the experiment. R_{min} and R_{max} are the ratio values under Ca^{2+}-free and Ca^{2+}-saturated conditions respectively as determined during the calibration procedure. S_{f2} and S_{b2} are the fluorescence values measured at an excitation of 380 nm under Ca^{2+}-free and Ca^{2+}-saturated conditions respectively.

For ratiometric measurements, a Ca^{2+} imaging system that allows for rapid switch of excitation wavelengths is essential. This can be achieved by a monochromator with a tuneable grating (e.g. Polychrome, Till Photonics), which switches between 340 and 380 nm within less than a millisecond. Pairs of images are taken by a CCD camera. The emission wavelength chosen for best recording of cytosolic Ca^{2+} changes is 510 nm, and a 490 nm longpass filter is used in front of the camera. Since glial Ca^{2+} signals are often rather slow as compared to neuronal Ca^{2+} signalling, an acquisition rate of 0.2–1 Hz per image pair is usually sufficient for most applications. Cover slips overgrown with cultured glial cells are mounted in an experimental chamber and continuously superfused with ACSF. We have studied the fundamental properties of astrocytic Ca^{2+} signalling such as receptor-mediated Ca^{2+} influx and Ca^{2+} release from intracellular stores as well as store-operated Ca^{2+} influx. (3, 16, 17)

2.4. Investigation of Ca^{2+} Oscillations and Ca^{2+} Waves

Cultured astrocytes have turned out to be ideally suited for studying Ca^{2+} oscillations and intra- and intercellular Ca^{2+} waves; this is due to the two-dimensional organisation of the cell culture, which allows the simultaneous imaging of the entire cell population in the field of view. Ca^{2+} oscillations may occur either spontaneously or can be evoked by agonists, which may critically depend on the agonist concentration used. Ca^{2+} oscillations can give rise to Ca^{2+} waves, which are propagated both *intra*cellularly along cellular processes as well as *inter*cellularly across cell boundaries within a syncytium (Fig. 12.4). The strong coupling of astrocytes by gap junctions plays a crucial role for the generation and propagation of these Ca^{2+} waves.

Spontaneous cytosolic Ca^{2+} oscillations are repetitive Ca^{2+} transients with variable amplitude, which occur either in bursts of often two to eight transients or are irregularly distributed in time with a frequency of usually 0.1–0.01 Hz. They have been reported in both neurons and glial cells in culture, in situ, and in vivo (18–20). Ca^{2+} oscillations in astrocytes have been associated with Ca^{2+}-dependent

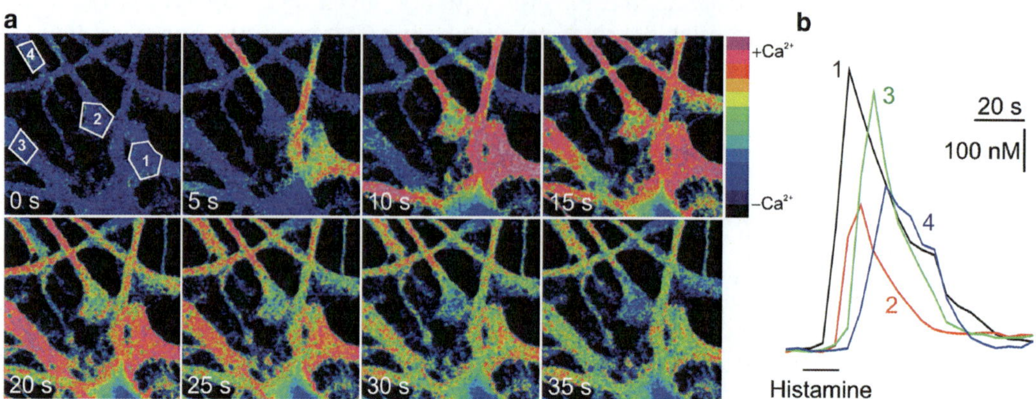

Fig. 12.4. Ca^{2+} wave in cultured astrocytes. (**a**) Time series of pseudo colour images of Fura-2-loaded cultured astrocytes stimulated briefly with histamine. Regions of interest (ROI) in the first image indicate areas in which the traces in (**b**) were analyzed. (**b**) Ca^{2+} traces measured in four astrocytes following application of 100 μM histamine. The delay in the onset of the Ca^{2+} transients from cell 1 to cell 4 reflects the propagation of the Ca^{2+} wave.

exocytosis of transmitters ("gliotransmitters"), gene regulation and modulation of regional blood flow.

Ca^{2+} waves are transmitted between astrocytes by two major pathways: First, intracellular Ca^{2+} mobilizing messengers such as $InsP_3$ diffuse from one cell to adjacent cells through gap junctions and secondly, ATP being released from one astrocyte in a Ca^{2+}-dependent manner can activate purinergic receptors of neighbouring cells thus initiating a new Ca^{2+} signal following release of Ca^{2+} from intracellular stores. The involvement of gap junctions in the propagation of Ca^{2+} waves has been shown by the use of gap junction blockers such as heptanol, octanol and carbenoxolone (21). However, since these substances also block gap junction hemichannels, anion channels and $P2X_7$ receptors (22, 23), which have been reported to mediate ATP release from astrocytes and may be included in the extracellular pathway of Ca^{2+} wave propagation (24–26), they do not allow for the discrimination between gap junction-dependent and gap junction-independent pathways of intercellular Ca^{2+} waves.

Evidence for an extracellular pathway of Ca^{2+} wave propagation was suggested by the finding that Ca^{2+} waves can cross cell-free areas of up to 120 μm width (27). Cell-free lanes can be produced in a confluent layer of astrocytes by scratching a razor blade or the tip of a glass micropipette through the cell layer few hours before the Ca^{2+} measurement. Ca^{2+} waves can be initiated by focal application of neurotransmitters by mechanical and by electrical stimulation of single astrocytes or a small group of astrocytes. The Ca^{2+} wave then spreads radially from the centre of stimulation. More directed propagation of Ca^{2+} waves in the direction of the bath perfusion has been observed both in astrocytes in culture and in *corpus callosum* brain slices, which has been taken as additional evidence for an extracellular messenger mediating intercellular Ca^{2+} waves (27, 28).

2.5. Ca^{2+} Measurements in Subcellular Compartments

Electron probe microanalysis revealed that vesicular or cisternal structures containing electron-dense material in frog ependymal glial cells contain deposits of calcium and phosphorus. The so-called "osmiophilic particles" in human astrocytes also contain calcium, this suggested that these organelles are storage sites of calcium was one of the first reports on Ca^{2+}-containing organelles in cells and in glial cells in particular (29).

Cell cultures are well suited for Ca^{2+} measurements in organelles such as the endoplasmic reticulum and mitochondria. Different methods have been employed to achieve dye accumulation in organelles (see Chaps. 9–11). Some Ca^{2+} indicators are preferentially taken up by organelles when incubated at temperatures above 30°C. Because the Ca^{2+} concentration in most organelles is much higher (in the order of 100 μM) than in the cytosol (in the order of 100 nM), Ca^{2+} indicators with K_D values of several μM should be chosen for monitoring organellar Ca^{2+}. The identity of the organelles can be verified by use of specific markers such as rhodamine 123 and mitotracker for mitochondria, ER tracker for endoplasmic reticulum, or lysotracker for lysosomes.

For Ca^{2+} measurements in mitochondria, rhod-2 and its derivatives can be used (30). The AM esters of these Ca^{2+} indicators hold a net positive charge and therefore accumulate in mitochondria due to the steep negative electrical gradient across the mitochondrial membrane. Uptake of rhod-2 AM can be enhanced by increasing the temperature above 30°C. Cytosolic Ca^{2+} buffering by mitochondria provides a potent mechanism to regulate the localized spread of astrocytic Ca^{2+} signals; dissipating the mitochondrial membrane potential using the mitochondrial uncoupler carbonyl cyanide *p*-trifluoromethoxy-phenyl-hydrazone (FCCP) with oligomycin prevented mitochondrial Ca^{2+} uptake and slowed the rate of decay of cytosolic (Ca^{2+}) transients, suggesting that mitochondrial Ca^{2+} uptake plays a significant role in the clearance of physiological cytosolic (Ca^{2+}) loads and for Ca^{2+} signalling in astrocytes (31, 32).

Golovina and Blaustein (33, 34) could selectively load astrocyte endoplasmic reticulum with the low-affinity Ca^{2+} indicator furaptra at 36°C while at 20°C most dye was trapped in the cytosol. The endoplasmic Ca^{2+} stores in mouse cortical astrocytes consist of ryanodine/caffeine-sensitive stores and ryanodine/caffeine-insensitive but cyclopiazonic acid/thapsigargin-sensitive stores, which can be visualized and analyzed at high magnification after selective loading with furaptra (34).

Changes in the submembrane Ca^{2+} concentration can be visualized by targeting genetically encoded Ca^{2+} indicators such as GCaMP2 to plasma membrane proteins. Lee et al. (35), e.g. measured submembrane Ca^{2+} changes in cultured astrocytes with GCaMP2 fused to the C-terminus of the α-subunit of the plasmamembrane Na^{+} pump facing the cytosol. Alternatively, total internal reflection fluorescence (TIRF) microscopy in combination with the Ca^{2+} indicator Fluo-4 has also be used to study submembrane Ca^{2+} signalling in cultured astrocytes (36).

3. Glial Cells in Brain Slices

Although cultured glial cells are well suited for investigation of fundamental properties of Ca^{2+} signalling, their usefulness to study complex interactions between glial cells and other glial cells, neurons or epithelial cells is limited. In addition, properties of cultured glial cells may differ from those of glial cells in brain tissue depending on the duration of culturing and on the culturing conditions. These constraints can only partly be circumvented by using acutely dissociated cells (37, 38). Therefore, cellular interactions and physiological processes involving glial cells are more appropriately studied in acute brain slices or if possible in vivo.

3.1. Preparation of Brain Slices

Particular care has to be taken during dissection and processing of brain tissue to minimize the traumatic impact and to maintain cellular functions. A juvenile mouse or rat is decapitated, the brain is dissected out of the skull and transferred into ice-cold CO_2/bicarbonate-buffered ACSF with reduced Ca^{2+} concentration and increased Mg^{2+} concentration to attenuate neural excitation (NaCl 125, KCl 2.5, $CaCl_2$ 0.5, $MgCl_2$ 2.5, d-glucose 25, $NaHCO_3$ 26, NaH_2PO_4 1.25, l-lactate 0.5, gassed with 95% O_2/5% CO_2 to adjust the pH to 7.4). Some groups also add 5–10 μM kynurenic acid in this saline and/or reduce the Na^+ concentration to reduce excitation mediated by ionotropic glutamate receptors and voltage-dependent channels respectively. A block of brain tissue containing the brain area of interest is dissected with a razor blade glued to the stage of a vibratome, and 250–300 μm thick brain slices are cut. The quality of the brain slices and hence largely the success of the experiments depends critically on this step of preparation. Therefore, clean blades must be used and the cutting edge must not be touched by scissors or forceps. The speeds of vibration and advance have to be carefully adjusted to keep shearing forces to a minimum. Brain slices are collected in 30°C warm, Ca^{2+}-reduced ACSF (see above), continuously gassed with carbogene, and allowed to equilibrate for 45 min. The temperature is then cooled down to room temperature for another 15 min before Ca^{2+} indicator loading in ACSF.

3.2. Ca^{2+} Indicator Loading

Like cultured glial cells, glial cells in brain slices are most often loaded with the Ca^{2+} indicator using bulk loading of AM esters. Fluo-4 AM, e.g. is preferentially taken up by glial cells and much less into neurons. Fluo-4 is a brightly fluorescent Ca^{2+} indicator with an increase in fluorescence of up to 50-fold upon binding Ca^{2+}. Its K_D of near 350 nM makes it ideal for Ca^{2+} measurements in the range from resting Ca^{2+} to 1–2 μM of free Ca^{2+}. Fluo-4 has an excitation maximum of 496 nm and can be applied to confocal microscopy using the 488 nm line of an argon laser. However, ratiometric Ca^{2+} measurements cannot be performed with Fluo-4 alone. For ratiometric Ca^{2+} imaging, Fura Red can be used alternating excitation at 458 nm (available with most argon ion lasers), which is the isosbestic point and 488 nm which shows a large decrease in fluorescence upon binding Ca^{2+} (39). Fura Red is less resistant to photobleaching and has a smaller quantum yield and dynamic range as compared to Fluo-4. Therefore, Fluo-4 is generally preferred over Fura Red.

Fluo-4 AM is prepared as a stock solution of 2 mM in 20% pluronic acid/80% DMSO. Brain slices are placed onto a nylon mesh in a dish (1.5–2 ml volume) and incubated in 2 μM Fluo-4 AM in Ca^{2+}-reduced ACSF for 60 min at room temperature. During incubation, the dish is placed in a carbogene and dark environment. After incubation, the slices are transferred into a beaker and can be stored in gassed ACSF at room temperature for up to 5 h.

Since bulk loading of Ca^{2+} indicator AM esters results in labelling of the whole tissue, the contrast is often not distinct enough to allow identification and separation of very small glial cell processes (<1 μM) and unspecific background labelling (Fig. 12.5). Therefore, single-cell loading might be necessary for high-resolution Ca^{2+} imaging. Single glial cells can be filled with the free salts of the high-affinity Ca^{2+} indicators Oregon Green BAPTA-1 and Fura-2 (Grosche et al. 1999) (40, 41) or the low-affinity Ca^{2+} indicator Fluo-5F (42) via a patch-pipette allowing for simultaneous recording of Ca^{2+} changes and membrane currents (Fig. 12.5). Ballistic loading of astrocytes in hippocampal slice cultures using Oregon Green BAPTA-1 dextran-coated tungsten particles and a "gene gun" has also been described (43). This method, however, has the disadvantage that the loading pattern of the Ca^{2+} indicator is less predictable.

3.3. Cell Identification

Though Fluo-4 AM is preferentially taken up by glial cells in brain slices, bulk loading cannot suppress some uptake of Ca^{2+} indicator by neurons. Hence, the identity of the cells to be investigated has to be confirmed. Antibodies against GFAP and S100b have often been used to identify astrocytes in brain slices that were fixed after the experiment (Fig. 12.6). This method is time consuming and does not always provide results of sufficient quality. For antibody studies of brain slices after Ca^{2+} imaging, Fluo-4 is first fixed in the cells with 1-ethyl-3-(3-dimethylaminopropyl)carbodiimide (EDAC; Sigma). Brain slices are then fixed using 4% formalin and labelled with an anti-GFAP antibody and fluorescent secondary antibody (44, 45). The antibody staining is compared with the Fluo-4 staining in the Ca^{2+} measurements to identify GFAP-positive cells, which are considered to be astrocytes. Since fixation frequently results in tissue shrinkage, retrieval of the original region of interest used during Ca^{2+} imaging is difficult and sometimes ambiguous. Therefore, other methods that allow identifying glial cells must be favoured especially those, which can identify the cell type during the Ca^{2+} imaging experiment.

One experiment of this kind is to counterstain astrocytes in brain slices with sulforhodamine 101 (46, 47). The red fluorescence can easily be separated from the green fluorescence of the most common Ca^{2+} indicators such as Fluo-4 and Oregon Green BAPTA-1 by appropriate mirrors and filters. Another possibility

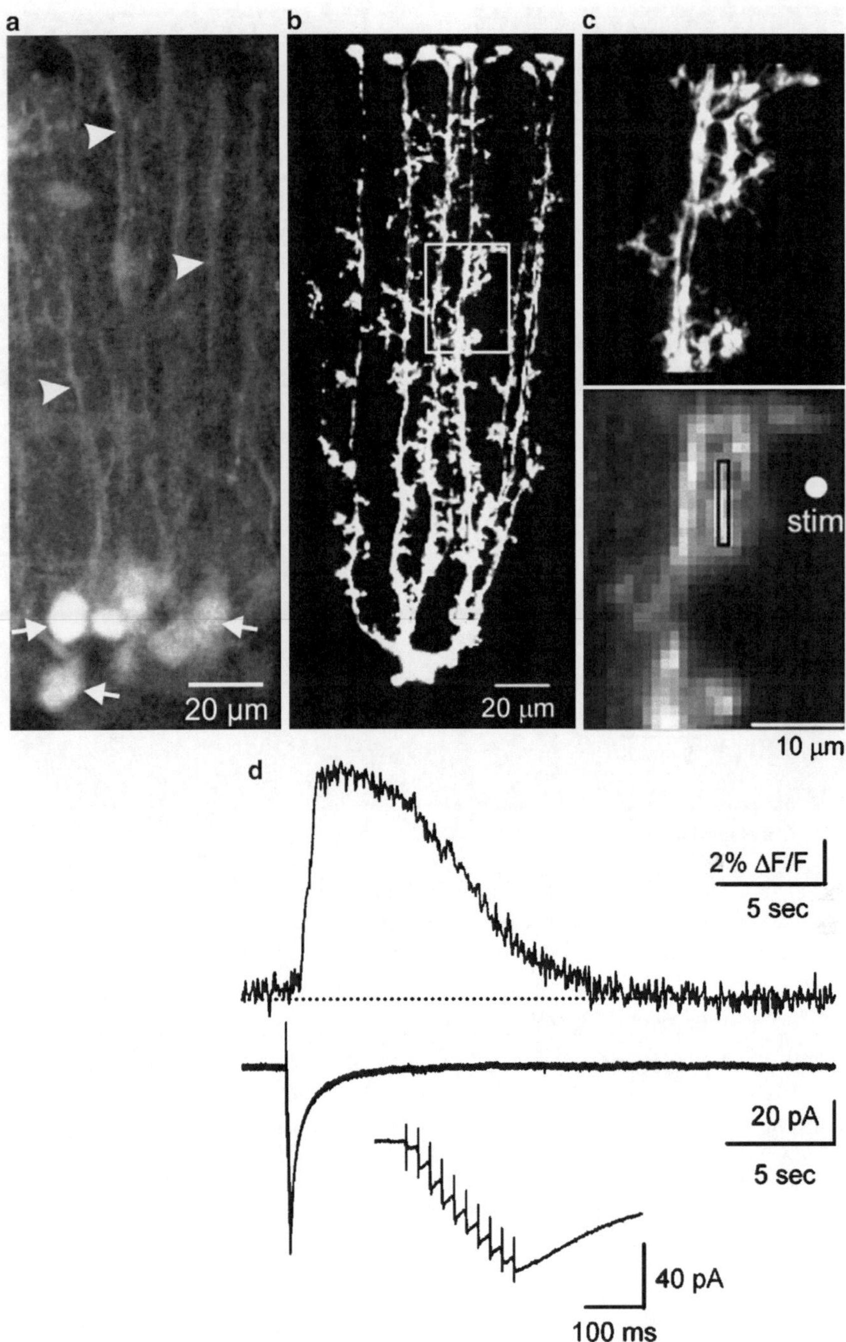

Fig. 12.5. Ca^{2+} measurements in Bergmann glia. (**a**) Bergmann glial cells in a rat cerebellar slice bulk-loaded with Fluo-4 AM. Cell somata (*arrows*) and large radial cell processes (*arrowheads*), but not microdomains can be visualized. (**b**) Bergmann glial cell filled with Alexa 594 and Fura-2 via a patch-clamp pipette. The box indicates the area in (), which was selected for Ca^{2+} measurements. (**c**) Microdomains as imaged with two-photon excitation of Alexa 594 (*upper panel*) and epifluorescence of Fura-2 (*lower panel*). The location of the stimulation pipette (stim) and the ROI for Ca^{2+} measurements (*rectangle*) are shown. (**d**) Ca^{2+} transient (*upper trace*) and whole-cell current (*lower trace*) in a Bergmann glial cell evoked by electrical stimulation of parallel fibres. (**b–d**), modified from Beierlein and Regehr (40).

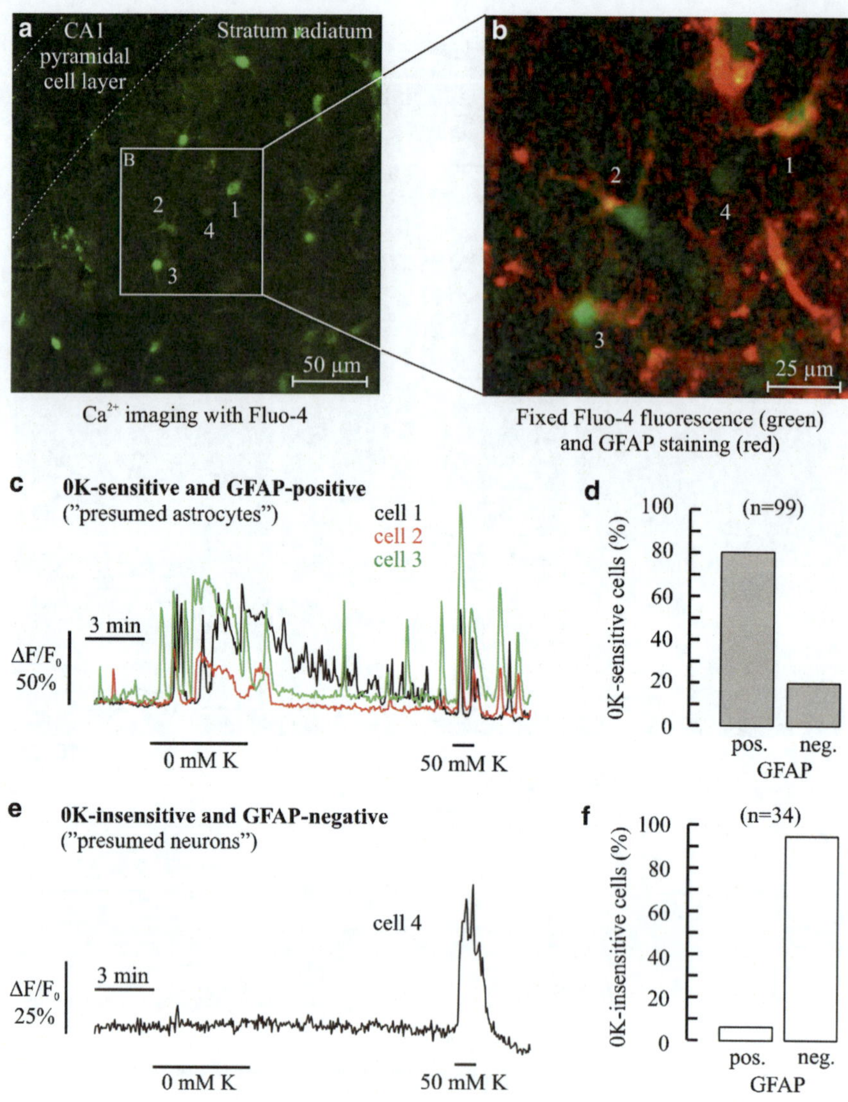

Fig. 12.6. Identification of astrocytes in brain slices. (**a**) Fluo-4-loaded acute hippocampal brain slice. (**b**) The area indicated by a *box* in (**a**) after fixation of Fluo-4 with EDAC (*green*) and GFAP antibody labelling (*red*) to identify GFAP-positive cells (cells 1–3; presumed astrocytes) and GFAP-negative cells (cell 4; presumed neuron). (**c**–**f**) GFAP-positive cells, but not GFAP-negative cells, respond to removal of external potassium (0 mM K). (**a**–**f**) from Beck et al. (44).

to counterstain glial cells is the expression of a fluorescent protein such as GFP, RFP, YFP or CFP under control of a glial cell-specific promoter in transgenic animals (Fig. 12.6; (48)). The expression of the fluorescent proteins, however, does not cover all cells of a given cell type resulting in an uncertain number of false-negative hits with respect of the cell identification. A physiological cell parameter that we frequently analyse to identify astrocytes in brain slices of immature animals (<postnatal day 21) is the Ca^{2+} response to the removal of external K^+ (Fig. 12.6; (44, 45, 49)). Immature

astrocytes express the inward rectifier K^+ channels Kir4.1, which becomes permeable to some divalent cations such as Ca^{2+}, when K^+ as the main ion permeating glial cell membranes is absent (50). Hence, Ca^{2+} signalling can be measured in astrocytes when K^+-free ACSF is superfused for 3–5 min, whereas neurons which lack the Kir4.1 channel do not show any Ca^{2+} signals in K^+-free solution. Astrocytes can be identified by this method with an accuracy of approximately 90% as verified by subsequent GFAP antibody staining (44, 45).

3.4. High-Resolution Imaging

3.4.1. Confocal Ca^{2+} Imaging

In conventional microscopy, out-of-focus fluorescence and scattered emission light blurs the image of deeper layers in brain slices, which can be circumvented by the use of confocal or two-photon microscopy. Confocal microscopy allows the visualization of and Ca^{2+} measurement in glial cells up to 100 μm deep in the tissue and cells even several hundreds of micrometers deep in the tissue can be imaged by two-photon microscopy (see Chap. 6). We routinely measure Ca^{2+} changes in Fluo-4-loaded astrocytes 50–100 μm deep in brain slices of the cerebellum, hippocampus and olfactory bulb (11, 20, 49).

The intensity of Fluo-4 fluorescence in confocal microscopy depends on several parameters that must be considered to obtain optimal results. These parameters are (without claim to be complete): (1) Type of Ca^{2+} indicator (see above) and amount of Ca^{2+} indicator in the cells, (2) laser power, (3) detector gain, (4) pinhole size, and, of course, (5) state of viability of the cells. We shall briefly comment on these points: (1) *Amount of Ca^{2+} indicator*: For bright fluorescence, it is desirable to load as much Ca^{2+} indicator into the cells as possible. However, since the Ca^{2+} indicator molecules bind Ca^{2+}, they act as Ca^{2+} buffer and a high intracellular concentration of Ca^{2+} indicator reduces the amplitude and slows down the kinetics of cytosolic Ca^{2+} transients. Therefore, it is preferable to use the lowest concentration of Ca^{2+} indicator that is sufficient to provide a decent signal-to-noise ratio. Experimentally, this Ca^{2+} indicator concentration can be established by loading acute brain slices with the Ca^{2+} indicator (see Sect. 3.2) for different times. Brain slices are removed from the incubation chamber after 30, 45, 60 and 90 min, and investigated with a confocal microscope to determine the shortest possible incubation time, and thus lowest Ca^{2+} indicator concentration resulting in sufficient staining quality.

(2, 3) *Laser power and detector gain*: In general, the laser power should be kept to a minimum to reduce photobleaching of the Ca^{2+} indicator. We reduce the laser output to 1% of the maximal output power of an 100-mW Argon laser at 488 nm using an acousto-optic tunable filter (AOTF) to excite Fluo-4 and Oregon Green BAPTA-1. A standard FITC filter set consisting of a 488-nm beam splitter and a 505–550-nm bandpass filter is used to detect Fluo-4 or Oregon Green BAPTA-1 fluorescence. When lasers are used for some time (usually 2–3 years or more), they loose power and the AOTF is adjusted accordingly to obtain the same results as with new lasers. In order to keep track of the laser performance, hours of laser use should be noted. The detector gain is set just below the limit where noticeable detector noise occurs. The gain should be set at the beginning of an experiment and should not be altered during an experiment to allow direct comparison of the different Ca^{2+} signals measured. When the signal-to-noise ratio needs to be improved, an increase in detector gain is preferable to increasing the laser power.

(4) *Pinhole size*: The pinhole size defines the axial resolution of confocal images, i.e. the thickness of the optical plane from which images are recorded, but also determines the amount of fluorescent light led to the detector and hence the brightness of the images. A pinhole size of approximately two "Airy units" (when using a 40×/NA 0.8 lens) is a good compromise to achieve sufficient axial resolution to investigate single cells with sufficient fluorescence yield. If Ca^{2+} indicator staining of cells is poor, detector gain and/or pinhole size should be increased rather than increasing the laser power. If a thin optical plane determining the "confocality" of the images is not needed, the pinhole should be increased in size so the detector gain and/or the laser power can be reduced.

(5) *State of cell viability*: Cells can be quickly tested for viability with established responses; e.g. brief exposure to 10–20 μM ADP or 100 μM glutamate for 10–30 s should elicit robust Ca^{2+} signals in most astrocytes. Cultures can be checked with propidium iodine for fraction of dead cells. In brain slices, the surfaces of the slice consist of traumatized cells, the depth of this layer of injured cells depending on the mode of slicing; injured cells take up non-esterized dye and/or exhibit large fluorescent indicative for high intracellular Ca^{2+}.

3.4.2. Two-Photon Ca^{2+} Imaging

In vivo two-photon calcium imaging is the only method that allows to record the activity of a dense neuronal and/or glial cell population with single-cell resolution. Schummers et al. (51) e.g. used two-photon imaging of calcium signals in the ferret visual cortex in vivo to discover that astrocytes like neurons respond to visual stimuli.

Fluo-4 can be used with two-photon or multi-photon excitation. The excitation spectrum of the two-photon excitation of Fluo-4 is very broad ranging from 750 to 900 nm. Thus, Fluo-4 can be combined with other fluorescent markers such as sulforhodamine 101 and genetically included fluorescent proteins (YFP, RFP) to identify cells. As in confocal microscopy, Fluo-4 has a large increase in fluorescence intensity upon binding of Ca^{2+} after two-photon excitation, rendering it as well suited to detect Ca^{2+}

signalling in glial cells in brain tissue including in vivo preparations. Hirase et al. (52) have topically applied Fluo-4 AM on the cerebral cortex of anesthetized rats and imaged cytosolic calcium fluctuation in astrocyte populations of superficial cortical layers in vivo using two-photon laser scanning microscopy. More recently, new probes have been reported, which are capable of monitoring calcium waves at a depth of 120–170 μm in live tissue for up to an hour using two-photon microscopy with no artefacts of photobleaching (53). 2-Acetyl-6-(dimethylamino)naphthalene-derived two-photon fluorescent Ca^{2+} probes (ACa1–ACa3) can be excited by a 780 nm laser beam, showing 23–50-fold enhancement in one- and two-photon excited fluorescence in response to Ca^{2+}.

In two-photon microscopy, excitation efficacy greatly depends on the numerical aperture of the lens. Most microscope companies sell low-magnification, high-numerical aperture lenses (between 16×, NA 0.75 and 20×, NA 0.9) with high IR (infra-red) transmission, which are particularly well suited for physiological two-photon experiments. Ca^{2+} signals in glial cells are usually slow as compared with neuronal Ca^{2+} signals. Hence, the image acquisition rate is set to values between 0.3 and 1 Hz. In experiments where the kinetics of glial Ca^{2+} transients might be faster, e.g. during synaptically evoked Ca^{2+} responses, the acquisition rate can be increased by decreasing the number of pixels of each image. This increase in temporal resolution, however, is achieved only at the expense of spatial resolution.

3.5. Ca^{2+} Signalling Evoked by Ligand Application

Glial Ca^{2+} signalling can be experimentally evoked by application of neurotransmitter receptor agonists. In addition, spontaneous Ca^{2+} signals often occurring as Ca^{2+} oscillations or Ca^{2+} waves that occur without any experimental stimulus have been studied (20, 54, 55). The exact mechanism of these spontaneous Ca^{2+} signals are not fully understood, but functional neurotransmitter receptors appear to mediate at least some of these spontaneous Ca^{2+} signals possibly by ongoing endogenous release or by some homeostatic level of neurotransmitter in the tissue (20, 55).

Glial cells in the central nervous system such as astrocytes, oligodendrocytes, Müller cells or olfactory ensheathing cells express a large set of functional neurotransmitter receptors linked to cytosolic Ca^{2+} increases (56–59). Most of the glial neurotransmitter receptors are G protein-coupled receptors, the activation of which results in Ca^{2+} release from intracellular stores mediated by inositol-trisphosphate. Well studied glial receptors linked to Ca^{2+} signalling are metabotropic glutamate receptors and P2Y purinergic receptors, which can be stimulated by bath application of DHPG and ADP respectively without stimulating ionotropic glutamatergic and purinergic receptors (Fig. 12.7). In addition, stimulation of receptors sensitive for acetylcholine, adrenaline, 5-hydroxytryptamine,

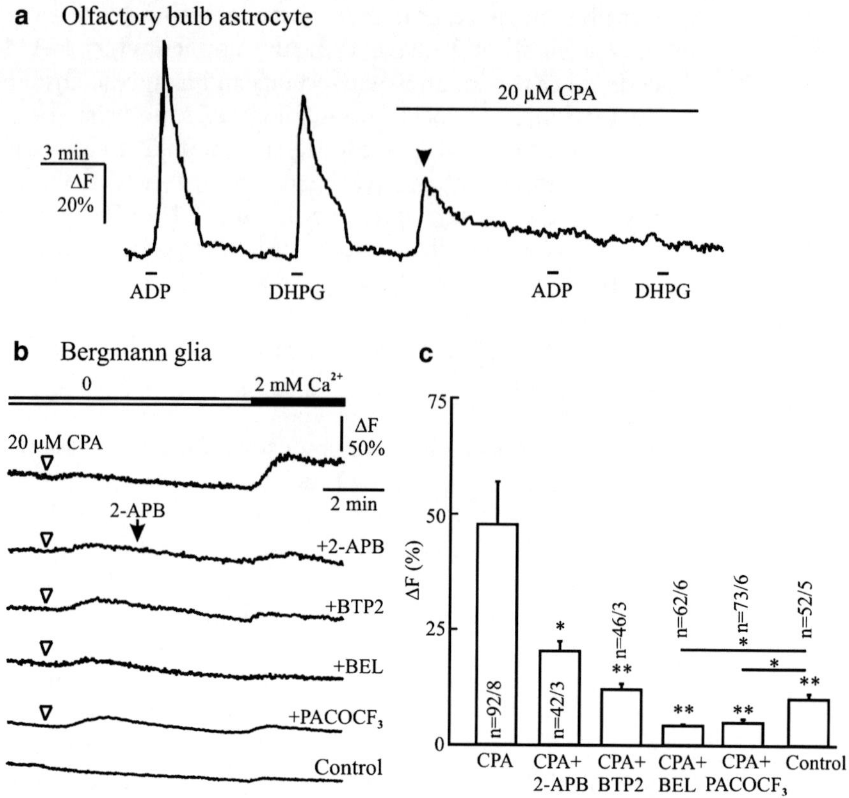

Fig. 12.7. Ca^{2+} release from internal stores and store-operated Ca^{2+} entry (SOCE). (**a**) Activation of $P2Y_1$ by 30 μM ADP and mGluR5 by 30 μM DHPG evokes Ca^{2+} transients in astrocytes in an olfactory bulb brain slice. Depletion of intracellular Ca^{2+} stores by cyclopiazonic acid (CPA) abolishes ADP- and DHPG-dependent Ca^{2+} transients. (**b**, **c**) Re-addition of 2 mM Ca^{2+} after Ca^{2+} store depletion with CPA induces a substantial increase in cytosolic Ca^{2+} (*upper trace*), indicative for SOCE, as recorded in astrocytes of acute cerebellar brain slices. This Ca^{2+} increase can be suppressed by the SOC channel blockers 2-APB and BTP2, and by inhibition of Ca^{2+}-independent phospholipase A_2 with BEL and $PACOCF_3$. When Ca^{2+} stores are not depleted by CPA, no SOCE is evoked (control, *lower trace*). (**a**) from M. Doengi, J.W. Deitmer and C. Lohr, unpublished, (**b**, **c**) from Singaravelu et al. (11).

gamma-aminobutyric acid (GABA), dopamine, histamine, adenosine, neuropeptides, growth factors and chemokines have been shown to elicit Ca^{2+} signalling in glial cells (57). Since activation of these receptors leads to intracellular Ca^{2+} release, depletion of intracellular Ca^{2+} stores abolishes receptor-mediated Ca^{2+} signalling in glial cells. Depletion of intracellular Ca^{2+} stores can be achieved by blockage of SERCA pumps with 1 μM thapsigargin or 10–30 μM cyclopiazonic acid for at least 10 min, which prevents the re-filling of the Ca^{2+} stores (Fig. 12.7).

To study the presence and the properties of neurotransmitter receptors in glial cells in situ, ligands of the corresponding receptor are applied with the bath perfusion. Since the drug-containing saline mixes with the drug-free saline in the experimental chamber, the onset of the drug application is delayed. In addition, bath application of drugs affects all cells in the tissue, which makes it impossible to investigate local Ca^{2+} signalling. To obtain local receptor-mediated Ca^{2+} transients with a prompt onset, ligands can be puff-applied from a focal micropipette. Puff pipettes with a tip diameter of a few micrometers can be pulled with a patch pipette puller and manipulated within a few micrometer close to a cell or region to be studied.

As a consequence of receptor-mediated intracellular Ca^{2+} release, and hence following emptying of the Ca^{2+} stores Ca^{2+} channels in the cytoplasmic membrane of the glial cells open to allow for Ca^{2+} entry and Ca^{2+} store replenishment (60) a mechanism known as store-operated Ca^{2+} entry (SOCE). Hence, the channels mediating Ca^{2+} influx upon Ca^{2+} store depletion are called store-operated Ca^{2+} channels (SOCCs). The molecular identity of store-operated Ca^{2+} channels is still unknown in glial cells. In cell lines, members of the TRP channel family or the Orai membrane protein family have been suggested to mediate store-operated Ca^{2+} entry (61, 62), but their involvement in glial Ca^{2+} signalling has not yet been confirmed.

The mechanism of store-operated Ca^{2+} entry in astrocytes has been investigated by depleting intracellular Ca^{2+} stores with cyclopiazonic acid in Ca^{2+}-free ACSF and subsequent re-addition of Ca^{2+}-containing ACSF. Ca^{2+} store depletion triggers the opening of store-operated Ca^{2+} channels and the Ca^{2+} increase upon Ca^{2+} re-addition reflects store-operated Ca^{2+} entry (Fig. 12.7; (11)).

In addition to Ca^{2+} signals induced by receptor activation, Ca^{2+} signals in glial cells can be evoked by photolysis of "caged Ca^{2+}" such as NP-EGTA (63). NP-EGTA is a photolabile EGTA derivative with a high Ca^{2+} affinity. Upon brief UV illumination (milliseconds to seconds), the molecule is cleaved resulting in a large decrease in its Ca^{2+} affinity and hence in the release of free Ca^{2+}. Alternatively, caged IP_3 can be used, which produces free IP_3 upon UV illumination and activates Ca^{2+} release from IP_3-sensitive stores (Fiacco and McCarthy 2004). However, caged IP_3 is membrane-impermeable and needs to be loaded into the cells via a patch-clamp pipette, whereas NP-EGTA is available as an acetoxymethyl ester (AM) and can easily be bulk-loaded into glial cells in culture or in brain slices.

3.6. Ca^{2+} Signalling Evoked by Endogenous Neurotransmitter Release

Endogenous neurotransmitter release can be evoked by electrical stimulation of neurons. Since glial cells are often in close contact to synapses, synaptic release of neurotransmitters can result in activation of glial neurotransmitter receptors and glial Ca^{2+} signalling (64, 65). In addition, extrasynaptic release of neurotransmitters has also been shown to evoke Ca^{2+} signalling in oligodendrocytes, NG2 cells and olfactory ensheathing cells (56, 59, 66). Neuron–glia interactions have been studied in detail in brain areas with distinct axon pathways and well established physiological properties of glial cells. Stimulation of Schaffer collaterals in hippocampal slices, e.g. induces Ca^{2+} transients in astrocytes of the CA1 region via metabotropic glutamate receptors (67). In addition, the same astrocytes can respond to acetylcholine release upon stimulation of axons from the alveus (68). Stimulation of olfactory nerve fibres results in the release of glutamate and ATP from olfactory receptor axon terminals in olfactory bulb slices (59); ATP then evokes Ca^{2+} transients in olfactory bulb astrocytes via activation of $P2Y_1$ receptors and after enzymatic degradation to adenosine of A_{2A} receptors (49). For stimulation of axons, either a patch-clamp pipette (1–4 Mohm resistance when filled with ACSF) or a bipolar stimulation electrode (70–200 μm distance between poles) are used. Stimulation pulses of 0.5–5 ms duration and 10–50 V (10–30 μA, respectively) amplitude reliably trigger action potentials in axons. Large Ca^{2+} transients can be induced in astrocytes by repetitive stimulation of axons at a frequency of 10–50 Hz for 1–15 s (Fig. 12.8).

The amplitude and duration of Ca^{2+} transients as well as the frequency of Ca^{2+} oscillations in astrocytes may vary depending on the level of synaptic activity. Moderate stimulation of parallel fibres has been shown to evoke Ca^{2+} increases that were restricted to only a few small processes of Bergmann glial cells closely associated with synaptic sites called microdomains (69), and that could spread within these processes (40) indicating that glial cells are able to respond locally to neuronal activity. Strong stimulation of parallel fibres in contrast evoked Ca^{2+} transients in all processes and the soma (41, 70). Studies aimed to investigate the synaptic control of astrocyte Ca^{2+} signalling have recently demonstrated that astrocytes display integrative properties for synaptic information processing as astrocytes can discriminate between

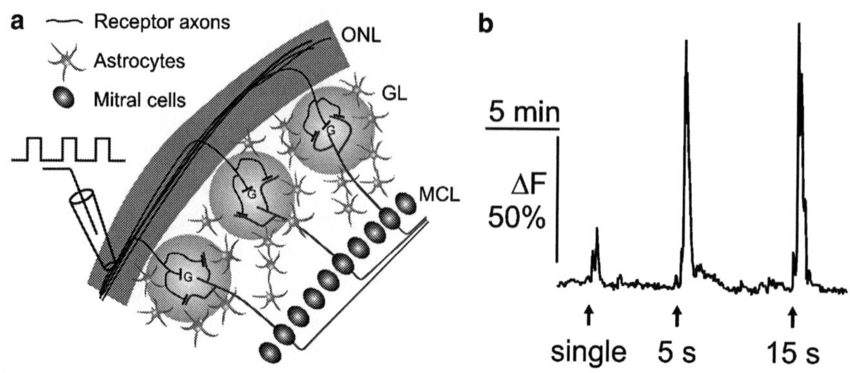

Fig. 12.8. Stimulation-induced Ca^{2+} transients in olfactory bulb astrocytes. (**a**) Schematic diagram of the cellular organization of the glomerular layer (GL) in the olfactory bulb. *ONL* olfactory nerve layer; *ML* mitral cell layer. (**b**) A single stimulation pulse as well as trains of stimuli (50 Hz) for 5 and 15 s evoked Ca^{2+} transients in periglomerular astrocytes. (**b**) modified from Doengi et al. (49).

the activity of synaptic terminals belonging to different axon pathways (68, 71).

Ca^{2+} signalling in glial cells has been shown to affect neuronal performance and blood capillaries (72–74). Ca^{2+} signals in astrocytes of hippocampal slices (75–77) and in glial cells of the retina (78) can lead to Ca^{2+} transients in adjacent neurons and increased neuronal activity due to the Ca^{2+}-dependent release of "gliotransmitters" such as glutamate or d-serine (79, 80). Ca^{2+} transients in astrocytic endfeet can also lead to vasodilation as well as vasoconstriction (81, 82).

In summary, to specifically trigger Ca^{2+} signalling in glial cells different techniques have been employed. Astrocytic Ca^{2+} transients can be evoked by bath application of agonists of glial metabotropic receptors (75, 77) by mechanical stimulation of a single astrocyte with a micropipette (78), by depolarization of current-clamped astrocytes (76) and by photorelease of caged Ca^{2+} (NP-EGTA) or caged IP_3, loaded into astrocytes via a patch-clamp pipette (71, 83), or by eliciting neuronal activity, which results in the release of neurotransmitters acting not only on post-synaptic neurons, but also on ambient glial cell processes.

4. Glial Cells In Vivo

In few cases, Ca^{2+} signalling in glial cells and neurons of living anaesthetized animals have been recorded using two-photon microscopy (see also Chap. 15). Animals are anaesthetized with ketamine/xylazine (0.1/0.01 mg/g), urethane (1.9 mg/g) or isofluorane (1.5% in O_2). A small craniotomy is made above the brain region to be investigated and the dura mater is carefully removed. To introduce the Ca^{2+} indicator into astrocytes in vivo, the AM ester of the Ca^{2+} indicator is dissolved in 20% pluronic acid/80% DMSO at a concentration of 10 mM and diluted in ACSF to a final concentration of 0.5–1 mM. Stosiek et al. (84) filled the Ca^{2+} indicator solution into a micropipette and pressure-injected it with 0.7 bar into the brain tissue for 1 min ("multicell bolus loading"). Alternatively, the Ca^{2+} indicator solution can be applied directly onto the exposed pia matter, where it is preferentially taken up by astrocyte endfeet and then distributes within the astrocytic syncytium (52, 85). A large number of astrocytes and neurons can be loaded simultaneously with the Ca^{2+} indicator by this way and visualized up to 200–300 μm deep in the tissue.

Astrocytes in vivo also generate spontaneous Ca^{2+} transients, Ca^{2+} waves and Ca^{2+} oscillations (47, 52, 86). In addition, Ca^{2+} signalling in astrocytes in vivo can also be evoked by sensory stimulation. Mechanical whisker stimulation results in Ca^{2+} transients in astrocytes of the barrel cortex (87) and visual stimulation induces Ca^{2+} signalling in astrocytes in the visual cortex (51).

5. Invertebrate Glial Cells

5.1. Leech Giant Glial Cells

The leech giant glial cells have long been used as a model system to study glial cell physiology including Ca^{2+} signalling (88, 89). Two giant glial cells are located in the central neuropil of each segmental ganglion and can easily be exposed by removing the collagen tissue and some larger neuronal cell bodies above the neuropil. Leech giant glial cells possess a large cell body of 80–100 μm diameter and widely ramified cell processes. The good accessibility and the large size make them ideal for Ca^{2+} imaging studies and hence local Ca^{2+} signalling evoked by neuronal activity have first been measured in this type of glial cell (90, 91). In addition, fluorescent Ca^{2+} measurements, ion measurements with ion-selective microelectrodes and electrophysiological recordings can be performed and even combined taking advantage of the size and robustness of the cells.

Since leech giant glial cells cannot be loaded with AM-esters of the Ca^{2+} indicators, the Ca^{2+} indicator is directly injected into the cell via a microelectrode (14). The potassium salt of either Ca^{2+} indicator, Fura-2, Fluo-3 or Oregon Green BAPTA-1 is dissolved in 100 mM KCl at a concentration of 5–10 mM and backfilled into a microelectrode pulled from a glass capillary. The microelectrode is fixed in a micromanipulator and connected to a bridge amplifier via chlorided silver wires. The microelectrode is impaled into a giant glial cell body and the Ca^{2+} indicator dye is iontophoretically injected into the cell with a current of −5 to −10 nA for 30 min. In order to maintain a constant dye concentration in the cells throughout the experiment, a current of −2 to −5 nA is injected continuously to replenish the dye which is transported out of the cell and/or diffused through gap junctions into neighbouring glial cells. Changes in the Ca^{2+} concentration in giant glial cells can be measured by conventional or confocal microscopy for up to several hours (14, 91–93).

Axons enter and leave the segmental ganglia through the connectives and the side nerves. Neuronal activity can be increased by electrical stimulation of axons in the side nerves. For that purpose, one side nerve is sucked into a glass pipette filled with saline and connected to one pole of a stimulation unit. The second pole is grounded in the bath. Electrical stimulation of axons in side nerves results in local Ca^{2+} signals in giant glial cell processes (91). In contrast to vertebrate glial cells, in which Ca^{2+} signalling is predominantly carried by Ca^{2+} release from intracellular stores, Ca^{2+} influx from the extracellular space mediates Ca^{2+} signalling in leech glial cells and the Ca^{2+} transients evoked by axonal stimulation can be blocked by ionotropic non-NMDA receptor antagonists (91, 94).

5.2. Antennal Lobe Glial Cells of Manduca sexta

In holometabolous insects such as the tabacco sphinx moth *Manduca sexta*, the central nervous system undergoes a severe morphological and functional reorganisation during metamorphosis, which resembles embryonic development of vertebrates. During that time, glial cells play a key role in axon guidance and morphogenesis in the insect antennal lobe the first central relay station of the olfactory pathway (95). Glial cell development is accompanied by changes in Ca^{2+} signalling, which has been studied in culture (96) and by two-photon Ca^{2+} imaging in situ and in vivo (89). For in situ studies, intact brains with antennae attached are dissected from *Manduca sexta* pupae and pinned into a Sylgard-lined experimental chamber. Brains are incubated in 2 μM Fluo-4-AM in insect saline (150 mM NaCl, 4 mM KCl, 2 mM $CaCl_2$, 1 mM $MgCl_2$, 10 mM HEPES buffer; pH was adjusted to 7.0 with NaOH; the osmolarity was adjusted to 390 mOsm with 50 mM mannitol and 25 mM glucose). Glial cells preferentially take up Fluo-4 and can easily be identified by their small cell bodies (approximately 10 μm in diameter) and their central position around glomeruli, while neuronal cell bodies are located in cell packets at the periphery of the antennal lobe (97, 98).

For in vivo Ca^{2+} imaging in *Manduca sexta*, pupae or adult moths are anaesthetized with CO_2 and embedded in 1% agarose in a plastic tube (Fig. 12.9; (99, 100)). The head capsule is opened to expose the brain. Glial cells are loaded with Fluo-4-AM dissolved in insect saline at a concentration of 5%. Olfactory receptor neurons are stimulated either by a stimulation pipette placed in the antennal nerve or by odorants puffed onto the antennae (Fig. 12.8). Stimulation of receptor neurons results in the release of acetylcholine (ACh) in the antennal lobe, which activates glial nicotinic ACh receptors and hence leads to depolarization of the glial membrane (99). The subsequent Ca^{2+} influx through voltage-gated Ca^{2+} channels is a prerequisite for the migration of these glial cells, and hence for the morphogenesis of the antennal lobe (89, 97). Insect brains and ganglia such as from *Manduca*, locust, flies and bees, have also been employed both for in vitro and in vivo Ca^{2+} imaging of neurons (101–105).

6. Conclusion

Different cells and tissues from vertebrate and invertebrate animals require some adjustments for optimal recording of cytosolic Ca^{2+}. Some, but by far not all, have been exemplarily discussed

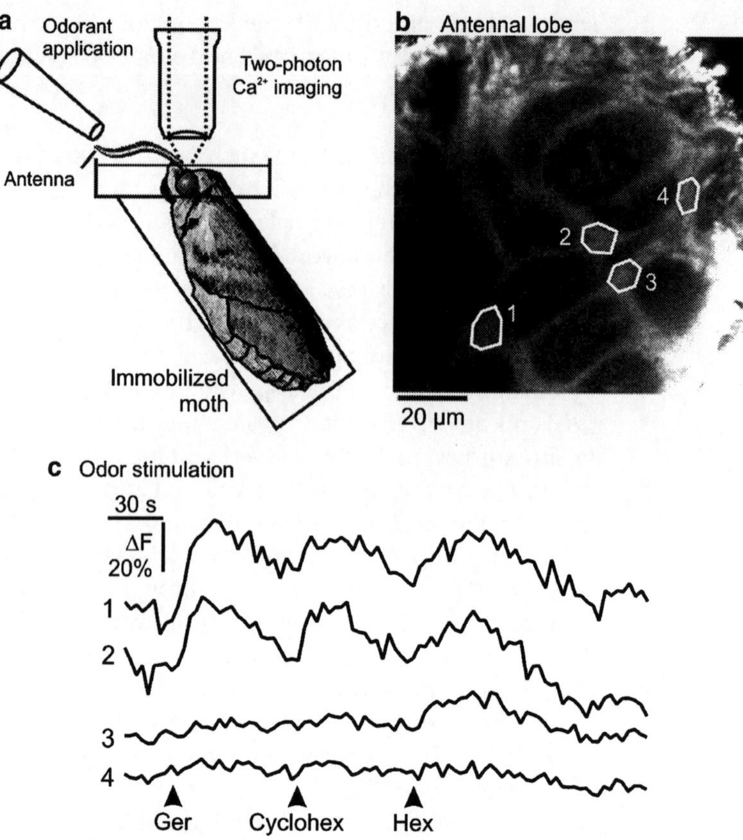

Fig. 12.9. In vivo two-photon Ca^{2+} imaging in the tabacco sphinx moth *Manduca sexta*. (**a**) Schematic diagram of the in vivo setup for imaging of Ca^{2+} signals in antennal lobe glial cells evoked by odorants puffed onto the antenna. (**b**) Fluo-4-loaded glial cells surrounding glomeruli in the antennal lobe of *Manduca*. ROIs indicate regions in which the traces as shown in (**c**) were recorded. (**c**) Calcium signalling in periglomerular antennal lobe glial cells evoked by stimulation of olfactory receptor neurons in the antenna by geraniol (Ger), cyclohexanol (Cyclohex) and 1-hexanal (Hex). (**b**, **c**) modified from Heil et al. (99).

here for various types of glial cells. Glial cell cultures are well suited to study basic features of Ca^{2+} signalling without the interference of other cell types as in tissue preparations. In addition, since cultured cells are not embedded deep in the tissue, conventional epifluorescence imaging systems can be employed to measure Ca^{2+} in the cytosol and even in organelles using ratiometric, UV-excitable Ca^{2+} indicators such as Fura-2. For studying complex interactions between glial cells and neurons, brain slices in combination with confocal microscopy or multi-photon microscopy are the techniques of choice. Recent developments in imaging instruments and sophisticated in vivo preparations have led to an increasing number of glial cell studies in living anaesthetized animals. Genetically encoded Ca^{2+} indicators could not be specifically expressed in glial cells in vivo so far, a deficit that surely will be overcome soon enabling in vivo studies in both major cell types in the brain even more efficiently. The combination of highly sophisticated imaging instruments and transgenic model organisms will likely lead to new insights into the role of glial cells in nervous system function.

Acknowledgements

We thank our collaborators, who contributed to our own studies on Ca^{2+} imaging in glial cells over many years. Financial support of our projects came primarily from the Deutsche Forschungsgemeinschaft and is gratefully acknowledged (LO 779/2, GK 845, SFB 530, TP B1, and DE 231/19-1,2).

References

1. Matthias K, Kirchhoff F, Seifert G et al (2003) Segregated expression of AMPA-type glutamate receptors and glutamate transporters defines distinct astrocyte populations in the mouse hippocampus. J Neurosci 23:1750–1758
2. Tabernero A, Medina JM, Giaume C (2006) Glucose metabolism and proliferation in glia: role of astrocytic gap junctions. J Neurochem 99:1049–1061
3. Brune T, Deitmer JW (1995) Intracellular acidification and Ca^{2+} transients in cultured rat cerebellar astrocytes evoked by glutamate agonists and noradrenaline. Glia 14:153–161
4. Ranscht B, Clapshaw PA, Price J, Noble M, Seifert W (1982) Development of oligodendrocytes and Schwann cells studied with a monoclonal antibody against galactocerebroside. Proc Natl Acad Sci USA 79: 2709–2713
5. Ito D, Imai Y, Ohsawa K, Nakajima K, Fukuuchi Y, Kohsaka S (1998) Microglia-specific localisation of a novel calcium binding protein, Iba1. Brain Res Mol Brain Res 57:1–9
6. Pfrieger FW, Barres BA (1997) Synaptic efficacy enhanced by glial cells in vitro. Science 277:1684–1687
7. Le Roux PD, Reh TA (1994) Regional differences in glial-derived factors that promote dendritic outgrowth from mouse cortical neurons in vitro. J Neurosci 14:4639–4655
8. Helmchen F, Imoto K, Sakmann B (1996) Ca2+ buffering and action potential-evoked Ca2+ signaling in dendrites of pyramidal neurons. Biophys J 70:1069–1081
9. Helmchen F, Borst JG, Sakmann B (1997) Calcium dynamics associated with a single action potential in a CNS presynaptic terminal. Biophys J 72:1458–1471
10. Regehr WG, Atluri PP (1995) Calcium transients in cerebellar granule cell presynaptic terminals. Biophys J 68:2156–2170
11. Singaravelu K, Lohr C, Deitmer JW (2006) Regulation of store-operated calcium entry by calcium-independent phospholipase A2 in rat cerebellar astrocytes. J Neurosci 26:9579–9592
12. Grynkiewicz G, Poenie M, Tsien RY (1985) A new generation of Ca^{2+} indicators with greatly improved fluorescence properties. J Biol Chem 260:3440–3450
13. Klein MG, Simon BJ, Szucs G, Schneider MF (1988) Simultaneous recording of calcium transients in skeletal muscle using high- and low-affinity calcium indicators. Biophys J 53:971–988
14. Munsch T, Deitmer JW (1995) Maintenance of Fura-2 fluorescence in glial cells and neurons of the leech central nervous system. J Neurosci Methods 57:195–204
15. Pet MJ, Wurster RD (1997) Determination of in situ dissociation constant for Fura-2 and quantitation of background fluorescence in astrocyte cell line U373-MG. Cell Calcium 21:233–240
16. Jung S, Pfeiffer F, Deitmer JW (2000) Histamine-induced calcium entry in rat cerebellar astrocytes: evidence for capacitative and non-capacitative mechanisms. J Physiol 527:549–561
17. Singaravelu K, Deitmer JW (2006) Calcium mobilization by nicotinic acid adenine

dinucleotide phosphate (NAADP) in rat astrocytes. Cell Calcium 39:143–153

18. Cornell-Bell AH, Finkbeiner SM, Cooper MS, Smith SJ (1990) Glutamate induces calcium waves in cultured astrocytes: long-range glial signaling. Science 247:470–473
19. Garaschuk O, Linn J, Eilers J, Konnerth A (2000) Large-scale oscillatory calcium waves in the immature cortex. Nat Neurosci 3:452–459
20. Zur Nieden R, Deitmer JW (2006) The role of metabotropic glutamate receptors for the generation of calcium oscillations in rat hippocampal astrocytes in situ. Cereb Cortex 16:676–687
21. Scemes E, Giaume C (2006) Astrocyte calcium waves: what they are and what they do. Glia 54:716–725
22. Suadicani SO, Brosnan CF, Scemes E (2006) P2X7 receptors mediate ATP release and amplification of astrocytic intercellular Ca^{2+} signaling. J Neurosci 26:1378–1385
23. Ye ZC, Oberheim N, Kettenmann H, Ransom BR (2009) Pharmacological "cross-inhibition" of connexin hemichannels and swelling activated anion channels. Glia 57:258–269
24. Cotrina ML, Lin JH, Alves-Rodrigues A et al (1998) Connexins regulate calcium signaling by controlling ATP release. Proc Natl Acad Sci USA 95:15735–15740
25. Duan S, Anderson CM, Keung EC, Chen Y, Swanson RA (2003) P2X7 receptor-mediated release of excitatory amino acids from astrocytes. J Neurosci 23:1320–1328
26. Wang Y, Roman R, Lidofsky SD, Fitz JG (1996) Autocrine signaling through ATP release represents a novel mechanism for cell volume regulation. Proc Natl Acad Sci USA 93:12020–12025
27. Hassinger TD, Guthrie PB, Atkinson PB, Bennett MV, Kater SB (1996) An extracellular signaling component in propagation of astrocytic calcium waves. Proc Natl Acad Sci USA 93:13268–13273
28. Schipke CG, Boucsein C, Ohlemeyer C, Kirchhoff F, Kettenmann H (2002) Astrocyte Ca2+ waves trigger responses in microglial cells in brain slices. FASEB J 16:255–257
29. Gambetti P, Erulkar SE, Somlyo AP, Gonatas NK (1975) Calcium-containing structures in vertebrate glial cells. Ultrastructural and microprobe analysis. J Cell Biol 64:322–330
30. Jou MJ, Peng TI, Sheu SS (1996) Histamine induces oscillations of mitochondrial free Ca2+ concentration in single cultured rat brain astrocytes. J Physiol 497:299–308
31. Boitier E, Rea R, Duchen MR (1999) Mitochondria exert a negative feedback on the propagation of intracellular Ca2+ waves in rat cortical astrocytes. J Cell Biol 145:795–808
32. Duchen MR, Verkhratsky A, Muallem S (2008) Mitochondria and calcium in health and disease. Cell Calcium 44:1–5
33. Golovina VA, Blaustein MP (2000) Unloading and refilling of two classes of spatially resolved endoplasmic reticulum Ca^{2+} stores in astrocytes. Glia 31:15–28
34. Golovina VA, Blaustein MP (1997) Spatially and functionally distinct Ca^{2+} stores in sarcoplasmic and endoplasmic reticulum. Science 275:1643–1648
35. Lee MY, Song H, Nakai J et al (2006) Local subplasma membrane Ca2+ signals detected by a tethered Ca2+ sensor. Proc Natl Acad Sci USA 103:13232–13237
36. Marchaland J, Cali C, Voglmaier SM et al (2008) Fast subplasma membrane Ca^{2+} transients control exo-endocytosis of synaptic-like microvesicles in astrocytes. J Neurosci 28:9122–9132
37. Bringmann A, Faude F, Reichenbach A (1997) Mammalian retinal glial (Muller) cells express large-conductance Ca(2+)-activated K+ channels that are modulated by Mg2+ and pH and activated by protein kinase A. Glia 19:311–323
38. Zhu Y, Kimelberg HK (2001) Developmental expression of metabotropic P2Y(1) and P2Y(2) receptors in freshly isolated astrocytes from rat hippocampus. J Neurochem 77:530–541
39. Lohr C (2003) Monitoring neuronal calcium signalling using a new method for ratiometric confocal calcium imaging. Cell Calcium 34:295–303
40. Beierlein M, Regehr WG (2006) Brief bursts of parallel fiber activity trigger calcium signals in bergmann glia. J Neurosci 26:6958–6967
41. Kulik A, Haentzsch A, Luckermann M, Reichelt W, Ballanyi K (1999) Neuron-glia signaling via $\alpha 1$ adrenoceptor-mediated Ca2+ release in Bergmann glial cells in situ. J Neurosci 19:8401–8408
42. Piet R, Jahr CE (2007) Glutamatergic and purinergic receptor-mediated calcium transients in Bergmann glial cells. J Neurosci 27:4027–4035
43. Benediktsson AM, Schachtele SJ, Green SH, Dailey ME (2005) Ballistic labeling and dynamic imaging of astrocytes in organotypic hippocampal slice cultures. J Neurosci Methods 141:41–53

44. Beck A, Nieden RZ, Schneider HP, Deitmer JW (2004) Calcium release from intracellular stores in rodent astrocytes and neurons in situ. Cell Calcium 35:47–58
45. Dallwig R, Deitmer JW (2002) Cell-type specific calcium responses in acute rat hippocampal slices. J Neurosci Methods 116:77–87
46. Kafitz KW, Meier SD, Stephan J, Rose CR (2008) Developmental profile and properties of sulforhodamine 101 - Labeled glial cells in acute brain slices of rat hippocampus. J Neurosci Methods 169:84–92
47. Nimmerjahn A, Kirchhoff F, Kerr JN, Helmchen F (2004) Sulforhodamine 101 as a specific marker of astroglia in the neocortex in vivo. Nat. Methods 1:31–37
48. Hirrlinger PG, Scheller A, Braun C et al (2005) Expression of reef coral fluorescent proteins in the central nervous system of transgenic mice. Mol Cell Neurosci 30:291–303
49. Doengi M, Deitmer JW, Lohr C (2008) New evidence for purinergic signaling in the olfactory bulb: $A2_A$ and $P2Y_1$ receptors mediate intracellular calcium release in astrocytes. FASEB J 22:2368–2378
50. Härtel K, Singaravelu K, Kaiser M, Neusch C, Hülsmann S, Deitmer JW (2007) Calcium influx mediated by the inwardly rectifying K+ channel Kir4.1 (KCNJ10) at low external K+ concentration. Cell Calcium 42:271–280
51. Schummers J, Yu H, Sur M (2008) Tuned responses of astrocytes and their influence on hemodynamic signals in the visual cortex. Science 320:1638–1643
52. Hirase H, Qian L, Bartho P, Buzsaki G (2004) Calcium dynamics of cortical astrocytic networks in vivo. PLoS Biol 2:E96
53. Kim HM, Kim BR, An MJ, Hong JH, Lee KJ, Cho BR (2008) Two-photon fluorescent probes for long-term imaging of calcium waves in live tissue. Chemistry 14:2075–2083
54. Nett WJ, Oloff SH, McCarthy KD (2002) Hippocampal astrocytes in situ exhibit calcium oscillations that occur independent of neuronal activity. J Neurophysiol 87:528–537
55. Weissman TA, Riquelme PA, Ivic L, Flint AC, Kriegstein AR (2004) Calcium waves propagate through radial glial cells and modulate proliferation in the developing neocortex. Neuron 43:647–661
56. Butt AM (2006) Neurotransmitter-mediated calcium signalling in oligodendrocyte physiology and pathology. Glia 54:666–675
57. Deitmer JW, Singaravelu K, Lohr C (2009) Calcium ion signaling in astrocytes. In: Haydon PG, Papura V (eds) Astrocytes in (patho)physiology of the nervous system. Springer, New York, pp 201–224
58. Metea MR, Newman EA (2006) Calcium signaling in specialized glial cells. Glia 54:650–655
59. Rieger A, Deitmer JW, Lohr C (2007) Axon-glia communication evokes calcium signaling in olfactory ensheathing cells of the developing olfactory bulb. Glia 55:352–359
60. Parekh AB, Putney JWJ (2005) Store-operated calcium channels. Physiol Rev 85:757–810
61. Lis A, Peinelt C, Beck A et al (2007) CRACM1, CRACM2, and CRACM3 are store-operated Ca2+ channels with distinct functional properties. Curr Biol 17: 794–800
62. Worley PF, Zeng W, Huang GN et al (2007) TRPC channels as STIM1-regulated store-operated channels. Cell Calcium 42:205–211
63. Parpura V, Haydon PG (1999) UV photolysis using a micromanipulated optical fiber to deliver UV energy directly to the sample. J Neurosci Methods 87:25–34
64. Agulhon C, Petravicz J, McMullen AB et al (2008) What is the role of astrocyte calcium in neurophysiology? Neuron 59:932–946
65. Schipke CG, Kettenmann H (2004) Astrocyte responses to neuronal activity. Glia 47:226–232
66. Gallo V, Mangin JM, Kukley M, Dietrich D (2008) Synapses on NG2-expressing progenitors in the brain: multiple functions? J Physiol 586:3767–3781
67. Porter JT, McCarthy KD (1996) Hippocampal astrocytes in situ respond to glutamate released from synaptic terminals. J Neurosci 16:5073–5081
68. Perea G, Araque A (2005) Properties of synaptically evoked astrocyte calcium signal reveal synaptic information processing by astrocytes. J Neurosci 25:2192–2203
69. Grosche J, Matyash V, Moller T, Verkhratsky A, Reichenbach A, Kettenmann H (1999) Microdomains for neuron-glia interaction: parallel fiber signaling to Bergmann glial cells. Nat Neurosci 2:139–143
70. Matyash V, Filippov V, Mohrhagen K, Kettenmann H (2001) Nitric oxide signals parallel fiber activity to Bergmann glial cells in the mouse cerebellar slice. Mol Cell Neurosci 18:664–670
71. Perea G, Araque A (2007) Astrocytes potentiate transmitter release at single hippocampal synapses. Science 317:1083–1086

72. Araque A, Sanzgiri RP, Parpura V, Haydon PG (1998) Calcium elevation in astrocytes causes an NMDA receptor-dependent increase in the frequency of miniature synaptic currents in cultured hippocampal neurons. J Neurosci 18:6822–6829
73. Mulligan SJ, MacVicar BA (2004) Calcium transients in astrocyte endfeet cause cerebrovascular constrictions. Nature 431:195–199
74. Zonta M, Angulo MC, Gobbo S et al (2003) Neuron-to-astrocyte signaling is central to the dynamic control of brain microcirculation. Nat Neurosci 6:43–50
75. Bezzi P, Carmignoto G, Pasti L et al (1998) Prostaglandins stimulate calcium-dependent glutamate release in astrocytes. Nature 391:281–285
76. Kang J, Jiang L, Goldman SA, Nedergaard M (1998) Astrocyte-mediated potentiation of inhibitory synaptic transmission. Nat Neurosci 1:683–692
77. Pasti L, Volterra A, Pozzan T, Carmignoto G (1997) Intracellular calcium oscillations in astrocytes: a highly plastic, bidirectional form of communication between neurons and astrocytes in situ. J Neurosci 17:7817–7830
78. Newman EA, Zahs KR (1998) Modulation of neuronal activity by glial cells in the retina. J Neurosci 18:4022–4028
79. Malarkey EB, Parpura V (2008) Mechanisms of glutamate release from astrocytes. Neurochem Int 52:142–154
80. Oliet SH, Mothet JP (2006) Molecular determinants of D-serine-mediated gliotransmission: from release to function. Glia 54:726–737
81. Gordon GR, Mulligan SJ, MacVicar BA (2007) Astrocyte control of the cerebrovasculature. Glia 55:1214–1221
82. Köhler RC, Roman RJ, Harder DR (2009) Astrocytes and the regulation of cerebral blood flow. Trends Neurosci 32(3): 160–169
83. Fiacco TA, McCarthy KD (2004) Intracellular astrocyte calcium waves in situ increase the frequency of spontaneous AMPA receptor currents in CA1 pyramidal neurons. J Neurosci 24:722–732
84. Stosiek C, Garaschuk O, Holthoff K, Konnerth A (2003) In vivo two-photon calcium imaging of neuronal networks. Proc Natl Acad Sci USA 100:7319–7324
85. Ding S, Wang T, Cui W, Haydon PG (2009) Photothrombosis ischemia stimulates a sustained astrocytic Ca^{2+} signaling in vivo. Glia 57(7):767–776
86. Hoogland TM, Kuhn B, Gobel W et al (2009) Radially expanding transglial calcium waves in the intact cerebellum. Proc Natl Acad Sci USA 106(9):3496–3501
87. Wang X, Lou N, Xu Q et al (2006) Astrocytic Ca^{2+} signaling evoked by sensory stimulation in vivo. Nat Neurosci 9:816–823
88. Deitmer JW, Rose CR, Munsch T et al (1999) Leech giant glial cell: functional role in a simple nervous system. Glia 28:175–182
89. Lohr C, Deitmer JW (2006) Calcium signaling in invertebrate glial cells. Glia 54:642–649
90. Lohr C (1998) Calciumsignale in den Riesengliazellen des Blutegels *Hirudo medicinalis* L. PhD thesis, University of Kaiserslautern
91. Lohr C, Deitmer JW (1999) Dendritic calcium transients in the leech giant glial cell in situ. Glia 26:109–118
92. Munsch T, Nett W, Deitmer JW (1994) Fura-2 signals evoked by kainate in leech glial cells in the presence of different divalent cations. Glia 11:345–353
93. Munsch T, Deitmer JW (1992) Calcium transients in identified leech glial cells in situ evoked by high potassium concentrations and 5-hydroxytryptamine. J Exp Biol 65:251–265
94. Rose CR, Lohr C, Deitmer JW (1995) Activity-induced Ca^{2+} transients in nerve and glial cells in the leech CNS. Neuroreport 6:642–644
95. Tolbert LP, Oland LA, Tucker ES, Gibson NJ, Higgins MR, Lipscomb BW (2004) Bidirectional influences between neurons and glial cells in the developing olfactory system. Prog Neurobiol 73:73–105
96. Lohr C, Tucker E, Oland LA, Tolbert LP (2002) Development of depolarization-induced calcium transients in insect glial cells is dependent on the presence of afferent axons. J Neurobiol 52:85–98
97. Lohr C, Heil JE, Deitmer JW (2005) Blockage of voltage-gated calcium signaling impairs migration of glial cells in vivo. Glia 50:198–211
98. Lohr C, Oland LA, Tolbert LP (2001) Olfactory receptor axons influence the development of glial potassium currents in the antennal lobe of the moth *Manduca sexta*. Glia 36:309–320
99. Heil JE, Oland LA, Lohr C (2007) Acetylcholine-mediated axon-glia signaling in the developing insect olfactory system. Eur J Neurosci 26:1227–1241
100. Hartl S, Heil JE, Hirsekorn A, Lohr C (2007) A novel neurotransmitter-independent communication pathway between axons and glial cells. Eur J Neurosci 25:945–956

101. Borst A, Egelhaaf M (1992) In vivo imaging of calcium accumulation in fly interneurons as elicited by visual motion stimulation. Proc Natl Acad Sci USA 89:4139–4143
102. Duch C, Levine RB (2002) Changes in calcium signaling during postembryonic dendritic growth in Manduca sexta. J Neurophysiol 87:1415–1425
103. Galizia CG, Sachse S, Rappert A, Menzel R (1999) The glomerular code for odor representation is species specific in the honeybee Apis mellifera. Nat Neurosci 2:473–478
104. Ryglewski S, Pflueger HJ, Duch C (2007) Expanding the neuron's calcium signaling repertoire: intracellular calcium release via voltage-induced PLC and IP3R activation. PLoS Biol 5:e66
105. Sachse S, Rueckert E, Keller A et al (2007) Activity-dependent plasticity in an olfactory circuit. Neuron 56:838–850
106. Galizia CG, Menzel R (2000) Odour perception in honeybees: coding information in glomerular patterns. Curr Opin Neurobiol 10:504–510

Index

T

U

W

Y